Analysis and Applications – ISAAC 2001

T0191626

International Society for Analysis, Applications and Computation

Volume 10

The titles published in this series are listed at the end of this volume.

Analysis and Applications – ISAAC 2001

Edited by

Heinrich G.W. Begehr
Freie Universität Berlin, Berlin, Germany

Robert P. Gilbert
University of Delaware, Newark, DE, U.S.A.

and

Man Wah Wong
York University, Toronto, Canada

KLUWER ACADEMIC PUBLISHERS
DORDRECHT / BOSTON / LONDON

A C.I.P. Catalogue record for this book is available from the Library of Congress.

ISBN 978-1-4419-5247-9

Published by Kluwer Academic Publishers,
P.O. Box 17, 3300 AA Dordrecht, The Netherlands.

Sold and distributed in North, Central and South America
by Kluwer Academic Publishers,
101 Philip Drive, Norwell, MA 02061, U.S.A.

In all other countries, sold and distributed
by Kluwer Academic Publishers,
P.O. Box 322, 3300 AH Dordrecht, The Netherlands.

Printed on acid-free paper

TABLE OF CONTENTS

Preface

The 3rd International ISAAC Congress took place from August 20 to 25, 2001 in Berlin, Germany, supported by the German Research Foundation (DFG), the city of Berlin through Investitionsbank Berlin and the Freie Universität Berlin. 10 ISAAC Awards were presented to young researchers in analysis its applications and computation from all over the world on the basis of financial support from Siemens, Daimler Crysler, Motorola and the Berlin Mathematical Society and book gifts from Birkhäuser Verlag, Elsevier, Kluwer Academic Publisher, Springer Verlag and World Scientific.

The ISAAC is grateful to all these institutions, firms and publishers for their support. Due to the support from DFG and from Investitionsbank Berlin many of the 362 registrated participants could be financially supported. Unfortunately the financial supports were granted too late to reach more people from former SU as the procedere for visa is still more than cumbersome and embassies are not at all flexible. Hence, a big part of the financial support could not be used and had to be returned.

The 10 plenary lectures were

I. Antoniou, I. Prigogine (Intern. Solvay Inst. Phys. Chem., Brussels): Irreversibility and the probabilistic description of unstable evolutions beyond the Hilbert space framework (read by I. Antoniou),

N.S. Bakhvalov, M.E. Eglit (Math. Mech. Dept., Lomonosov State Univ. Moscow): Homogenization of some multiparametric problems (read by M.E. Eglit),

A.D. Bruno (Keldysh Inst. Appl. Math., Moscow): Power geometry as a new calculus,

M. Essén (Dept. Math., Uppsala Univ.): Some results on Q spaces,

P. Kuchment (Texas A & M Univ., College Station): Differential and pseudo-differential operators and graphs as models of mesoscopic systems,

V. Maz'ya (Dept. Math., Linköping Univ.): The Schrödinger and the relativistic Schrödinger operators on the energy space: boundedness and compactness criteria,

S. Nikols'kii (Steklov Inst. Math., RAS Moscow): Boundary problems and algebraic polynomials,

T. Nishitani (Dept. Math., Osaka Univ): Hyperbolicity for systems
S. Saitoh (Dept. Math., Gunma Univ.): Theory of reproducing kernels,
M. Singer (Dept. Math., North Carolina State Univ., Raleigh, NC): Galois theory of linear differential equations with applications to Hamiltonian mechanics,
D. Tataru (Dept. Math., Univ. Cal., Berkely): The nonlinear wave equation.

In 38 sessions from real, complex and hypercomplex analysis, from potential theory, differential geometry, differential equations and operator theory, from mathematical physics, engineering and life sciences, and from numerical analysis and computation nearly 400 scientists from altogether 41 countries participated in the congress.

The abstracts of the plenary, the main and the regular talks can be found at the webpage of the "Atlas Mathematical Conference Abstracts" under **http://at.yorku.ca/cgi-bin/amca/cahk-01**
ISAAC and the local organizers highly appreciated this service, on the basis of which the abstract booklet was easily produced. Thanks are due to Elliott Pearl from York University for his perfect arrangement of the abstracts.

The ISAAC Awards for young scientists in analysis its applications and computation were given to

Dr. Jin Chen, Professor at Fudan University in Shanghai, China: Inverse Problems,
Dr. Young-Bok Chung, Professor at the Chonnam National University in Kwangju, Korea: Complex Analysis of Several Variables,
Dr. Michael Dreher, Research Associate at the University of Tsukuba, Japan: Weakly Hyperbolic and Degenerate Parabolic Differential Equations,
Dr. Misolav Engliš, Researcher at the Academy of Sciences in Prague, Czech Republic: Quantization Methods on Manifolds and Invariant Differential Operators on Hermitean Spaces,
Dr. Vassili G. Gelfreich, Researcher at the Freie Universität Berlin, Germany: Dynamical Systems,
Dr. Maarten de Hoop, Professor at Colorado School of Mines in Golden, Colorado, USA: Wave Propagation and Scattering,
Dr. Frank Jochmann, Postdoc at the University of Leipzig, Germany: Partial Differential Equations of Mathematical Physics,
Dr. Alexey Karapetyants, Professor at the Autonomous University of Mexico D.F.: Bergman Spaces, Toeplitz Operators, and Singular Integral Operators,

Dr. Vladislav V. Kravchenko, Professor at the National Polytechnic Institute in Mexico D.F.: Quaternionic Analysis and Partial Differential Equations of Mathematical Physics,

Dr. Ya-Yuang Wang, Professor at the Jiao Tong University in Shanghai, China: Nonlinear Partial Differential Equations, Shock Waves, and Geometric Optics.

The prices were equipped with DM 800 each and with several books. Two awardees not participating on the congress did only get the certificates. These can be viewed on the ISAAC homepage at FU Berlin under

http://www.math.fu-berlin.de/rd/ag/isaac/publication.html

The award committee consisted of H. Begehr, E. Brüning, Ju. Dubinskii, M. Essén, R.P. Gilbert, J. Kajiwara, I. Laine, I. Netuka, D.A. Tarzia, M.W. Wong, C.C. Yang. The ISAAC society will continue in encouraging young researchers in analysis, its applications and computation by awarding prices for high quality research at its international congresses.

The proceedings of the congress will be published in several volumes. This present volume contains some of the plenary (1 hour) and main (45 minutes) lectures Nonlinear problems are the main target of mathematical analysis and its applications today. Survey articles report on new methods either for quite general problems as e.g. the power geometry for as well algebraic as differential equations or homogenization methods to describe inhomogeneous material or for special involved physical processes like electromagnetic wave propagation in thin (molecular) structures the theory of differential equations in graph-structures. Decomposition of the Hilbert space of an arbitrary Hamiltonian into subspaces of decaying and of non decaying states is explained. There are two main areas within the contributions : complex analysis and hyperbolic partial differential equations. In complex analysis Q-spaces and $Q^{\#}$-spaces are reported on, hyperbolically convex functions forming a non-linear space are studied, applying value distribution methods the proximity of preimages of polygons are investigated. Moreover, value distribution is generalized to algebraically closed fields of characteristic p. Necessary and sufficient conditions for the validity of Riemann's hypothesis are given and the important theory of reproducing kernels are shown to be widely applicable. Analytic functions and functionals in several complex variable are investigated on the basis of a series of new norms.

Hyperbolicity and strong hyperpolicity is surveyed on for first order systems together with their Cauchy problem. This is also in the center of research on second order strictly hyperbolic operators, on the Schrödinger equation and the wave equation in different situations even without boundary conditions. Also inverse scattering problems are considered. Other problems come from potential theory, harmonic analysis and dynamical systems.

The camera-ready manuscript of the volume in the Kluwer style was prepared by Barbara G. Wengel. She together with Daria Schymura, Astrid Begehr and Andreas Krausz were heavily involved with the congress organization. Thanks are due to them and to all the other helpers from both mathematical institutes of the FU Berlin among them in particular the dynamical systems group around Professor B. Fiedler (FU) and Dr. K.R. Schneider (WIAS).

The next international ISAAC Congress will take place at York University in Toronto from August 11 to 16, 2003.

DECAY AND COMPUTABILITY

I. Antoniou[1,2] and Z. Suchanecki[1,2,3]
[1] *International Solvay Institute for Physics and Chemistry*
CP 231, ULB Campus Plaine
Bd. du Triomphe, 1050 Brussels, Belgium
[2] *Theoretische Natuurkunde*
Free University of Brussels
[3] *Institute of Mathematics University of Opole*

Abstract   We study Hamiltonians with singular spectra of Cantor type with a constant ratio of dissection and present strict connections between the decay properties of states and the algebraic number theory as well as the computability of decaying and non decaying states.

1. Introduction

The decay and the decay rate of a Hamiltonian system depends on the nature of its spectrum. If the Hamiltonian has point spectrum then each state is non decaying. On the other hand an easy consequence of the Radon-Nikodym theorem is that if the Hamiltonian has an absolutely continuous spectrum then each state is decaying. If the Hamiltonian consists of both point and absolutely continuous spectrum then the underlying Hilbert space can be decomposed as a direct sum of two Hilbert spaces. Each of these spaces reduces the Hamiltonian and one of them consists only of decaying states while the other only of non decaying. However, the spectrum of an arbitrary Hamiltonian does not necessarily consist of these two parts only. Hamiltonian systems with singular continuous spectra may have both decaying and non decaying states.

We have shown recently (see [1, 2] and the references therein) that the Hilbert space of an arbitrary Hamiltonian H can also be decomposed on two parts, which reduce H, in such a way that one Hilbert space consists of decaying states and the other of non decaying. The division line between decaying and non decaying states goes through the singular part of the spectrum, which means that singular spectra may behave like point spectra but may also behave as absolutely continuous spectra.

1

H.G.W. Begehr et al. (eds.), Analysis and Applications - ISAAC 2001, 1–6.
© *2003 Kluwer Academic Publishers.*

This article is devoted to study one of the simplest class of singular spectrum, namely a Cantor type set with a constant ratio of dissection [3]. Such spectra in spite of their simplicity exhibit all the above mention complexity of the behavior of Hamiltonian systems. They provide constructive examples of fractal spectra and show strict connections with algebraic number theory.

We show the one-to-one correspondence between decaying states from systems with Cantor type spectra and the algebraic S-numbers. We also analyze the question of separability of decaying and non decaying states and the related question of computability of decaying and non decaying states of Hamiltonians with singular continuous spectra.

2. Decay in Hamiltonian systems with fractal spectra

By the states of a Hamiltonian system we understand the elements of a separable Hilbert space $\mathcal{H}$ on which a Hamiltonian, i.e. a selfadjoint operator H acts. We do not assume that the states are normalized. The time evolution of a state $\psi \in \mathcal{H}$ is governed by the unitary group

$$U_t = e^{-itH} , \quad t \in \mathbb{R} ,$$

on $\mathcal{H}$. In particular, in a quantum mechanical system states are wavefunctions and the evolution group is the solution of the Schroedinger equation

$$\partial_t \psi = -iH\psi , \quad \hbar = 1 .$$

Since the Hamiltonian H is a selfadjoint operator on $\mathcal{H}$ it admits the spectral resolution

$$H = \int_{\mathbb{R}} \lambda \, dE_\lambda ,$$

where $\{E_\lambda\}$ is the spectral family of H. Consequently

$$U_t = \int_{\mathbb{R}} e^{-i\lambda t} \, dE_\lambda .$$

A state $\psi \in \mathcal{H}$ is called a *decaying state* if its *survival amplitude* decays asymptotically, $t \to \infty$

$$\langle \psi, U_t \psi \rangle = \int_{\mathbb{R}} e^{-i\lambda t} d\langle \psi, E_\lambda \psi \rangle \longrightarrow 0 .$$

For any given state $\psi \in \mathcal{H}$ we denote by $F(\lambda)$ the spectral distribution function:

$$F(\lambda) = \langle \psi, E_\lambda \psi \rangle , \quad \text{for} \ \lambda \in \mathbb{R} .$$

It can be shown [4] that each state ψ can be uniquely decomposed as

$$\psi = \psi_\mathrm{p} + \psi_\mathrm{sc} + \psi_\mathrm{ac}$$

in such a way that the corresponding spectral distribution function F_p, F_sc and F_ac is discrete, singular and absolutely continuous respectively. Obviously ψ_p is non decaying and ψ_ac is decaying. It turns out that the singular continuous state ψ_sc can be uniquely decomposed further as a sum of two states such that one is non decaying and the other decaying [1, 2]. In fact the whole Hilbert space can be decomposed as a direct sum of two closed subspaces

$$\mathcal{H} = \mathcal{H}^\mathrm{D} \oplus \mathcal{H}^\mathrm{ND},$$

called decaying and no-decaying space respectively. Moreover the spaces $\mathcal{H}^\mathrm{D}$ and $\mathcal{H}^\mathrm{ND}$ reduce the Hamiltonian operator H, which means that the spectral properties of H can be studied separately and independently on these subspaces.

Singular spectra may appear in Hamiltonian systems of the form $-\frac{d^2}{dx^2} + V$, where the potential V is, for example, an almost periodic function [5]. In this article, however, we shall not study perturbed Hamiltonians. Instead, given a Hamiltonian with a specified singular spectrum, we shall study its properties. Let us, therefore, show first how to correspond a Hamiltonian to a given spectrum. Suppose that the required spectrum σ is a Borel subset of $\mathbb{R}$ and a measure μ, which we would like to regard as a spectral measure, is a Borel measure on σ. In the spectral representation, the Hamiltonian with spectral measure μ is the multiplication operator

$$Hf(\lambda) = \lambda f(\lambda) \tag{1}$$

on the Hilbert space $L^2(\sigma, \mu)$. The spectral projectors E_λ of H are:

$$E_\lambda f(\lambda') = \begin{cases} f(\lambda), & 0 \le \lambda' < \lambda \\ 0, & \text{otherwise.} \end{cases}$$

The spectral measure $\mu(d\lambda) = dF(\lambda)$ corresponds to the distribution function F associated with the cyclic vector $\psi = 1$:

$$F(\lambda) = \langle \psi, E_\lambda \psi \rangle. \tag{2}$$

We introduce now an important class of Cantor type sets which will serve as the supports of singular measures. In order to simplify the notation we restrict our considerations to the interval $[a, b]$. Let r be a real number, $0 < r < \frac{1}{2}$. In the first step divide the interval $[a, b]$ into three parts of the lengths proportional to r, $1 - 2r$ and r respectively. Then remove the middle open interval. In the second step divide each of two remaining intervals into three parts of lengths also proportional to r, $1 - 2r$ and r respectively. Then remove the middle open intervals, and so on. In this way we obtain in the k-th step a closed set σ_k consisting of 2^k disjoint intervals, each one of the length $(b - a)r^k$. Denote $\sigma = \bigcap_k \sigma_k$ and observe that σ is a closed set with points of the form

$$x = a + (b - a)(1 - r) \sum_{k=1}^{\infty} \varepsilon_k r^{k-1}, \tag{3}$$

where $\varepsilon_k = 0$ or 1. Cantor's ternary set on the interval $[0, 1]$ with points $x = \sum_{k=1}^{\infty} \frac{\varepsilon_k}{3^k}$, is obtained by putting $a = 0$, $b = 1$ and $r = \frac{1}{3}$.

Let us focus our attention on Cantor type sets on the interval $[0, 2\pi]$ with a constant ratio of dissection. Therefore each point $x \in \sigma$ has the form

$$x = 2\pi(1 - r) \sum_{k=1}^{\infty} \varepsilon_k r^{k-1}. \tag{4}$$

On the set σ define the distribution function F putting for the points x of the form (4)

$$F(x) = \sum_{k=1}^{\infty} \frac{\varepsilon_k}{2^k}.$$

We extend F on $[0, 2\pi]$ putting

$$F(x) = \sup_{\substack{y \in \sigma \\ y \leq x}} F(y).$$

The function F is non decreasing and continuous with $F'(x) = 0$ for almost all $x \in [0, 2\pi]$, therefore singular.

According to the above prescription we can define a Hamiltonian system on the Hilbert space $\mathcal{H}$ of all functions f such that $\int_{\mathbb{R}} |f(x)|^2 dF(x) < \infty$ putting as H the multiplication operator.

It is easy to see that the function (state) $\psi \equiv 1$ is a cyclic vector for H, i.e. the set of all finite linear combinations of $H^n \psi$, $n = 0, 1, 2, \ldots$, is dense in $\mathcal{H}$. Therefore, the operator H, considered as a Hamiltonian on $\mathcal{H}$, has purely singular continuous spectrum.

We would like to know whether the constructed in the previous section cyclic state $\psi = \psi(r)$ is decaying. It turns out that the decay of ψ

depends on the ratio of dissection. In Ref. [6] we have given the full answer to this question describing the algebraic properties of the ratio of dissection r that decide about the decay properties of the corresponding state ψ. We present below the algebraic characterization of decaying states associated with the Cantor type sets with a constant ratio of dissection. First, however, recall some basic facts from algebraic number theory (see, for example, [7]).

An algebraic integer is a root of an equation of the form

$$a_n x^n + a_{n-1} x^{n-1} + \ldots + a_0 = 0 , \tag{5}$$

where a_k are integer numbers and $a_n = 1$. If α is a root of the polynomial (5) which is irreducible, i.e. there is no polynomial of degree $m < n$ with the integer coefficients and the leading coefficient having α as a root, then the other roots of (5) are called the conjugates of α. An algebraic integer $\alpha > 1$ such that each its conjugate α', $\alpha' \neq \alpha$, satisfies $|\alpha'| < 1$ is called an S-number.

We have [6]

Proposition 1 *Let* $\psi(r)$ *be the cyclic state of the Hamiltonian H with the spectrum of Cantor type with a constant ratio of dissection* r. *The state* $\psi(r)$ *is non decaying if and only if* $1/r$ *is an S-number. Correspondingly,* $\psi(r)$ *is decaying if and only if* $1/r$ *is not an S-number.*

The S-numbers include all integers $n > 1$. It is also easy to verify that any number of the form $\frac{1}{2}(p + \sqrt{p^2 + 4q})$, where $p, q \in \mathbb{N}$, is an S-number (its only conjugate is $\frac{1}{2}(p - \sqrt{p^2 + 4q}) < 1$). For example the golden number $\frac{\sqrt{5}+1}{2}$ is an S-number. On the other hand, none of the irreducible rationals $\frac{p}{q}$ with $p, q \in \mathbb{N}\backslash\{1\}$ is an S-number. In fact such $\frac{p}{q}$ is not even an algebraic integer [7]. Therefore if the ratio of dissection is any irreducible rational number $k/n < 1/2$, where k and n are integers different from 1, then the corresponding cyclic state is decaying.

A natural question is whether decaying and non decaying states can be in some sense separated. It is surprising that the answer to this question is negative. This follows from the fact that since the irreducible rationals are dense in $\mathbb{R}$, they are also arbitrary close to S-numbers. In other words, for any dissection rate r determining a non decaying state $\psi(r)$ and any $\varepsilon > 0$ one can find a dissection rate r' with $|r - r'| < \varepsilon$, such that the state $\psi(r')$ is non decaying.

It follows from the above considerations that it is impossible to isolate non decaying states associated with Cantor type sets. On the other hand the decaying states can be separated from non decaying. In fact we have [6]

6

Proposition 2 *For each ratio of dissection* r, *which determines a decaying state* $\psi(r)$, *there is* $\delta > 0$ *such that the nearest ratio of dissection* r', *which determines a non decaying state* $\psi(r')$ *is at the distant larger than* δ.

Concluding remark

Proposition 1 shows how inappropriate the computational modeling of decaying and non decaying states can be. In the case of Hamiltonians with fractal spectra the construction of decaying states amounts to the construction of a Cantor type set with a given ratio of dissection. According to Proposition 2 it is possible to construct such decaying states $\psi(r)$ for which the distance δ of r from the nearest inverse of an S-number is within the computing accuracy. However any construction of a non decaying state $\psi(r)$ for which r has an infinite dyadic expansion is completely unreliable. The reason is that we can not perform computations on numbers with infinite dyadic expansion. Therefore any truncation of the dissection rate give us, in general, a decaying state instead. Physically speaking any finite approximation of such non decaying state is a decaying state. Only in the infinite limit we obtain non decay. Moreover the possibility of construction of decaying states is also rather theoretical because very little is known about the localization of S-numbers.

References

[1] Antoniou, I., Suchanecki, Z.: Spectral characterization of decay in quantum mechanics. Trends in Quantum Mechanics, eds. H.-D. Doebner, S.T. Ali, M. Keyl and R.F. Werner, 158–166, World Scientific, Singapore 2000.

[2] Antoniou, I., Shkarin, S.A.: Decay spectrum and decay subspace of normal operators. J. Edinburg Math. Soc. (in press).

[3] Salem, R.: Algebraic numbers and Fourier analysis. Reprint. Orig. publ. 1963 by Heath, Boston. The Wadsworth Mathematics Series, Belmont, California 1983.

[4] Weidmann, H.: Linear Operators in Hilbert space. Springer Verlag 1980.

[5] Bellisard, J.: Schrödinger operators with almost periodic potential: an overview. Lect. Notes on Phys. 153 (1982), 356–363.

[6] Antoniou, I., Suchanecki, Z.: Quantum systems with fractal spectra. Chaos Solitons and Fractals – Special Volume (submitted).

[7] Stewart, I.N., Tall, D.O.: Algebraic Number Theory. Chapman and Hall, London 1979.

DIFFERENTIAL AND PSEUDO-DIFFERENTIAL OPERATORS ON GRAPHS AS MODELS OF MESOSCOPIC SYSTEMS

Peter Kuchment

Texas A & M University

College Station, TX, USA 77843

kuchment@math.tamu.edu

Abstract   The lecture contains a brief survey on graph models for wave propagation in thin structures.

Keywords: Spectrum, mesoscopic physics, quantum wire, pgotonic crystal.

1.   Introduction

This lecture[1] intends to present a brief survey of a field that has been emerging in the last couple of decades. It is a short version of an article that will be published elsewhere.

What do we mean by a mesoscopic system? Those who are familiar with this notion, will notice that we will stretch it here to use in the situations where normally the system is not considered to be mesoscopic. In the narrow meaning *mesoscopic systems* (see e.g. [21, 39, 57, 77]) are those that have some dimensions too small to be treated using purely classical physics, while too large to be conveniently considered on the quantum level only. More precisely, these are physical systems whose one, two, or all three dimensions are reduced to a few nanometers. They hence look as surfaces, wires, or dots and are called correspondingly *quantum walls*, *quantum wires*, and *quantum dots*. While quantum dots are probably most familiar to general scientific public, in this lecture we will be concerned with circuits of quantum wires only. We will also

---

[1]This paper was presented as a plenary talk at the 3rd Congress of ISAAC in Berlin in August 2001.

H.G.W. Begehr et al. (eds.), Analysis and Applications - ISAAC 2001, 7–30.
© 2003 *Kluwer Academic Publishers.*

include into our considerations the cases of thin graph like acoustic or optical structures. In this case the word mesoscopic will be abused, since the characteristic dimensions of such systems will be normally much larger than nanometers. So, we will use the word mesoscopic as an equivalent of *thin graph-like* systems. One can expect that transport of electronic, electromagnetic, or acoustic waves in such media could be studied using some approximate models on graphs (when the "thin" dimensions are ignored). This is exactly the direction that we choose in this lecture. It is practically impossible to provide a brief survey of all existing studies of the kind described above, so according to the author's own interests, we will concentrate on spectral problems only. In particular, we will be interested in existence and location of spectral gaps.

Our main plan is to present first some motivations for studying such systems, then to indicate what is known about the validity of the graph models, and finally to give some examples (which will be far from exhausting all known cases) of effects one can observe using such models.

2. Main mathematical objects

2.1 Metric graphs

Let us first define the graphs that we will deal with. We will consider graphs as metric rather than purely combinatorial objects. In other words, a graph will be a one-dimensional variety with singularities at its vertices. More precisely, a graph Γ with finite valences of all its vertices v_i will be assumed to be embedded into $\mathbb{R}^2$ (or $\mathbb{R}^3$) in such a way that all edges e_j are sufficiently smooth (usually C^2 suffices) finite length curves with transversal intersections at vertices. We also assume that every compact domain contains only a finite number of edges and vertices. Our graphs of interest will be assumed to be finite or periodic with respect to a lattice in $\mathbb{R}^2$ (albeit higher dimensional analogs also exist).

We equip each edge e_j with the arc length coordinate x_j that will usually be denoted by x, which should not lead to any confusion. The functions $f(x)$ of interest on Γ are defined along the edges, rather than at the vertices only like in discrete models. One can for instance define in a natural way the space $L_2(\Gamma)$ of square integrable measurable functions on Γ. Our spectral problems will be introduced for some operators in this space.

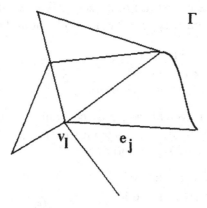

Γ

Figure 1.   Graph Γ.

2.2   Operators

The operators of interest in the simplest cases are the second arc length derivative

$$f(x) \to -\frac{d^2 f}{dx_j^2},$$ (1)

a more general Schrödinger operator

$$f(x) \to -\frac{d^2 f}{dx_j^2} + V(x)f(x),$$

or a magnetic Schrödinger operator

$$f(x) \to \left(i\frac{d}{dx_j} + a(x)\right)^2 + V(x)f(x).$$

The real potentials a and V will be assumed smooth enough and in the case of infinite periodic graphs also periodic. Precise conditions can be found in the cited literature. Certainly, the operators are not defined until one describes their domains. Although it is rather clear that the natural conditions should require that f belongs to the Sobolev space H^2 on each edge e_j, one also needs to impose boundary value conditions at the vertices. It is possible to find all such boundary value conditions that make these operators self-adjoint (see [24] and references therein and [45, 46]). One standard type of "Kirchhoff" boundary conditions is

$$
\begin{cases}
f(x) \text{ is continuous on } \Gamma \\
\text{at each vertex } v_l \text{ one has } \sum_{\{j \mid v_l \in e_j\}} \frac{df}{dx_j}(v_l) = \alpha_l f(v_l) \ ,
\end{cases}
\tag{2}
$$

where the sum is taken over all edges e_j containing the vertex v_j, and derivatives are taken in the directions from the vertex. Here α_l are some fixed real numbers. The most common case is

$$
\begin{cases}
f(x) \text{ is continuous on } \Gamma \\
\text{at each vertex } v_l \text{ one has } \sum_{\{j \mid v_l \in e_j\}} \frac{df}{dx_j}(v_l) = 0 \ .
\end{cases}
\tag{3}
$$

There are many other conditions, among which are those where the roles of the function and the derivatives in (2) are switched. One of the questions to consider will be which one of these conditions (if any) arise in the asymptotic limit of a problem of interest in a thin domain. We will also have to face more general operators, including those of higher order, the ones with more general boundary conditions, and also "pseudo-differential" operators (whatever such an operator on a graph could mean).

2.3   Spectral problem

As it has already been indicated, for any operator H of the type described above we will be interested in its spectrum. In particular, in the case of an infinite graph presence of spectral gaps is importanct. Existence of such a gap means that waves of certain frequencies (electrons of certain energies) cannot propagate through the system. This is known to be one of the major considerations in the solid state theory [2] and in the theory of photonic crystals [43, 49, 70]. The reason for studying these spectral problems on graphs is that as it is explained below, they arise as natural asymptotic models for finding spectra of much more complex problems in thin ("mesoscopic") domains surrounding the graph.

3.   The origins of the problem

We will describe here some of the manifold reasons for studying graph models. They naturally arise in chemistry, physics, and mathematics. We will address at some length only a few of them, leaving it to the audience to look up the rest in the literature. Thin circuit type structures for which graph models are natural candidates for simplified (due to reduced dimensionality) considerations have arisen in the past few decades in many areas of science and engineering, among which one can mention in particular the free-electron theory of conjugated molecules, quantum chaos, quantum wires, atomic and molecular wires, and nanowires (one can refer for instance to [1, 3, 5, 7, 20, 21, 24, 35, 36, 37, 39, 42, 47, 55, 57, 66, 67, 68, 73, 77] for descriptions of these fields and related mathematical considerations).

Let us address briefly some of such examples.

3.1 A free-electron theory of conjugated molecules

We will follow the formulation of the graph model given in [68].

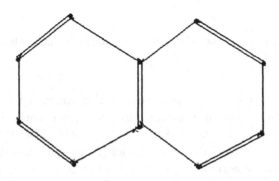

Figure 2. Naphtalene molecule.

Some organic molecules like the one of naphthalene contain systems of conjugated bonds, i.e., a system of alternating single and double bonds (Fig. 2). Every atom contributes three electrons for chemical binding. In the first approximation one thinks that two of those (so called σ-electrons) form σ bonds that maintain the "skeleton" of the molecule, i.e., the graph obtained by eliminating the doubling of bonds (Fig. 3).

This "skeleton" creates a potential which influences all remaining so called π-electrons (one per each atom). So, the π-electrons can be thought of as confined to the "skeleton" graph by the potential. It was suggested in [68] that one can obtain a simplified approximate model for π-electrons using a second order (ordinary) differential Hamiltonian on the skeleton Γ. In order to obtain such a model the authors of

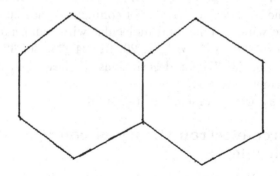

Figure 3. The "skeleton "of the naphtalen molecule.

[68] assumed first that the single particle Hamiltonian for a π-electron is the Laplace operator in a narrow tube around Γ with zero Dirichlet condition on its boundary. Then the width of the tube was allowed to tend to zero. In this case the ground energy increases to infinity, so one needs to shift the spectrum back to zero. The authors of [68] provided heuristic arguments for the claim that after this shift the asymptotic limit of the spectrum is given by the spectrum of the second arch length derivative on Γ (1) with the boundary conditions (3). Then this much simpler model was used for studying the orbitals of π-electrons.

We will argue later on in this talk that this claim about the limiting behavior of the spectrum of the Dirichlet Laplacian in a narrow tube around Γ seems to be incorrect.

3.2 Circuits of quantum wires

As it has been mentioned already, quantum wires are quasi-one-dimensional objects whose other two dimensions are reduced to a few nanometers. So one can envision a graph Γ with a "fattened graph" domain Ω_d around Γ of thickness $d \ll 1$ (see the figure below).

Electron propagation in Ω_d is assumed to be governed by the Laplace operator $-\Delta u = -\sum \frac{\partial^2 u}{dx_j^2}$ with either Dirichlet, or Neumann conditions on $\partial\Omega_d$ (i.e., either the function, or its normal derivative vanish at the boundary). The Neumann case arises in the case of thin superconducting structures ([1], [64] – [67]).

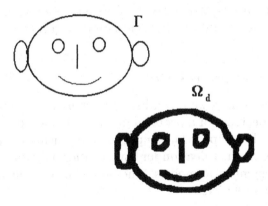

Figure 4. A 'fattened' graph.

So, now the problem is: How do the spectra of the Neumann $-\Delta_N$ and Dirichlet $-\Delta_D$ Laplacians in Ω_d behave when $d \to 0$ (i.e. in the thin domain limit)? Do they converge to the spectrum of a differential operator on the graph?

3.3 Photonic crystals

Here the domain Ω_d of the previous section is assumed to be filled with an optically dense dielectric, while the rest is filled with air. One wonders whether the propagation of electromagnetic waves governed when $d \to 0$ by an operator on Γ? It is rather clear that in general this is not the case, since the waves do penetrate the air, but for some "dielectric" modes a pseudodifferential operator on Γ does arise (see details below).

3.4 Other applications

There are quite a few other cases when one wants to use a graph model. First of all, one can think of acoustic and electromagnetic waveguides (see e.g., [14, 15, 56]). Another option is to use graph models as test grounds for studying the features that depend upon or are influenced by multiple connectedness of the material, for instance Aharonov-Bohm effect [7], quantum chaos [47], and scattering [5, 34, 45, 46, 55, 59]. Yet another source of such models is averaging in dynamical systems, when one has slow motion in graph directions and fast one across the graph. Then averaging naturally leads to the models of the kind described above. One can find a very interesting discussion and results on such situations in [31, 32]. There is a variety of other mathematical topics that also lead to differential operators on graphs (e.g., [16]-[19], [22, 23, 58, 69]).

4.   Convergence results

By convergence results we mean here statements that guarantee convergence of the spectrum of a problem in a thin domain to the spectrum of a problem on the graph. The importance of such theorems lies not in providing rigorous justification for asymptotics only, but also in finding the correct asymptotic models, since the choice of such a model is sometimes far from obvious.

4.1   Neumann Laplacian

The case of the Neumann Laplacian is probably the simplest one. We assume that a finite graph Γ is C^2 embedded into $\mathbb{R}^2$ (multidimensional generalizations also hold). The "fattened graph" domain Ω_d consists of narrow tubes along the edges joined by some neighborhoods of vertices. The tubes have width $d \times p(x)$, where $p(x) > 0$ is a C^1 function on the edge and x is the arc length coordinate. Each vertex neighborhood is contained in a ball of radius of order $\sim d$ and starshaped with respect to a smaller ball of a radius of the same order of smallness.

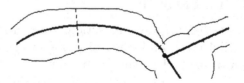

Figure 5.   Local structure of Ω_d.

Consider the Schrödinger operator Ω_d

$$H_d(\mathbf{A}, q) = (\frac{1}{i}\nabla + \mathbf{A}(x))^2 + q(x),$$

where the scalar electric $q(x)$ and vector magnetic $\mathbf{A}(x)$ potentials are defined in a fixed neighborhood of Γ, q is of the Lipschitz class, and $\mathbf{A}$ belongs to C^1. We impose Neumann conditions on $\partial\Omega_d$.

We also consider the following operator $H(\mathbf{A}, q)$ on Γ (exact definition that we skip requires quadratic forms):

$$H(\mathbf{A}, q)f(x_j) = -\frac{1}{p}(\frac{d}{dx_j} + iA_j^\tau(x))p(\frac{d}{dx_j} + iA_j^\tau(x))f + qf,$$

where we use $q(x)$ to denote the restriction of the potential q to Γ and A_j^T is the tangential component of the field $\mathbf{A}$ to the edge e_j of Γ. In order to complete the description of the operator we need to impose some boundary conditions at vertices. These are:

1. f is continuous through each vertex.
2. at each vertex v

$$\sum_{\{j|v\in e_j\}} p_j(v) \frac{df_j}{dx_j}(v) = 0.$$

Here p_j denotes the function that provides the width of the tube along e_j (see the description of the domain above). So, the values $p_j(v)$ at the same vertex can be different for different edges e_j adjacent to v.

The next theorem summarizes some of the results of [53, 54], [64]-[67], [71]:

Theorem 1 *For any $n = 1, 2, \ldots$*

$$\lim_{d\to 0} \lambda_n(H_d(\mathbf{A}, q)) = \lambda_n(H(\mathbf{A}, q)),$$

where λ_n is the n-th eigenvalue counted in increasing order taking into account multiplicities.

This result shows that the asymptotic behavior of the spectrum of $H_d(\mathbf{A}, q)$ when $d \to 0$ is dictated by the spectrum of the graph operator $H(\mathbf{A}, q)$. Convergence of solutions of the corresponding heat equations in absence of potentials was shown in [31, 32].

4.1.1   Large protrusions at the vertices.   We address here
the case when the vertex neighborhoods are of radii decaying slower than the width d of the tubes. One can expect that for sufficiently large protrusions at vertices the coupling of different edges might decay and an additional "life at vertices" can arise. In fact, the result (and the proof provided in [53]) of Theorem 1 still holds while the exterior and interior sizes of the vertex neighborhoods decay as d^α, where $\alpha \in (0.5, 1]$. The situation changes however, when $\alpha \leq 0.5$. In this case the limit operator does not act in the space $L_2(\Gamma)$ (or any space of functions on Γ) anymore, but rather in some its finite-dimensional extension that corresponds to vertex states. Let us formulate the corresponding results of [54]. We assume for simplicity that the potentials are equal to zero, the edges are straight, the tubes have thickness $2d$, and the vertex neighborhoods are balls of radii d^α.

Assume that the total number of vertices is m. We will use the notation H_d for the negative Neumann Laplacian $-\Delta_{N,d}$ in Ω_d. Consider

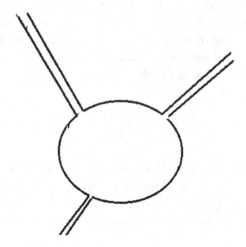

Figure 6.   A large protrusion at a vertex.

the space $\mathcal{H} = L_2(\Gamma) \oplus \mathbb{C}^m$ and the operator H on this space that can be described as follows: it acts on $L_2(\Gamma) \oplus \{0\}$ as the negative second derivative along each edge with zero Dirichlet conditions at each vertex, and the whole space $\{0\} \oplus \mathbb{C}^m$ is in the kernel of H (precise definition needs to be a little bit more accurate).

Theorem 2 ([54, 79]) *Let $0 < \alpha < 0.5$, then when $d \to 0$, the spectrum of the operator H_d converges to the spectrum of H.*

This theorem shows in particular that in the limit the edges completely decouple, while some additional vertex states arise.

In order to tackle the borderline case $\alpha = 0.5$ one needs to introduce a little bit different operator in the same extended space $\mathcal{H} = L_2(\Gamma) \oplus \mathbb{C}^m$. We will avoid giving the precise description of the quadratic form or the operator, since the corresponding spectral problem can be rewritten onto Γ itself, which results in a problem that involves the spectral parameter in the boundary conditions (see the theorem below). This is not anything unusual, for instance in transmission type spectral problems that involve two domains with a common boundary, the effect of one of the domains can be represented as an energy-dependent boundary condition imposed in the second domain.

Theorem 3 ([54, 79]) *Let $\alpha = 0.5$, then when $d \to 0$, the spectrum of the operator H_d converges to the spectrum of the following problem on*

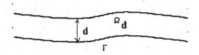

Figure 7. A trip of width d.

the graph:

$$\begin{cases} -\frac{d^2 f}{dx_j^2} & = \lambda f & \text{on each edge } e_j \\ f & \text{is continuous} & \text{through each vertex } v_l \\ \sum_{\{j|\, v_l \in e_j\}} \frac{df}{dx_j}(v_l) & = \frac{\lambda \pi}{2} f(v_l) & \text{at each vertex } v_l \end{cases}.$$

4.2   Dirichlet Laplacian

The more difficult problem of convergence of spectra of Dirichlet Laplacians arises in mesoscopic physics [24]. Consider the domain Ω_d which is the parallel strip of width d along a smooth curve Γ (see the picture below).

One is interested in the behavior of the spectrum of the (negative) Dirichlet Laplacian $-\Delta_{D,d}$ in Ω_d when $d \to 0$. It is easy to see (for instance, on the example of a linear strip) that due to the transversal modes, the spectrum of $-\Delta_{D,d}$ blows up (i.e., its bottom goes to infinity) when $d \to 0$. It is natural hence to shift the spectrum down by the first transversal eigenvalue $\lambda_1 = \left(\frac{\pi}{d}\right)^2$. Changing the coordinates appropriately in order to flatten the strip, one can derive then the following result:

Theorem 4 (*see* [24] *and references therein*) *The spectrum of* $-\Delta_{D,d} - \lambda_1$ *converges to the spectrum of the operator*

$$-\frac{d^2}{dx^2} - \frac{\gamma(x)^2}{4}, \tag{4}$$

where x *is the arc length coordinate on* Γ *and* $\gamma(x)$ *is the curvature of* Γ.

This result, however, does not translate to the case of a graph Γ easily. Indeed, let us assume for simplicity that the edges are straight. Then one expects to obtain the second arc length derivative along the edges. The question remains whether one can impose some boundary conditions at vertices such that the spectrum of the resulting operator on Γ provides the asymptotics of the (appropriately shifted) spectrum of the Dirichlet Laplacian in a thin neighborhood. As the reader might remember, we quoted before chemistry studies [68] where it was suggested that the right boundary conditions are (3). This conclusion, however, does not seem correct. The first indication that one can expect trouble comes from the formula (4). Indeed. trying to force a smooth curve Γ to turn sharply in order to approximate an angle formed by two rays, one observes that the square of curvature term would lead to the "square of the delta function" at the corner. Here is a more convincing argument: the results of [4] show that if Γ is an angle with straight sides, then for small d the Dirichlet Laplacian has a bound ground state, whose energy and simultaneously its distance from the rest of the spectrum grow to infinity when d tends to zero. This immediately shows that no choice of boundary conditions at vertices can ever make convergence results analogous to Theorems 1 or 4 possible. A close look at heuristic arguments of [68] shows that they implicitly assumed the absence of states confined to vertices, which is exactly what does occur. One can imagine, however, that the limit operator could live on an extension of the space $L_2(\Gamma)$ rather than on Γ itself, the additional components being responsible for the vertex states. In this case, though, the convergence must have different meaning for different components of the extension. This program has not been implemented.

4.3   Photonic crystals

We will survey here some results on the so called photonic crystals that are of the nature addressed in this lecture. A photonic crystal first suggested in [44, 78] (see also [43, 49, 62, 70] for surveys of this topic) is a periodic dielectric medium, whose properties with regard to light should resemble properties of semi-conductors with respect to electron propagation. The governing equation is the Maxwell system in a periodic medium that serves here as an analog of the Schrödinger operator with periodic potential in the solid state theory. One of the issues in particular interest is the structure of the spectrum of the stationary Maxwell operator in a periodic medium, in particular existence of spectral gaps (which mean existence of frequency regions in which electromagnetic waves are not allowed to propagate in the medium). We will concentrate here on the case of $2D$ photonic crystals, i.e. the ones that are periodic with respect to two variables and homogeneous with respect to the third one. The figure below shows an example of the cross-section of such a medium. Here the dark areas are assumed to be filled with an optically

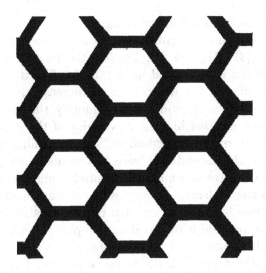

Figure 8. The crossection of a $2D$ photonic crystal.

dense dielectric, while the rest is filled with the air (or another dielectric of low optical density).

The dielectric constant is assumed to be $\varepsilon(x) = \varepsilon_0 > 1$ in the dark domains of thickness d and $\varepsilon = 1$ (air) in the white ones. The material is assumed to have no magnetic properties, so the magnetic permeability μ equals to 1. We will be interested in this paper in the thin high contrast structures, i.e. those where d is small and ε_0 is large. Neither of these two conditions can be easily satisfied in practice with optical meterials, but acoustic analogs of photonic crystals, which enjoy much the same properties, allow for very high contrast materials. Besides, it has been discovered that in some instances the thin high contrast approximation gives interesting hints to the properties of more realistic media.

It is well known [41, 43] that in the $2D$ case there are two polarizations of the electromagnetic waves: the one where the magnetic field is orthogonal to the plane of periodicity (i.e., the plane of the picture above), and the one where the electric field is directed this way. For the monochromatic waves of frequency ω the Maxwell equation for these two polarizations boils down correspondingly to the following spectral problems:

$$-\nabla \cdot \frac{1}{\varepsilon}\nabla H = \lambda H \tag{5}$$

and

$$-\Delta E = \lambda \varepsilon E, \tag{6}$$

where $\lambda = (\omega/c)^2$ and c is the speed of light.

High contrast and thin structure asymptotics: $d \to 0$, $\varepsilon_0 d \to \infty$ were considered in [8], [31]-[31], [49, 51, 52, 74] (the more realistic cases of the finite limit of $\varepsilon_0 d$ were treated in [31, 32, 51, 9]). It was discovered that for the H-mode (5) the waves become increasingly air waves, i.e. concentrate in the air only, and correspondingly the spectrum of (5) asymptotically concentrates in a small vicinity of the spectrum of the Neumann Laplacian on one "air bubble," thus opening large gaps in between. However, the E-mode (6) leads to two distinct types of waves: the air waves that behave in a manner essentially similar to the one just described, while the dielectric waves prefer to stay inside the narrow dielectric tubes (due to the total internal reflection) and are evanescent in the air. These provide a much more complicated spectrum with a very narrow bands separated by narrow gaps of approximately the same size. One can suspect that the dielectric waves could be governed by an operator living on the graph Γ obtained when the dielectric tubes shrink. This happens to be true.

Theorem 5 [31] *After appropriate rescaling (one needs to zoom in to make the small bands and gaps observable), the spectrum of dielectric modes converges to the spectrum of the problem*

$$-\Delta u = \lambda \delta_\Gamma u, \tag{7}$$

where δ_Γ is the Dirac's delta-function of the graph Γ (i.e. $< \delta_\Gamma, \phi >= \int_\Gamma \phi dx$).

Although the problem (7) seems to involve the whole plane, its spectrum in fact can also be described as the spectrum of the Dirichlet-to-Neumann operator Λ_Γ on the graph Γ. Let us briefly remark how this operator is constructed. Starting with a function $\phi(x)$ on Γ one uses it as the Dirichlet boundary value to find a harmonic function u on each face of the planar graph Γ (see Fig. 9).

When such a function u is found, it is automatically continuous through Γ, while its normal derivatives do not have any reason to be continuous. Now one takes the jump of the normal derivatives of u to get a function $\psi(x)$ on Γ. The "jump" here means the sum of outward normal derivatives of u from the two faces adjacent to a given edge. Now the Dirichlet-to-Neumann operator Λ_Γ is the operator transforming ϕ into ψ:

$$\Lambda_\Gamma \phi = \psi. \tag{8}$$

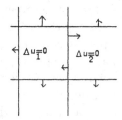

Figure 9. Dirichlet – to – Neuman map.

In the case when Γ is smooth (i.e., when no vertices are present), Λ_Γ coincides up to a smoothing operator with $2\sqrt{-\Delta_\Gamma}$, where $-\Delta_\Gamma$ is the Laplace-Beltrami operator on Γ (i.e., just the second arc length derivative). In particular, the spectra $\sigma(\Lambda_\Gamma)$ and $\sigma(2\sqrt{-\Delta_\Gamma})$ asymptotically coincide for high energies. This also holds in higher dimensions [51]. Numerical calculations show that in sufficiently smooth cases the spectra converge very rapidly (see [51] for this and other related discussions). Since the spectrum of $2\sqrt{-\Delta_\Gamma}$ can be immediately calculated, this gives a fast method of approximate calculations of the spectrum of $\sigma(\Lambda_\Gamma)$ and hence of dielectric modes.

The question arises: is there any analog of this approximation in the more realistic case of presence of vertices, i.e. of non-smooth graph Γ? Analogously to the smooth case, one can think of Λ_Γ as of a "pseudo-differential" operator of first order on Γ. In other words, does there exist a differential spectral problem A of order $2m$ on the graph such that the $2m$th power of Λ_Γ is in some sense close to A, and hence $\sigma(\Lambda_\Gamma) \approx \sigma(\sqrt[2m]{A})$? There are not many indications why this should be true, besides that it would be helpful to have such a relation. This study was attempted in [52] with some heuristic analysis. It was discovered that in the case of symmetric junctions at vertices one can sometimes write some reasonable differential operators as candidates for A. Although no theorem about convergence of spectra was proven, the numerical experiments conducted show amazing agreement of spectra. Take for example the case of symmetric triple junctions at the vertices (e.g., honeycomb lattice). Then analysis of [52] shows that the spectrum of the following problem is a good candidate for the approximation to the spectrum of Λ_Γ:

$$\begin{cases} -\dfrac{d^2 u_j}{dx_j^2} = \lambda^2 u & \text{on each edge } e_j \\ u \text{ is continuous} & \text{at each vertex } v \\ \displaystyle\sum_{j \in J(v)} \dfrac{du}{dx_j}(v) = -\left(\dfrac{3\lambda}{2}\right) \cot \dfrac{\pi}{3} u(v) & \text{at each vertex } v \end{cases} \tag{9}$$

Numerics show amazing agreement between the spectra of the two problems for the case of honeycomb lattice. This is especially interesting since the nature of the two spectra can be significantly different: the spectrum of the problem (9) has a pure point part of infinite multiplicity [51, 52], while the spectrum of Λ_Γ was conjectured [49, 51], and then proven [12] to be absolutely continuous.

Another observation of [52] was that in the case of a symmetric quadruple junction at each vertex (square lattice) one needs to employ a fourth order differential problem on the graph Γ, which is also responsible for some observed differences between the spectral behavior of the square and honeycomb Dirichlet-to-Neuman operators [51].

It is not clear at the moment how to make the analysis of [52] rigorous and whether there exists any its analog for asymmetric vertex junctions.

5.   Some applications

The graph models described above have been used in many different ways in order to understand the properties of systems that they approximate. Without trying to survey these applications, we will just provide a few nuggets of such nature.

5.1   Curved wires can bind electrons

Looking at (4) one understands that if the quantum wire Γ is essentially flat except of a sharp bent, the operator (4) is the one-dimensional Schrödinger operator with a deep potential well potential $-\gamma^2/4$. This means that some bound sates will arise. One can show (see [24] and references therein) that a bound state survives also for small non-zero values of the width d of the wire. In other words, **bent quantum wires can bind electrons**. Analogously, **bent quantum walls can create currents confined to their edges** (see the figure below).

Figure 10.   Electring current along a bent quantum wall.

5.2   Photonic band gaps

The first analytic proof of possibility of opening spectral gaps in photonic crystallic materials was obtained in [27, 28] using asymptotic analysis of the type described in Section 4.3 above. See also [33, 38] for similar considerations.

5.3   Opening spectral gaps by "decorating" the graph

Existence of spectral gaps is a favorite question in many areas, most prominently in solid state physics and photonic crystals theory (e.g., [2, 43, 49, 70]). The most common situation when the gaps might arise is of a periodic medium, since according to the Floquet theory (e.g., [48, 49, 63]) the spectrum of an elliptic periodic operator has a natural band-gap structure. There have been indications of a different mechanism that leads to spectral gaps, namely, proliferation of small scatterers with an internal structure in the medium. This has been noticed for systems modeled in $\mathbb{R}^n$ [60, 61] as well as on graphs [6]. However, the simplest and clearest model was probably delivered in the recent paper [72]. It deals with a discrete problem, i.e. graphs are considered as combinatorial objects, functions on graphs take their values at vertices, and operators of interest are discrete Laplacians or some their generalizations. Let us have a graph Γ and Laplacian on it. Consider an auxiliary graph Σ with a distinguished vertex v. Let us now attach a copy of Σ to each vertex of Γ identifying the vertex with the distinguished vertex v of Σ. In other words, we "decorate" the graph Γ with copies of Σ - "flowers" grown out of each and every vertex of Γ. In Fig. 11 below, the underlying graph Γ is drawn in the plane and the vertically attached pieces are copies of Σ.

The main result of [72] is the following

Theorem 6 *Decoration creates gaps in the spectrum of the discrete Laplacian at locations that depend on the decoration Σ only.*

One can find precise details and extensions of this result in [72]. We formulated it for the simplest case of the Laplace operators, while it holds for a wide class of Hamiltonians [72].

At the first glance the techniques of [72] do not apply to the case of differential (rather than difference) operators on graphs considered in this lecture. It is possible, however, to show that the result still holds:

Theorem 7 [50] *An analog of Theorem 6 holds for differential operators on graphs considered above.*

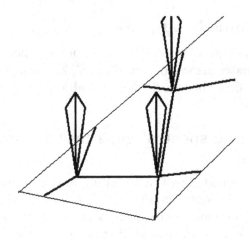

Figure 11. Graph decoration.

It is interesting to mention that there are instances when without analysis similar to the one done in [72] the effect of decoration was effectively used in physics and engineering. We will quote here just one example of this kind, namely the ground plane for cellular phone antennas developed in the UCLA photonic crystals group headed by E. Yablonovitch [75]. Fig. 12 shows this ground plane, which is essentially a metallic plate with little metallic "mushrooms" grown on it.

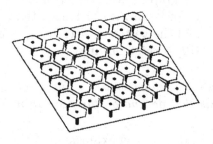

Figure 12. UCLA ground plane.

The main feature is that at certain frequencies of electromagnetic waves the plate turns into an insulator, i.e. a frequency gap arises. It seems that this effect is in direct relation with the decoration mechanism described in this section.

5.4   Confinement of waves in purely periodic media

There is a general understanding that probably came first from solid state physics (e.g., [2]) that one cannot confine a wave in a periodic medium, i.e. that the corresponding periodic operators of mathematical physics have no pure point spectrum. In physics terms this means absence of bound states. To be precise, one is talking here about elliptic operators with periodic coefficients, with one of the main examples being the Schrödinger operator with a periodic potential $-\Delta + V(x)$. Other periodic operators of interest include for instance Maxwell operator or divergence type second order operators (it is known that bound states do occur for operators of higher orders [48]). Starting with [76], the problem of absence of bound states in periodic media has been attracting intense interest of researchers, with major advancements occurring in the last several years (see for instance [11] for a survey and references).

One can also consider from this point of view differential and pseudo-differential operators of the type discussed above on periodic graphs and ask the same question: can a pure point spectrum (i.e., a bound state) arise? Let us look first at second order differential problems on graphs with conditions (2). It is easy to observe that there are resonant situations when one can have compactly supported eigenfunctions that look like sinusoidal waves running around a cycle in the graph [51]. This happens when natural commensurability conditions on the lengths of edges in the cycle is satisfied. What about the Dirichlet-to-Neumann operator (8) on a periodic graph that arises in the study of photonic crystals? Numerical analysis for certain geometries (for instance, for the honeycomb structure) suggests existence of bound states. Some analysis of those was done in [51]. It was, however, conjectured in [49, 51] that these are not actual bound states, but just strong resonances, and that the spectrum of the Dirichlet-to-Neumann operator on a periodic graph is in fact absolutely continuous. This conjecture was proven in [12]. Still, existence of strong resonances could possibly be used in applications. We remind the reader that bound states are usually created by introducing impurities into an otherwise purely periodic medium. Their existence is crucial for applications like enhancement of spontaneous emission and lasing. One can think that the "leaky" states that are strong resonances (and hence leak slowly) could be used for similar purposes. This is exactly what was done in experimental studies [13] for spontaneous emission enhancement and in [40] for lasing. In both cases photonic crystals with no impurities were used and the leaky modes were employed in the ways impurity states usually are. It needs to be noted that this was done with no relation to the mathematical analysis of the problem indicated in this section.

5.5 Opening spectral gaps for long waves

The general rule of thumb (which is also the basis of homogenization theory, see [10]) is that long waves in a medium periodic with a small period "do not notice" the periodic variations and essentially behave as in a "homogenized" medium. In particular, no spectral gaps should open in the long wave (i.e., low frequency) region. Analysis done in [27, 28] of high contrast thin photonic crystal structures shows that one can open spectral gaps for arbitrarily long waves without increasing the characteristic sizes of the structure. In fact, similar observation was used in creating the UCLA ground plane mentioned above [75].

6. Acknowledgment

This research was partly sponsored by the NSF through the Grant DMS 9610444 and by the Department of Army, Army Research Office, and state of Kansas through a DEPSCoR Grant. The author expresses his gratitude to NSF, ARO and the state of Kansas for this support. The content of this paper does not necessarily reflect the position or the policy of the federal government, and no official endorsement should be inferred.

References

[1] Alexander, S.: Superconductivity of networks. A percolation approach to the effects of disorder. Phys. Rev. B, 27 (1983), 1541-1557.

[2] Ashcroft, N.W., Mermin, N.D.: (1976) Solid State Physics. New York-London: Holt, Rinehart and Winston.

[3] Aviram, A., Ratner, M. (Eds.): (1998). Molecular Electronics: Science and Technology. Ann. New York Acad. Sci., Vol. 852.

[4] Avishai, Y., Bessis, D., Giraud, B.G., Mantica, G.: Quantum bound states in open geometries. Phys. Rev. B 44(1991), no. 15, 8028-8034.

[5] Avishai, Y., Luck, J.M.: Quantum percolation and ballistic conductance on a lattice of wires. Phys. Rew. B 45(1992), no. 3, 1074-1095.

[6] Avron, J., Exner, P., Last, Y.: Periodic Schrödinger operators with large gaps and Wannier-Stark ladders. Phys. Rev. Lett. 72(1994), 869-899.

[7] Avron, J., Raveh, A., Zur, B.: Adiabatic quantum transport in multiply connected systems. Rev. Mod. Phys. 60(1988), no. 4, 873-915.

[8] Axmann, W.,Kuchment, P., Kunyansky, L.: Asymptotic methods for thin high contrast 2D PBG materials. Journal of Lightwave Technology, 17(1999), no. 11, 1996- 2007.

[9] Benchama, N., Kuchment, P.: Asymptotics models for acoustic waves in high contrast media, in preparation.

[10] Bensoussan, A., Lions, J.L., Papanicolaou, G. (1980): Asymptotic Analysis of Periodic Structures. Amsterdam: North-Holland.

[11] Birman, M.Sh., Suslina, T.A.: Periodic magnetic Hamiltonian with a variable metric. The problem of absolute continuity. Algebra i Analiz 11(1999), no.2. English translation in St. Petersburg Math J. 11(2000), no. 2, 203-232.

[12] Birman, M.Sh., Suslina, T.A., Shterenberg, R.G.: Absolute continuity of the spectrum of a two-dimensional Schrödinger operator with potential supported on a periodic system of curves. Preprint ESI no. 934, http://www.esi.ac.at, 2000, to appear in Algebra i Analiz (St. Petersburg Math. J.).

[13] Boroditsky, M., Vrijen, R., Krauss, T.F., Coccioli, R., Bhat, R., Yablonovitch, E.: Spontaneous emission extraction and Purcell enhancement from thin-film 2D photonic crystals. Journal of Lightwave Technology, 17(1999), no. 11, 2096-2112.

[14] Carini, J.P., Londergan, J.T., Murdock, D.P., Trinkle, D., Yung, C.S.: Bound states in waveguides and bent quantum wires. I. Applications to waveguide systems. Phys. Rev. B 55(1997), 9842-9851.

[15] Carini, J.P., Londergan, J.T., Murdock, D.P.: Bound states in waveguides and bent quantum wires. II. Electrons in quantum wires. Phys. Rev. B 55(1997), 9852-9859.

[16] Carlson, R.: Hill's equation for a homogeneous tree. Electronic J. Diff. Equations (1997), no. 23, 1-30

[17] Carlson, R.: Adjoint and self-adjoint operators on graphs. Electronic J. Diff. Equations (1998), no. 6, 1-10

[18] Carlson, R.: Inverse eigenvalue problems on directed graphs. Trans. Amer. Math. Soc. 351 (1999), no. 10, 4069-4088.

[19] Carlson, R.: Nonclassical Sturm-Liouville problems and Schrödinger Operators on Radial Trees. Preprint 2000.

[20] de Gennes, P.-G.: Champ critique d'une boucle supraconductrice ramefiee. C. R. Acad. Sc. Paris 292B (1981), 279-282.

[21] Datta, S. (1999): Electronic Transport in Mesoscopic Systems. Cambridge: Cambridge Univ. Press.

[22] Evans, W.D., Harris, D.J.: Fractals, trees and the Neumann Laplacian. Math. Ann. 296(1993), 493-527.

[23] Evans, W.D., Saito, Y.: Neumann Laplacians on domains and operators on associated trees, to appear in Quart. J. Math. Oxford.

[24] Exner, P., Seba, P.: Electrons in semiconductor microstructures: a challenge to operator theorists. Schrödinger Operators, Standard and Nonstandard (Dubna 1988). Singapore: World Scientific, Singapore 1989; pp. 79-100.

[25] Figotin, A.: High-contrast photonic crystals. Diffuse waves in complex media (Les Houches, 1998), 109–136, NATO Sci. Ser. C Math. Phys. Sci., 531. Dordrecht: Kluwer Acad. Publ. 1999.

[26] Figotin, A., Godin, Yu.: Spectral properties of thin-film photonic crystals. SIAM J. Appl. Math., 61(2001), no. 6, 1959-1979.

[27] Figotin, A., Kuchment, P.: Band-Gap Structure of the Spectrum of Periodic and Acoustic Media. I. Scalar Model. SIAM J. Applied Math. 56(1996), no. 1, 68-88.

[28] Figotin, A., Kuchment, P.: Band-Gap Structure of the Spectrum of Periodic and Acoustic Media. II. 2D Photonic Crystals. SIAM J. Applied Math. 56(1996), 1561-1620.

[29] Figotin, A., Kuchment, P.: Spectral properties of classical waves in high contrast periodic media. SIAM J. Appl. Math. 58(1998), no. 2, 683-702.

[30] Figotin, A., Kuchment, P.: Asymptotic models of high contrast periodic photonic and acoustic media (tentative title). Parts I and II, in preparation.

[31] Freidlin, M. (1996): Markov Processes and Differential Equations: Asymptotic Problems, Lectures in Mathematics ETH Zürich. Basel: Birkhäuser Verlag.

[32] Freidlin, M., Wentzell, A.: Diffusion processes on graphs and the averaging principle. Annals of Probability, 21(1993), no. 4, 2215-2245.

[33] Friedlander, L.: On the density of states in periodic media in the large coupling limit. To appear in Comm. Partial Diff. Equations.

[34] Gerasimenko, N., Pavlov, B.: Scattering problems on non-compact graphs. Theor. Math. Phys., 75(1988), 230-240.

[35] Giovannella, C., Lambert, C.J. (Editors) (1998): Lectures on Superconductivity in Networks and Mesoscopic Systems: Pontignano, Italy September 1997 (Aip Conference Proceedings, Vol 427). Amer Inst of Physics.

[36] Griffith, J.S.: A free-electron theory of conjugated molecules. I. Polycyclic Hydrocarbons. Trans. Faraday Soc., 49 (1953), 345-351.

[37] Griffith, J.S.: A free-electron theory of conjugated molecules. II. A derived algebraic scheme. Proc. Camb. Philos. Soc., 49 (1953), 650-658.

[38] Hempel, R., Lienau, K.: Spectral properties of periodic media in the large coupling limit. Comm. Partial Diff. Equations 25(2000), 1445-1470.

[39] Imry, Y. (1997): Introduction to Mesoscopic Physics. Oxford: OxfordUniversity Press.

[40] Inoue, K., Sasada, M., Kuwamata, J., Sakoda, K., Haus, J.W.: A two-dimensional photonic crystal laser. Japan J. Appl. Phys. 85(1999), no. 8, 5768-5770.

[41] Jackson, J.D. (1962): Classical Electrodynamics. New York: John Wiley & Sons.

[42] Joachim, C., Roth, S. (Eds.) (1997): Atomic and Molecular Wires. NATO Adv. Ser.: E Appl. Sci., Vol. 341. Dordrecht: Kluwer.

[43] Joannopoulos, J.D., Meade, R.D., Winn, J.N. (1995): Photonic Crystals, Molding the Flow of Light. Princeton: Princeton Univ. Press.

[44] John, S.: Strong localization of photons in certain disordered dielectric superlattices. Phys. Rev. Lett. 58 (1987), 2486.

[45] Kostrykin, V., Schrader, R.: Kirchgoff's rule for quantum wires. J. Phys. A 32(1999), 595-630.

[46] Kostrykin, V., Schrader, R.: Kirchgoff's rule for quantum wires. II: The inverse problem with possible applications to quantum computers. Preprint 2000.

[47] Kottos, T., Smilansky, U.: Quantum chaos on graphs. Phys. Rev. Lett. 79(1997), 4794-4797.

[48] Kuchment, P. (1993): Floquet Theory for Partial Differential Equations. Basel: Birkhäuser Verlag.

[49] Kuchment, P.: The Mathematics of Photonics Crystals. Ch. 7 in Mathematical Modeling in Optical Science. Bao, G., Cowsar, L. and Masters, W. (Editors). (2001). Philadelphia: SIAM.

[50] Kuchment, P.: Opening spectral gaps for differential operators on graphs using Schenker-Aizenman decoration procedure. In preparation

[51] Kuchment, P., Kunyansky, L.: Spectral Properties of High Contrast Band-Gap Materials and Operators on Graphs. Experimental Mathematics, 8(1999), no. 1, 1-28.

[52] Kuchment, P., Kunyansky, L.: Differential operators on graphs and photonic crystals. To appear in Adv. Comput. Math.

[53] Kuchment, P., Zeng, H.: Convergence of spectra of mesoscopic systems collapsing onto a graph. J. Math. Anal. Appl. 258(2001), 671-700.

[54] Kuchment, P., Zeng, H.: Convergence of spectra of mesoscopic systems collapsing onto a graph II. Large protrusions at vertices. In preparation.

[55] Melnikov, Yu.B., Pavlov, B.S.: Two-body scattering on a graph and application to simple nanoelectronic devices. J. Math. Phys. 36(1995), 2813–2825.

[56] Mittra, R., Lee, S.W. (1971): Analytic Techniques in the Theory of Guided Waves. London: Collier - Macmillan.

[57] Murayama, Y. (2001): Mesoscopic Systems. John Wiley & Sons.:

[58] Naimark, K., Solomyak, M.: Eigenvalue estimates for the weighted Laplacian on metric trees. Proc. London Math. Soc., 80(2000), 690–724.

[59] Novikov, S.: Schrödinger operators on graphs and symplectic geometry. The Arnoldfest (Toronto, ON, 1997), 397–413, Fields Inst. Commun., 24, Amer. Math. Soc., Providence, RI, 1999.

[60] Pavlov, B.S.: A model of zero-radius potential with internal structure. Theor. Math. Phys. 59(1984), 544–580.

[61] Pavlov, B.S.: The theory of extensions and explicitly solvable models. Russian. Math. Surveys 42(1987), 127–168.

[62] Photonic & Acoustic Band-Gap Bibliography, http://home.earthlink.net/ jp-dowling/pbgbib.html

[63] Reed, M., Simon, B. (1978): Methods of Modern Mathematical Physics, IV: Analysis of Operators. New York: Academic Press.

[64] Rubinstein, J., Schatzman, M.: Spectral and variational problems on multiconnected strips. C. R. Acad. Sci. Paris Ser. I math. 325(1997), no. 4, 377–382.

[65] Rubinstein, J., Schatzman, M.: Asymptotics for thin superconducting rings. J. Math. Pures Appl. (9) 77 (1998), no. 8, 801–820.

[66] Rubinstein, J., Schatzman, M.: On multiply connected mesoscopic superconducting structures. Semin. Theor. Spectr. Geom., no. 15, Univ. Grenoble I, Saint-Martin-d'Heres, 1998, 207–220.

[67] Rubinstein, J., Schatzman, M.: Variational problems on multiply connected thin strips I: Basic estimates and convergence of the Laplacian spectrum. Preprint 1999.

[68] Ruedenberg, K., Scherr, C.W.: Free-electron network model for conjugated systems. I. Theory. J. Chem. Physics, 21(1953), no. 9, 1565–1581.

[69] Saito, Y.: The limiting equation of the Neumann Laplacians on shrinking domains. Preprint 1999.

[70] Sakoda, K. (2001): Optical Properties of Photonic Crystals. Berlin: Springer Verlag.

[71] Schatzman, M.: On the eigenvalues of the Laplace operator on a thin set with Neumann boundary conditions. Applicable Anal. 61(1996), 293–306.

[72] Schenker, J., Aizenman, M.: The creation of spectral gaps by graph decoration. Lett. Math. Phys. 53 (2000), no. 3, 253.

[73] Serena, P.A., Garcia, N. (Eds.) (1997): Nanowires. NATO Adv. Ser.: E Appl. Sci., Vol. 340, Dordrecht: Kluwer.

[74] Shepherd, T.J., Roberts, P.J.: Soluble two-dimensional photonic-crystal model. Phys. rev. E 55(1997), no. 5, 6024–6038.

[75] Sievenpiper, D., Zhang, L., Jimenez Broas, R.F., Alexópoulos, N.A., Yablonovitch, E.: High-impedance electromagnetic surfaces with a forbidden frequency band. IEEE Trans. Microwave Theory and Tech. 47(1999), no. 11, 2059–2074.

[76] Thomas, L.E.: Time dependent approach to scattering from impurities in a crystal. Comm. Math. Phys. 33(1973), 335–343.

[77] Thornton, T.J.: Mesoscopic devices. Rep. Prog. Phys. 57(1994), 311–364.

[78] Yablonovitch, E.: Inhibited spontaneous emission in solid-state physics and electronics. Phys. Rev. Lett. 58: 2059 (1987).

[79] Zeng, H.: Convergence of spectra of mesoscopic systems collapsing onto a graph. PhD Thesis, Wichita State University, Wichita, KS 2001.

HOMOGENIZATION OF SOME MULTIPARAMETRIC PROBLEMS *

N.S. Bakhvalov, M.E. Eglit
Mech. and Math. Department
Lomonosov Moscow State University
Moscow 119992, Russia
wbakh@math.msu.su,eglit@mech.math.msu.su

Abstract   Inhomogeneous media are considered in this paper, e.g., composites and mixtures. Let L be the length scale of the problem and d be the typical length of inhomogeneities. The ratio $\varepsilon = d/L$ is supposed to be small. Then the averaged equations can often be obtained that describe a certain homogeneous medium and have solutions close in some sense to solutions of original equations. For periodic media an asymptotic homogenization method to obtain the averaged equations is developed making use of the presence of a small parameter ε.

In many problems there are additional small parameters γ_i, besides ε. For example the following parameters can be small: the ratios of different phases moduli, the ratios of coefficients determining different properties of a phase, the ratios of inhomogeneity scales in different directions. The averaged equations essentially depend on the relations between small parameters ε and γ_i. In some cases the homogenized equations are of another type than equations describing the process in original medium. For example, instead of differential equations we obtain integro-differential equations.

Construction of averaged equations for periodic media includes solution of the so-called cell-problems. They are boundary-value problems for partial differential equations. As a rule they can be solved only numerically. In some cases analytical approximate solutions to cell-problems and explicit formulae for effective coefficients can be obtained due to presence of additional small parameters. The explicit formulae for effective moduli are very useful, especially in optimal design of materials and constructions.

The paper is a brief review of some author's results concerning the effect of different small parameters.

*The work was supported by Russian Foundation for Basic research (Projects No 02-01-00490, 02-01-00613)

H.G.W. Begehr et al. (eds.), Analysis and Applications - ISAAC 2001, 31–49.

1. Introduction

Our world is very inhomogeneous. Often we deal with objects which contain a great amount of inhomogeneities: mixtures, composite materials, porous media, technical and biological structures. Two examples are presented in Figures 1 and 2. Figure 1 is a view of a large building. There are more than 1000 windows in a front wall, which make the construction strongly inhomogeneous.

Figure 1.   Main building of Lomonosov Moscow State University (http://www.msu.ru)

Figure 2 is a photo of a heart muscle cell made by electronic microscopy. The cell contain about 200 deep invaginations filled by an intercellar fluid.

Usually it is impossible and not needed to know all the details of the process in each inhomogeneity cell. Often we need to know the global response of the system only and consider the so-called effective properties of an inhomogeneous body: effective heat conductivity, effective diffusion coefficient, effective elastic moduli, and so on. In these cases we may replace in our study the original inhomogeneous body by a certain homogeneous body with the effective properties. This approach is called homogenization or averaging.

How one can find the effective properties of inhomogeneous bodies? One way is to use experiments and physical hypotheses. The other way is the mathematical one. Let we know the equations and boundary conditions for each phase The aim is to find the equations that describe a homogeneous body and whose solutions are close in some sense to solutions of the original problem.

For periodic structures an asymptotic homogenization method [1] has been proposed, developed and applied to different problems. The foun-

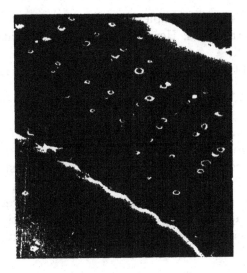

Figure 2. A heart muscle cell of a rat (*Courtesy of S.A. Regirer*)

dations of this method can be read, e.g., in the books by Bensoussan, Lions, Papanicolaou [3]; Sanchez-Palencia [4]; Bakhvalov, Panasenko [2]. This branch of partial differential equations theory was named homogenization theory.

Let we have a medium with periodic structure (Fig. 3)

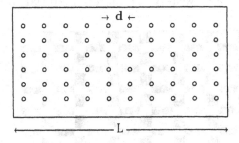

Figure 3. A periodic medium

Let the global length scale of the problem L be much larger than the period d of the medium properties, so that

$$\varepsilon = \frac{d}{L} << 1.$$

The homogenization algorithm includes the following steps.

1) Dimensionless slow variables $x_i = \bar{x}_1/L$ and fast variables $y_i = \bar{x}_i/d = x_i/\varepsilon$ are introduced. Here $\bar{x}_i$ are spatial coordinates, d is the size of a periodicity cell.

2) Solution u of the original problem is tried in the form of asymptotic series

$$u = u(t, x_i, y_i, \varepsilon) = \sum \varepsilon^n u_n, \qquad u_n \in M,$$

where M is the space of 1-periodic functions of y_i.

3) the series are substituted into original equations, boundary and initial conditions; coefficients at all powers of ε in the resulting equations are equated to 0. Thus the so-called cell-problems appear that should be solved to calculate the coefficients of the series. Averaging over the periodicity cell is applied to obtain the averaged (effective) equations.

4) The closeness of the averaged and original equations solutions is proved.

Below we deal with this algorithm and take into account the presence of additional small parameters besides ε.

2. Composites reinforced by periodic system of plates and bars

Let us consider constructions and composites reinforced by a periodic system of mutually orthogonal plates and bars [10]. One can imagine these constructions as large buildings consisting of identical rectangular rooms filled by soft and weakly conducting material (matrix) (Fig. 4).

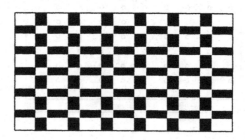

Figure 4. A reinforced composite

Plates (walls) and bars (intersections of walls) are supposed to be relatively thin. Materials can be anisotropic. The properties of matrix, plates, bars and nodes (i.e., intersections of the bars) can be different. Moduli of plates, bars and nodes are assumed to be larger then the matrix moduli.

Denote the dimensionless period along x_j by ε_j, the thicknesses of plates by d_j, $\nu_j = d_j/\varepsilon_j$. It is supposed that

$$d_j \ll \varepsilon_j \ll 1, \quad \bar{\nu} = \frac{\max(d_1, d_2, d_3)}{\min(\varepsilon_1, \varepsilon_2, \varepsilon_3)} \ll 1, \quad \theta \ll 1. \qquad (1)$$

Here

$$\theta = \max_{j_1, j_2, j_3} \theta^{j_1 j_2 j_3}, \quad \theta^{j_1 j_2 j_3} = \max_{\substack{i_k \geq j_k \\ i_1 + i_2 + i_3 > j_1 + j_2 + j_3}} \frac{\pi^{i_1 i_2 i_3}}{\pi^{j_1 j_2 j_3}}, \quad \pi^{ijk} = \frac{V^{ijk}}{Q^{ijk}}$$

V^{ijk} are volumes of components and Q^{ijk} are typical values of components elastic moduli. The upper indices i, j, k mark different parts of a cell: 000 are for matrix, 100, 010 and 001 are for plates, 110, 101 and 011 are for bars, 111 is for a nod.

The equations

$$\frac{\partial}{\partial x_i} a_{kilj} \frac{\partial u_l}{\partial x_j} = f_k, \quad k = 1, \ldots, n \qquad (2)$$

describe, for example, displacements in an elastic composite or stationary heat conduction obeying the Fourier law; a_{kilj} are supposed to be 1-periodic functions of $y_j = x_j/\varepsilon_j$; n is the dimension of the vector $\mathbf{u}$ with components u_l ($l, k \equiv 1$ if u is a scalar, e.g., a temperature). At the condition $\varepsilon_j \ll 1$ the equations with constant coefficients $\bar{a}_{kilj}$ can be constructed so that their solutions are close to the solutions of original equations (2). Coefficients $\bar{a}_{kilj}$ are called the effective moduli of periodic medium.

To calculate the effective moduli by the standard homogenization algorithm one should find vectors $\mathbf{m}^{qp}$ with components m_l^{qp} ($q = 1, 2, 3$; $p, l = 1, ..., n$) satisfying the equations

$$\frac{\partial}{\partial x_i} a_{kilj} \frac{\partial m_l^{qp}}{\partial x_j} = 0, \quad k = 1, \ldots, n \qquad (3)$$

and conditions

$$\mathbf{m}^{qp} - x_q \mathbf{e}_p \in \mathbf{M}. \qquad (4)$$

Here $\mathbf{e}_p$ is a unit vector of coordinate axis x_p ($\mathbf{e}_p$ should be replaced by 1 if $n = 1$), $\mathbf{M}$ is a subspace of functions from $H^{1,loc}(R_s)$ that are periodic over x_j with a period ε_j.

Let

$$U(\mathbf{u}, \mathbf{v}) = a_{kilj} \frac{\partial u_l}{\partial x_j} \frac{\partial v_k}{\partial x_i}, \quad \Lambda(\mathbf{u}, \mathbf{v}) = \langle U(\mathbf{u}, \mathbf{v}) \rangle. \qquad (5)$$

Here $< \cdot >$ means averaging over a periodicity cell

$$< f >= \frac{1}{\varepsilon_1\varepsilon_2\varepsilon_3} \int\limits_0^{\varepsilon_1} (\int\limits_0^{\varepsilon_2} (\int\limits_0^{\varepsilon_3} f(x_1,x_2,x_3)dx_3)dx_2)dx_1.$$

Coefficients $\bar{a}_{kilj}$ are calculated by the formulae

$$\bar{a}_{kilj} = \Lambda(\mathbf{m}^{jl}, \mathbf{m}^{ik}) \qquad (6)$$

At the condition (1) approximate solutions $\hat{m}^{qp}$ to the problem (3), (4) can be constructed in an explicit form by the following way. Note that at $p = q$ the problems (3), (4) describe a unit extension of a periodicity cell along the x_q - axis at the following conditions: body forces are absent; $\mathbf{u}$ and its derivatives are periodic over the other space variables. At $p \neq q$ they correspond to a shear of a periodicity cell along the x_p-axis in the planes that are orthogonal to the x_q - axis.

For composites under consideration at the first step the displacements in the plates and bars orthogonal to x_q-axis are certain constants. In the plates parallel to x_q-axis they are solutions of the extension or shear problems at the conditions for surfaces of plates to be free of forces and for displacements and their derivatives to satisfy the periodicity conditions. In the bars parallel to x_q-axis the displacements are those at simple extension or simple shear of isolated bars. At the second step the displacements are spliced specially to satisfy the conditions (4).

Approximate values of the effective coefficients $\hat{a}_{kilj}$ are calculated by the formulae (6) with $\hat{m}^{qp}$ instead of m^{qp}. An example of explicit formulae for $\hat{a}_{kilj}$, which have been obtained by this way, is written below

$$\hat{a}_{k1l1} = (A_{11}^{000}\mathbf{e}_k, \mathbf{e}_l) \left((A_{11}^{010} - A_{12}^{010}(A_{22}^{010})^{-1}A_{21}^{010})\mathbf{e}_k, \mathbf{e}_l\right) \nu_2$$

$$+ \left((A_{11}^{001} - A_{13}^{001}(A_{33}^{001})^{-1}A_{31}^{001})\mathbf{e}_k, \mathbf{e}_l\right) \nu_3$$

$$+ (A_{11}^{011}\mathbf{e}_k + A_{12}^{011}\mathbf{e}_{1k2}^{011} + A_{13}^{011}\mathbf{e}_{1k3}^{011}, \mathbf{e}_l)\nu_2\nu_3,$$

$$\hat{a}_{k1l2} = (A_{12}^{000}\mathbf{e}_k, \mathbf{e}_l) + \left((A_{12}^{001} - A_{23}^{001}(A_{33}^{001})^{-1}A_{31}^{001})\mathbf{e}_k, \mathbf{e}_l\right)\nu_1\nu_2.$$

Vectors $\mathbf{e}_{1kj}^{011}$, $\mathbf{e}_{1kl}^{011}$ are determined by relations

$$\sum_{j=1}^{3} A_{lj}^{011}\mathbf{e}_{1kj}^{011} = 0, \quad \mathbf{e}_{1kl}^{011} = \mathbf{e}_k.$$

Here A_{ij} – matrices with elements a_{kilj}, and the upper indices mark the values of elastic moduli for matrix, plates and bars, respectively.

For small ν and θ the values of $\hat{a}_{kilj}$ are close to a_{kilj}. The following estimates are proved

$$|\overline{a}_{kilj} - \hat{a}_{kilj}| \leq r_{kilj}$$

where

$$r_{kiki} = O(q_i(\nu + \theta)), \quad r_{kili} = O(q_i(\sqrt{\nu} + \sqrt{\theta})) \quad \text{at} \quad k \neq l,$$

$$r_{kilj} = O(\sqrt{q_1 q_2}(\sqrt{\nu} + \sqrt{\theta})) \quad \text{at} \quad i \neq j,$$

$$q_1 \equiv Q^{000} + Q^{010}\nu_2 + Q^{001}\nu_3 + Q^{011}\nu_2\nu_3,$$

$$q_2 \equiv Q^{000} + Q^{100}\nu_1 + Q^{001}\nu_3 + Q^{101}\nu_1\nu_3,$$

ν_i are relative thicknesses of plates.

3. Composites reinforced by an arbitrary system of high moduli fibres

Let us consider composites reinforced by fibres [8] The fibres can be inhomogeneous and curvilinear; they can be located not periodically; small breaks of fibres are permitted (Fig. 5).

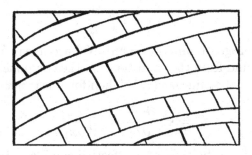

Figure 5. A fibrous material

At certain conditions effective characteristics of such composites are additive functions of characteristics of fibres. These conditions are the relations that contain ratios of fibres and matrix moduli, the fibres diameters and distances between fibres.

We consider the boundary value problem for the equations (2) in the domain G at the condition $\mathbf{u}|_{\partial G} = \gamma$. As usual the solution is understood in the sense of integral identity

$$\mathbf{u} \in H^1(G), \quad \Lambda_G(\mathbf{u}, \varphi) = 0 \quad \forall \quad \varphi \in H_0^1(G).$$

Here

$$\Lambda_G(\mathbf{u}, \varphi) = \int_G U \, dx,$$

U is defined by (5).

Let V_m ($m = 1, ..., M$) be fibres, a_{kilj} be their moduli; $V_0 = G \setminus (\bigcup_{m=1}^{M} V_m)$ be matrix and $dist(V_i, V_j) \geq \alpha_0 > 0$ for $i, j > 0$. We consider the case of small ratios ε and α of the typical fibre diameter and distance α_0 to the macroscale L. Besides we assume that the ratio ω of the fibres moduli (thermal conductivity, elasticity) to matrix moduli is large. Let γ be a volume concentration of fibres.

The main point is an assumption that the influence of matrix on heat conduction and elastic behavior of the composite in the first approximation can be neglected. Then, e.g., heat can propagate only along fibres and the total heat flux equals the sum of fluxes through fibres. On this way it is easy to calculate approximately the effective moduli.

To calculate the effective thermoconductivity coefficient we should at first find the relation between the heat flux $\mathbf{q}^m$ along the fibre V_m and the temperature gradient along it. If V_m is homogeneous, straight and oriented along the x_1-axis then the temperature $v^{(1)}$ in V_m is given by the relation

$$v^{(1)} = v + N_j^m \frac{\partial v}{\partial x_j},$$

where v is the averaged temperature and N_j^m are linear functions of x_2, x_3 that can be found from the conditions on the components of the heat flux: $q_1^m = 0$, $q_2^m = 0$ at linear $v(x_1)$. Then for linear $v(x_1)$

$$q_1^m = k_{1j} \frac{\partial v^{(1)}}{\partial x_j} = c_{11}^m \frac{\partial v}{\partial x_1}.$$

The coefficient c_{11}^m can be called the effective conductivity of V_m.

Let the fibre V_m be inclined to x_1-axis and directed along x_1'. Then its effective conductivity tensor components c_{ij}^m equal $c_{11}^m g_{i1} g_{j1}$ where g_{ij} are components of the orthogonal matrix transforming coordinates x_j' into x_i. If fibres are curvilinear (e.g., circular) they can be transformed into straight ones by a certain nonlinear transformation of coordinates. After that we apply the described algorithm and return to old variables.

Introduce $\hat{k}_{ij}$ by the relations $\hat{k}_{ij} = 0$ in V_0, $\hat{k}_{ij} = \sum_{i=1}^{M} c_{ij}^m$ where $c_{ij}^m = 0$ outside V_m. Suppose the existence of $\bar{k}_{ij}(x)$ such that

$$\max_{i,j} \left\| \frac{\hat{k}_{ij}}{\gamma} - \bar{k}_{ij} \right\|_{(-1)} \leq \delta_A \ll 1,$$

As usual, here

$$\|\mathbf{f}\|_{(-1)} = \sup_{\varphi \in H_0^1(G)} \frac{(\mathbf{f}, \varphi)_G}{\|\nabla \varphi\|_{L_2(G)}}.$$

If the material is homogeneous in average, then $\bar{k}_{ij}$ are the average values of $\widehat{k}_{ij}/\gamma$. The solution of original problem is close to the solution of the averaged problem for the homogeneous material with moduli $\bar{k}_{ij}$. The following inequality is proved

$$\int_{G \setminus V_0} |\nabla(\mathbf{u} - \mathbf{v}^{(1)})|^2 \, dx \leq b(R_1 + R_2), \tag{7}$$

where

$$R_1 = \varepsilon^2 + \omega^{-1} + \omega \gamma^2 \delta_A^2, \qquad R_2 = \varepsilon/(\omega \alpha).$$

There is an additional term R_3 in the right part of (7) if the fibres have "breaks"; it is equal to a common measure of all breaks.

So the effective moduli $\bar{k}_{ij}$ are additive functions of c_{ij}^m if R_1, R_2 and R_3 are small.

As an example let us consider a material that can be obtained from a material reinforced by periodic system of homogeneous straight fibres by the following way. Original fibres are replaced by fibres with different diameters not exceeding $O(\varepsilon)$ and located at a distance not larger than $O(\varepsilon)$. Let the ratios of the thermal conductivities of the new and old fibres are inversely proportional to the ratios of their cross-section areas. Then

$$\delta_A^2 = O\left(\varepsilon^2 \ln \frac{1}{\varepsilon} + \frac{1}{\omega^2}\right)$$

and the values of $\bar{k}_{ij}$ for the new material equal those for original one and therefore can be calculated by the formulae given in [2].

4. Media with periods of different order in different directions

For media with essential difference in values of the periods in different directions explicit formulae for effective coefficients can be obtained. Consider first two-dimensional problems for media with long thin pores [9], [17] (Fig. 6).

The dimensionless period is ε_1 in the x_1 direction and ε_2 in the x_2 direction and

$$\varepsilon_2 \ll \varepsilon_1 \ll 1, \quad \text{i.e.} \quad \omega = \frac{\varepsilon_1}{\varepsilon_2} \gg 1.$$

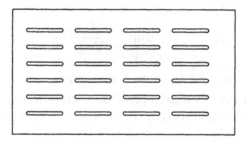

Figure 6.   A material with long thin pores

Denote the domain occupied by the material by G, its solid part by G_s, the external boundary of G by Γ and the boundary of pores by S. The following boundary value problem is studied

$$\frac{\partial}{\partial x_i}\left(k\,\frac{\partial u}{\partial x_i}\right) = f \quad \text{in } G_s, \quad k = const,$$

$$u\big|_\Gamma = g, \qquad \frac{\partial u}{\partial n}\Big|_S = 0.$$

Here $\mathbf{n}$ is the normal to the boundary of pores.

It has been proved by Bakhvalov and Saint Jean Paulin [9] for materials with symmetric pores and by Yakubenko [17] for arbitrary long pores, that if $\varepsilon_1 + \varepsilon_2/\varepsilon_1 \to 0$ then the solution u is well approximated by solution v of the averaged problem

$$\hat{k}_{ij}\frac{\partial^2 v}{\partial x_i \partial x_j} = f \quad \text{in } G, \quad v\big|_{\partial G} = g,$$

where

$$\hat{k}_{11}^0 = k\left\langle \frac{1}{1-\bar{g}(x_1)}\right\rangle_1^{-1}; \quad \hat{k}_{22}^0 = k(1-a); \quad \hat{k}_{12}^0 = \hat{k}_{21}^0 = 0; \quad (8)$$

$$\hat{k}_{ij} = \hat{k}_{ij}^0/\mu_0, \quad \langle f\rangle_1 = \int_0^1 f\,dy_1.$$

Here μ_0 is the area of the solid part of a periodicity cell, a is the minimum distance between pores in x_1 direction, $g(x_1)$ is a pore thickness in x_2 direction, $\bar{g}(x_1) = 0$ outside of pores, $\bar{g}(x_1) = g(x_1)$ in pores. Formulae (8) give approximate values of effective thermoconductivity coefficients at small ε_1 and small $\varepsilon_2/\varepsilon_1$.

Now let us consider a periodic material without pores at the condition

$$\varepsilon_2 << \varepsilon_1 << 1.$$

The effective equation of an infinite accuracy for thermoconductivity process in such materials has been considered in [14] at the assumption of smoothness of the coefficients of the original equation. Explicit approximate formulae for effective elastic coefficients $\bar{A}_{ij}$ have been derived and justified for media with measurable coefficients, e.g., composite materials in [18]. An example of the formulae for matrices $\hat{A}_{ij}$ is given below [18].

$$\hat{A}_{11} = C_2, \ \hat{A}_{12} = C_2 \langle A_4 A_3^{-1} \rangle_1, \ \hat{A}_{21} = \langle A_3^{-1} A_4 \rangle_1 C_2$$

$$\hat{A}_{22} = \langle A_3^{-1} \rangle_1 + \langle A_3^{-1} A_4 \rangle_1 C_2 \langle A_4 A_3^{-1} \rangle_1$$

Here matrices A_1, A_2, A_3, A_4 and C_2 are defined by the formulae

$$A_1 = \langle A_{11} - A_{12} A_{22}^{-1} A_{21} \rangle_2^{-1}, \quad A_2 = \langle A_{12} A_{22}^{-1} \rangle_2,$$

$$A_3 = \langle A_{22}^{-1}(1 + A_{21} A_1 A_2) \rangle_2, \quad A_4 = \langle A_{22}^{-1} A_{21} A_1 \rangle_2,$$

$$C_2 = \langle A_1 - A_1 A_2 A_3^{-1} A_4 \rangle_1^{-1}, \quad \langle f \rangle_k = \int_0^1 f \, dy_k.$$

It has been proved that the following estimates are valid

$$\|\hat{A}_{ij} - \bar{A}_{ij}\| \le \beta(\omega), \quad \|\mathbf{u} - \mathbf{v}\|_{L^2(G_\varepsilon)} \le Q V_3 \left(\sqrt{\varepsilon_1} + \beta(\omega)\right),$$

$$\beta(\omega) \to 0, \quad \text{at} \quad \omega \to \infty$$

Here $\bar{A}_{ij}$ are the effective moduli obtained by the exact solution of the cell-problems, $\mathbf{u}$ is a solution of the original problem, $\mathbf{v}$ is the solution of the problem with the coefficients $\hat{A}_{ij}$, Q is a constant independent of ε_1, ω, and $V_3 = \|\mathbf{v}\|_{H_3}$.

5. Dispersion of waves in linear stratified media

Elastic wave propagation in stratified media (Fig. 7) have been studied by many authors. Large amount of the papers deal with the media, which consist of homogeneous isotropic layers. Here we study media with arbitrary anisotropy and with arbitrary periodic distribution of the density and elastic properties along the x_1-axis and apply the homogenization theory algorithm to derive the homogenized equations [11], [12], [13]. To describe dispersion of waves the homogenized equations should include terms with higher derivatives of displacements.

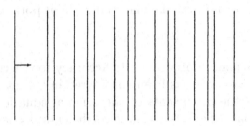

Figure 7. Waves in a stratified medium

Let us consider the equation

$$L(\mathbf{u}) = -\rho \frac{\partial^2 \mathbf{u}}{\partial t^2} + \frac{\partial}{\partial x_i}\left(A_{ij}\frac{\partial \mathbf{u}}{\partial x_j}\right) = 0$$

with ρ and matrices A_{ij} being periodic functions of fast variables $y_i = x_i/\varepsilon$ and satisfying conditions usual for elasticity theory. The following parameters are supposed to be small

$$\varepsilon \ll 1, \quad \frac{1}{\tau} \ll 1, \quad \tau = \frac{cT}{L}$$

where L is a typical wave length, ε is the ratio of the medium period to L, c is a typical value of the wave velocity, and T is the observation time.

It is known [2] that at a small ε a solution $\mathbf{u}$ can be written as

$$\mathbf{u} \sim \sum_{q,l_1,l_2,l_3=0}^{\infty} \varepsilon^{q+l_1+l_2+l_3} N^q_{l_1 l_2 l_3} \frac{\partial^{q+l_1+l_2+l_3}\mathbf{v}}{\partial t^q \partial x_1^{l_1} \partial x_2^{l_2} \partial x_3^{l_3}}$$

with $\mathbf{v}$ not depending on fast variables y_i and $N^q_{l_1 l_2 l_3}$ being periodic functions of y_i. Then

$$L\mathbf{u} \sim \sum_{q,l_1,l_2,l_3=0}^{\infty} \varepsilon^{q+l_1+l_2+l_3-2} H^q_{l_1 l_2 l_3}(y_1, y_2, y_3) \frac{\partial^{q+l_1+l_2+l_3}\mathbf{v}}{\partial t^q \partial x_1^{l_1} \partial x_2^{l_2} \partial x_3^{l_3}},$$

where

$$H^q_{l_1 l_2 l_3} = -\rho N^{q-2}_{l_1 l_2 l_3} + \frac{\partial}{\partial y_i}\left(A_{ij}\frac{\partial N^q_{l_1 l_2 l_3}}{\partial y_j}\right) + A_{ij}\frac{\partial N^q_{l_1-\delta_{i1},l_2-\delta_{i2},l_3-\delta_{i3}}}{\partial y_j}$$

$$+ \frac{\partial}{\partial y_i}(A_{ij}N^q_{l_1-\delta_{j1},l_2-\delta_{j2},l_3-\delta_{j3}}) + A_{ij}N^q_{l_1-\delta_{i1}-\delta_{j1},l_2-\delta_{i2}-\delta_{j2},l_3-\delta_{i3}-\delta_{j3}}.$$

To obtain the homogenized equation one should find $N^q_{l_1 l_2 l_3}$ to satisfy the relations

$$H^q_{l_1 l_2 l_3}(y_1, y_2, y_3) = 0 \quad \text{at } q + l_1 + l_2 + l_3 < 2,$$

$$H^q_{l_1 l_2 l_3}(y_1, y_2, y_3) = h^q_{l_1 l_2 l_3} = \text{const at } q + l_1 + l_2 + l_3 \geq 2, \qquad \left\langle N^q_{l_1 l_2 l_3} \right\rangle = 0.$$

The homogenized equation of infinite order on ε has the form

$$\sum_{m \geq 2} \varepsilon^{m-2} h^q_{l_1 l_2 l_3} \frac{\partial^m \mathbf{v}}{\partial t^q \partial x_1^{l_1} \partial x_2^{l_2} \partial x_3^{l_3}} \sim 0, \quad m = q + l_1 + l_2 + l_3 \qquad (9)$$

where

$$h^q_{l_1 l_2 l_3} = \left\langle H^q_{l_1 l_2 l_3} \right\rangle, \qquad \langle f \rangle = \int_0^1 \int_0^1 \int_0^1 f(y_1, y_2, y_3) \, dy_1 \, dy_2 \, dy_3$$

Properties of homogenized equation (9) for arbitrary periodic media have been investigated in [5], [11] It was proved that it is an Euler equation for a certain functional. Therefore, the matrices $h^q_{l_1 l_2 l_3}$ satisfy the conditions

$$(h^q_{l_1 l_2 l_3})^T = (-1)^{l_1 + l_2 + l_3} h^q_{l_1 l_2 l_3}.$$

It is possible to exclude derivatives of higher order over t in the equation (9). Two canonical forms of the equation, A and B, can be obtained. They differ in terms with derivatives of higher order. The form A contains mixed derivatives over t, x_i; coefficients at derivatives over x_i coincide with those obtained in statics ("static elastic moduli of higher order"). The form B contains no mixed derivatives over t, x_i; its coefficients at higher derivatives over x_i differ from static ones ("dynamic elastic moduli of higher order"). We are interested in major terms of the homogenized equation, which are responsible for dispersion of waves. Further the equations up to ε^2 only will be studied.

For waves propagating normally to the layers the equation in the form A reads

$$L^A \mathbf{v} = -\widehat{\rho} \frac{\partial^2 \mathbf{v}}{\partial t^2} + \widehat{A} \frac{\partial^2 \mathbf{v}}{\partial x^2} + \varepsilon h_1^2 \frac{\partial^3 \mathbf{v}}{\partial t^2 \partial x} + \varepsilon^2 \widetilde{h}_2^2 \frac{\partial^4 \mathbf{v}}{\partial t^2 \partial x^2} = 0$$

where

$$x = x_1, \quad h_l^q = h^q_{l00}, \quad \widehat{\rho} = \langle \rho \rangle, \quad \widehat{A} = \left\langle A_{11}^{-1} \right\rangle^{-1}.$$

So in general the effective equation contains the third and fourth derivatives of displacements. In some cases however the coefficients at the third derivatives are equal to zero. Specifically, $h_1^2 = 0$ in the following cases:) if v is a scalar; b) if the material is locally orthotropic with a plane of symmetry parallel to the layers, in particular, for locally isotropic materials; c) for media with arbitrary local anisotropy if $\rho(y)$ $A_{11}(y_1)$ are even functions of y_1. The latter is true, e.g., for arbitrary two-layer media.

The following important statement is proved in [11], [12]: if $h_1^2 = 0$, then $\tilde{h}_2^2 \geq 0$.

Note that the equations with the higher derivatives of displacements are introduced phenomenologically, e.g., in the so-called couple-stress elasticity [15]. The matrix $\tilde{h}_2^2$ is usually assumed to be negative in this theory though couple-stresses are supposed to appear due to microinhomogeneity of the material.

Consider harmonic waves propagating normally to the layers, $\mathbf{v} = e^{i(kx-\omega t)}\mathbf{e}$. If $h_1^2 = 0$ then the dispersion relation is

$$(c^2 E - \widehat{A} + p^2\tilde{h}_2^2)\mathbf{e} = 0, \quad c = \frac{\omega}{k}, \quad p = \varepsilon\omega.$$

Therefore the velocity of harmonic waves that are solutions of equation $L^A\mathbf{v} = 0$ up to ε^4 does not increase as frequency increases. This effect is referred to as a negative dispersion.

Consider now waves propagating in arbitrary direction. An analytical study have been conducted in [13] for a scalar case at the conditions $A_{12} = 0, A_{13} = 0$. This is true, for example, for acoustic waves in stratified locally isotropic fluid. Explicit formulae for all coefficients responsible for dispersion (up to $O(\varepsilon^4)$) have been obtained and investigated. It has been proved that in this case the velocity of harmonic waves up to $O(\varepsilon^4)$ does not increase as frequency increases.

Elastic waves along the layers have been investigated for media consisting of a periodic system of isotropic or anisotropic homogeneous layers. Homogenized equations in the form B for waves propagating along the x_2-axis is

$$L^B\mathbf{v} = -\widehat{\rho}\frac{\partial^2\mathbf{v}}{\partial t^2} + \widehat{A}_2\frac{\partial^2\mathbf{v}}{\partial x_2^2} + \varepsilon H_3\frac{\partial^3\mathbf{v}}{\partial x_2^3} + \varepsilon^2 H_4\frac{\partial^4\mathbf{v}}{\partial x_2^4} = 0$$

with

$$\widehat{\rho} = \langle\rho\rangle, \quad \widehat{A}_2 = \langle A_{22}\rangle.$$

Matrices $\widehat{A}_2$, H_3, H_4 have been calculated numerically for different media. For two-layer periodic media with arbitrary local anisotropy H_3 was found to be equal to zero. In three-layer periodic media $H_3 \neq 0$ even for isotropic layers. In all cases two of the three eigenvalues of the matrix H_4 are positive.

6. Soft moduli inclusions

Let us compare solution to two problems, A and B (Fig.8).

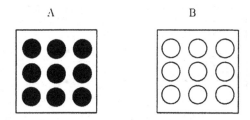

A   B

Figure 8. Media with low moduli inclusions (A) and with empty pores (B)

The problem A deals with a body containing a periodic system of inclusions. The matrix material moduli are A_{ij}^1. The moduli of the inclusions are supposed to have an order η in comparison to the matrix moduli. The problem B deals with the same geometrical structure but the inclusions are replaced by empty pores.

Let $\mathbf{u}_\eta^\varepsilon$ be a solution of the boundary value problem A in G

$$\frac{\partial}{\partial x_i}\left(A_{ij}\frac{\partial \mathbf{u}_\eta^\varepsilon}{\partial x_j}\right) = \mathbf{f}, \quad (\mathbf{u}_\eta^\varepsilon - \mathbf{g})|_{\partial G} = 0,$$

$$A_{ij} = A_{ij}^1, \quad x \in G_\varepsilon, \quad A_{ij} = A_{ij}^2 = \eta A_{ij}^0, \quad x \in G/G_\varepsilon,$$

and $\mathbf{u}_0^\varepsilon$ be a solution of the boundary value problem B in G_ε

$$\frac{\partial}{\partial x_i}\left(A_{ij}^1\left(\frac{x}{\varepsilon}\right)\frac{\partial \mathbf{u}_0^\varepsilon}{\partial x_j}\right) = \mathbf{f}, \quad x \in G_\varepsilon,$$

$$(\mathbf{u}_0^\varepsilon - \mathbf{g})|_{\Gamma_\varepsilon} = 0, \quad \left(A_{ij}^1\frac{\partial \mathbf{u}_0^\varepsilon}{\partial x_j}\nu_i^x\right)\Big|_{S_\varepsilon} = 0.$$

It has been proved in [7] that in general

$$\lim_{\eta\to 0}\lim_{\varepsilon\to 0}\mathbf{u}_\eta^\varepsilon \neq \lim_{\varepsilon\to 0}\mathbf{u}_0^\varepsilon.$$

Limit behavior of solutions to dynamic problems for media with inclusions at the condition $\eta \to 0$ have been studied in [16].

7. Sound propagation in mixtures. Effect of heat conductivity

The effective equations for acoustic waves propagation in mixtures can be obtained by homogenization theory at the assumption of periodicity of the structure with small ratio ε of the period d to the typical wave length L.

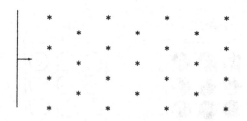

Figure 9.   Sound propagation in a mixture

Usually viscosity, shear elasticity and heat conduction are neglected in study of acoustic waves in homogeneous fluids. Still in mixtures these phenomena can influence the effective equations essentially at certain conditions [6]. Here the effect of heat conduction is described. The dimensionless forms of the momentum end energy conservation equations for small disturbances are

$$-\rho\frac{\partial^2 \mathbf{u}}{\partial t^2} + \operatorname{grad}\left(k_{11}\operatorname{div}\mathbf{u} - k_{12}T\right) + \rho f = 0,$$

$$-\rho c\frac{\partial T}{\partial t} + \gamma^2 \operatorname{div}\left(\kappa\operatorname{grad}T\right) - \frac{\partial}{\partial t}\left(k_{21}\operatorname{div}\mathbf{u}\right) = 0,$$

where

$$\gamma = \frac{l}{L}, \qquad l = \sqrt{\chi^*\tau}, \qquad \chi^* = \frac{\kappa^*}{\rho^* c^*}, \qquad \tau = \frac{L}{\sqrt{k_{11}^*/\rho^*}}.$$

Here L is a typical wave length, T is a temperature, c is a specific heat capacity, κ is a thermoconductivity coefficient, and a star marks their typical values; τ is a time scale, and l is known to have an order of the distance, which a "heat wave" covers during the time τ. Usually

$$l \ll L, \quad \text{i.e.,} \quad \gamma \ll 1$$

so γ is an additional small parameter besides $\varepsilon = d/L$.

The homogenized equation depends essentially on the relation between l and d. Introduce

$$a = \frac{\gamma}{\varepsilon} = \frac{l}{d} = \sqrt{\frac{\kappa^* L}{d^2 c^* \sqrt{k_{11}^* \rho^*}}}$$

Consider 3 cases: I. $a \ll 1, l \ll d$; II. $a \gg 1, l \gg d$; III. $a \sim 1$, $l \sim d$.

The homogenized equations in the cases I and II have the form

$$R\frac{\partial^2 \mathbf{v}}{\partial t^2} = \operatorname{grad}(\lambda \operatorname{div} \mathbf{v}) + R\mathbf{f},$$

$$C\frac{\partial T_0}{\partial t} = \frac{\partial}{\partial t}\bar{k}_{21}\operatorname{div}\mathbf{w},$$

where R is matrix calculated by the solution of the cell-problems and therefore depending on the form of the inclusions. The effective compressibility coefficient λ is different for the cases I and II. In the case I

$$\lambda = \bar{\lambda}_1 = \left\langle \left(k_{11} + \frac{k_{12}k_{21}}{\rho c} \right)^{-1} \right\rangle^{-1}.$$

In the case II $\lambda = \bar{\lambda}_2 \le \bar{\lambda}_1$,

$$\bar{\lambda}_2 = \bar{k}_{11} + \frac{\bar{k}_{12}\bar{k}_{21}}{\overline{\rho c} + \bar{k}_{22}}$$

$$\bar{k}_{22} = \left\langle \frac{k_{12}k_{21}}{k_{11}} \right\rangle - \frac{\bar{k}_{12}\bar{k}_{21}}{\bar{k}_{11}} = \langle k_{11}\alpha^2 \rangle - \bar{k}_{11}\bar{\alpha}^2 \ge 0$$

$$\alpha = \frac{k_{12}}{k_{11}}, \quad \alpha_s = \frac{\rho c}{k_{21}}, \quad \bar{\alpha} = \langle \alpha \rangle, \quad \overline{\rho c} = \langle \rho c \rangle$$

$$\bar{k}_{11} = \left\langle \frac{1}{k_{11}} \right\rangle^{-1}, \quad \bar{k}_{12} = \bar{k}_{11}\left\langle \frac{k_{12}}{k_{11}} \right\rangle, \quad \bar{k}_{21} = \bar{k}_{11}\left\langle \frac{k_{21}}{k_{11}} \right\rangle$$

The effective heat capacity per unit volume C in the case I is

$$C = C_1 = \overline{\rho c},$$

and in the case II

$$C = C_2 = (\overline{\rho c} + \bar{k}_{22}) \ge C_1.$$

So, for a mixture with a fine structure $(d \ll l)$ the effective compressibility and sound velocities are less than those for a mixture with more rough structure $(d \gg l)$. The effective heat capacity is larger in a fine structure. This property together with the lower value of the effective heat conductivity can be used to prevent heat conduction by mixtures.

Now consider the case III $(l \sim d)$. In this case the effective medium is the medium with memory. The effective equation has the form

$$R\frac{\partial^2 \mathbf{v}}{\partial t^2} = \operatorname{grad}\Omega(x, t) + R\mathbf{f},$$

48

where

$$\Omega(x,t) = \bar{\lambda}_1 \mathrm{div} \mathbf{v} + \int\limits_0^t \bar{K}(t-\tau) \mathrm{div} \mathbf{v} d\tau.$$

The zeroth term T_0 in the asymptotic series for temperature is

$$T_0 = -\frac{1}{\lambda_1 \alpha_s} \Omega(x,t) + \int\limits_0^t \Gamma(t-\tau, y) \Omega(x,\tau) d\tau$$

where $\Gamma(t,y)$ is a solution of the following cell-problem

$$\frac{\partial \Gamma}{\partial t} = \frac{a^2}{\lambda_1 \alpha \alpha_s} \mathrm{div}_y(\kappa \mathrm{grad}_y \Gamma), \quad \Gamma(0,y) = -\frac{a^2}{\lambda_1 \alpha \alpha_s} \mathrm{div}_y \left(\kappa \mathrm{grad}_y \frac{1}{\lambda_1 \alpha_s} \right)$$

8. Conclusion

Homogenization theory provides a method to obtain a homogenized model to describe processes in microinhomogeneous media. The method is based on asymptotic expansions of solutions on a small parameter ε, which is the ratio of typical length of an inhomogeneity to the global length scale of the problem.

Presence of additional small parameters a) sometimes gives a possibility to obtain explicit formulae for effective moduli; b) sometimes change the type of homogenized equations depending on the relations between different small parameters.

References

[1] Bakhvalov, N.S.: Averaging of partial differential equations with the rapidly oscillating coefficients. Doklady Akademii Nauk SSSR 221, 3, (1975), 516–519.

[2] Bakhvalov, N.S., Panasenko, G.P.: Homogenization. Averaging Processes in Periodic Media. Mathematical Problems in Mechanics of Composite Materials, Nauka, Moscow 1984. Kluwer Academic Publishers, Dordrecht-Boston-London 1989.

[3] Bensoussan, A., Lions, J.L., Papanicolaou, G.: Asymptotic Methods in Periodic Structures. North Holland, Amsterdam 1978.

[4] Sanchez-Palencia, E.: Non Homogeneous Media and Vibration Theory. Lecture Notes in Physics 127, Springer, Berlin 1980.

[5] Bakhvalov, N.S., Eglit, M.E.: Variational properties of averaged models for periodic media. Trudy MIAN 192 (1990), 5–19.

[6] Bakhvalov, N.S., Eglit, M.E.: Homogenization of dynamic problems singularly depending on small parameters. Proceedings of Second Workshop on Composite Media and Homogenization Theory (Trieste, 1993) World Scientific, Singapore 1995, 17–35.

[7] Bakhvalov, N.S., Eglit, M.E.: The limiting behavior of periodic media with soft media inclusions. Comp. Maths Math. Phys. 35, no 6 (1995), 719–730.

[8] Bakhvalov, N.S., Eglit, M.E.: Explicit Calculation of Effective Moduli for Composites Reinforced by an Irregular System of Fibres. Doklady Mathematics 51 (1995), 46–50.

[9] Bakhvalov, N.S., Saint Jean Paulin, J.: Homogenization for thermoconductivity in a porous medium with periods of different orders in the different directions. Asymptotic Analysis 13 (1996), 253–276 .

[10] Bakhvalov, N.S., Eglit, M.E.: Effective moduli of composites reinforced by systems of plates and bars. Comp. Maths Math. Phys. 38, no 5 (1998), 783–804.

[11] Bakhvalov, N.S., Eglit, M.E.: Effective equations with dispersion for waves propagation in periodic media. Doklady Math. 370 (2000), 1–4.

[12] Bakhvalov, N.S., Eglit, M.E.: Long-waves asymptotics with dispersion for the waves propagation in stratified media. Part 1. Waves orthogonal to the layers. Russian J. Numer. Analys. and Math. Modelling 15 (2000), 3–14.

[13] Bakhvalov, N.S., Eglit, M.E.: Long-waves asymptotics with dispersion for waves propagation in stratified media. Part 2. Waves in arbitrary direction. Russian J. Numer. Analys. And Math. Modelling 15, no 3 (2000).

[14] Dubinskaya, V.Yu.: Asymptotic expansion for a solution of a stationary heat conduction problem in a medium with two small parameters. Doklady RAN 333(5) (1993), 571–574.

[15] Mindlin, R.D., Tiersten, H.F.: Effects of couple-stresses in linear elasticity. Arch. Ration. Mech. and Analysis 11 (1962), 415–448.

[16] Sandrakov, G.V.: The homogenization of nonstationary problems the theory of strong nonuniform elastic media. Doklady Mathematics 355, no 5 (1997), 605–608.

[17] Yakubenko, T.A.: Averaging a periodic porous medium with periods of different orders in different directions. Russian J. Numer. Analys. and Math. Modelling 13, no 2 (1998), 149–157.

[18] Yakubenko, T.A.: Averaging of periodic structures with nonsmooth data. Mosscow State University, Mech. and Math. Dept., Preprint no 2, 1999, 30 pp.

POWER GEOMETRY AS A NEW CALCULUS

Alexander D. BRUNO

Keldysh Institute of Applied Mathematics,
Moscow 125047 Russia

bruno@spp.keldysh.ru

Abstract   Power Geometry develops Differential Calculus and aims at nonlinear
problems. The algorithms of Power Geometry allow to simplify equations,to resolve their singularities, to isolate their first approximations,
and to find either their solutions or the asymptotics of the solutions.
This approach allows to compute also the asymptotic and the local expansions of solutions. Algorithms of Power Geometry are applicable to
equations of various types: algebraic, ordinary differential and partial
differential, and also to systems of such equations. Power Geometry
is an alternative to Algebraic Geometry, Group Analysis, Nonstandard
Analysis, Microlocal Analysis etc.

Introduction

Power Geometry develops Differential Calculus and aims at the nonlinear problems. Its main concept consists in the study of nonlinear
problems not in the original coordinates, but in the logarithms of these
coordinates. Then to many properties and relations, which are nonlinear
in the original coordinates, some linear relations can be put in correspondence. The algorithms of Power Geometry are based on these linear relations. They allow to simplify equations, to resolve their singularities,
to isolate their first approximations, and to find either their solutions
or the asymptotics of the solutions. This approach allows to compute
also the asymptotic and the local expansions of solutions. Algorithms of
Power Geometry are applicable to equations of various types: algebraic,
ordinary differential and partial differential, and also to systems of such
equations. Power Geometry is an alternative to Algebraic Geometry (a
resolution of complicated singularities of an algebraic manifold and finding asymptotics of its branches [3, Ch. 2]), to Group Analysis (finding
the group of symmetries of a system of equations and its self-similar

H.G.W. Begehr et al. (eds.), Analysis and Applications - ISAAC 2001, 51–71.

solutions [3, Ch. 7; 4]), to Nonstandard Analysis (a regularization of singular perturbations [3, Ch. 6, Sect. 1; 7]), to Microlocal Analysis [8] and to other disciplines.

First elements of the plane Power Geometry were proposed by I. Newton for one algebraic equation. The space Power Geometry was first proposed in [1] for the autonomous ODE system. An introduction to Power Geometry for algebraic equations and for autonomous ODE systems see in [2]. The more advanced presentation for all types of equations see in [3–6,9,10].

Applications to some complicated problems from Robotics, Mechanics, Celestial Mechanics, Hydrodynamics and Thermodynamics showed the effectiveness of this approach (see [3, Ch. 8, Sect. 5]). In particular, the strictly mathematical justification of the boundary layer theory was given [3, Ch. 6, Sect. 6] (see also [12]). Recent applications to motions of a rigid body see in [13–16].

Here in Sections 1–6 methods of Power Geometry are explained for the simplest case when we have an equation and we look for its solutions in the form of power expansions; in Section 7 all power expansions of solutions to the fourth Painleve equation are found; in Section 8 a system of equations is shortly considered.

1.   Support [3,4]

Let $X' = (x_1,\ldots,x_{n-1})$ be independent variables, x_n be the dependent variable and $X = (X',x_n) \in \mathbb{C}^n$. For $R = (r_1,\ldots,r_n) \in \mathbb{R}^n$, we denote $X^R = x_1^{r_1}\ldots x_n^{r_n}$. The *power–differential monomial* $a(X)$ is a product of the usual monomial $\text{const}\, X^R$ and of a finite number of derivatives $\partial^k x_n/\partial X'^{K'}$, where $K' = (k_1,\ldots,k_{n-1})$ and $k = \|K'\| \overset{\text{def}}{=} k_1+\ldots+k_{n-1}$. To each power–differential monomial $a(X)$ we put in correspondence its vectorial *power exponent* $Q(a) \in \mathbb{R}^n$ by the following rules:

$$Q(\text{const}\ X^R) = R,$$

$$Q(\partial^k x_n/\partial X'^{K'}) = (-K',1),$$

$$Q(ab) = Q(a) + Q(b),$$

where $b(X)$ is a power–differential monomial. The sum $f(X) \overset{\text{def}}{=} \sum\limits_{i=1}^{s} a_i(X)$, were $a_i(X)$ are power–differential monomials, is called the *power–differential sum*. Its *support* $\mathbf{S}(f) \overset{\text{def}}{=} \{Q(a_i), i = 1,...,s\} \subset \mathbb{R}^n$ is formed by power exponents of its monomials.

Example 1 Let us consider the fourth Painleve equation

$$f(z,w) \overset{\text{def}}{=} -2ww'' + w'^2 + 3w^4 + 8zw^3 + 4(z^2 - a)w^2 + 2b = 0, \quad (1)$$

where constants $a, b \in \mathbb{C}$. We assume that $a, b \neq 0$. Here $n = 2$, $x_1 = z$, $x_2 = w$ and the support $\mathbf{S}(f)$ consists of 6 points

$$Q_1 = (-2, 2), \quad Q_2 = (0, 4), \quad Q_3 = (1, 3), \quad Q_4 = (2, 2),$$

$$Q_5 = (0, 2), \quad Q_6 = (0, 0) \quad (2)$$

(see Fig. 1).

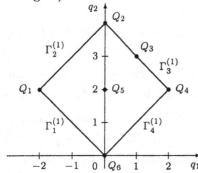

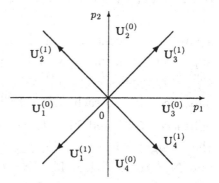

Figure 1. Support and polyhedron for Equation (1)

Figure 2. Normal cones to faces of the polyhedron in Fig 1.

2. Polyhedron and truncations [3]

Let us write the power–differential sum $f(X)$ in the form

$$f(X) \overset{\text{def}}{=} \sum f_Q(X) \text{ over } Q \in \mathbf{S}(f), \quad (3)$$

where $f_Q(X)$ are power–differential sums with point supports: $\mathbf{S}(f_Q) = \{Q\}$. Let $\Gamma(f)$ denote the closure of the convex hull of the support $\mathbf{S}(f)$. For the usual polynomial $f(X)$, it is called *the Newton polyhedron*. Its boundary $\partial\Gamma$ consists of *faces* $\Gamma_j^{(d)}$, where $d = \dim \Gamma_j^{(d)} < n$. Thus, $\Gamma_j^{(0)}$ is a vertex, $\Gamma_j^{(1)}$ is an edge and so on. Let the space $\mathbb{R}_*^n$ be dual to $\mathbb{R}^n$, i.e. we have the *scalar product*

$$\langle P, Q \rangle = p_1 q_1 + \ldots + p_n q_n$$

for $P = (p_1, \ldots, p_n) \in \mathbb{R}_*^n$ and $Q \in \mathbb{R}^n$. To each face $\Gamma_j^{(d)}$, in the space $\mathbb{R}_*^n$ there correspond its *normal subspace* $\mathbf{N}_j^{(d)}$, formed by all vectors P orthogonal to the face $\Gamma_j^{(d)}$, and its *normal cone* $\mathbf{U}_j^{(d)}$, formed by all such vectors P that the intersection of the polyhedron Γ with its supporting hyperplane $\langle P, Q \rangle = \text{const} = \max \langle P, Q_i \rangle$ over $Q_i \in \mathbf{S}(f)$ is exactly the face $\Gamma_j^{(d)}$. Evidently, $\mathbf{N}_j^{(d)} \supset \mathbf{U}_j^{(d)}$. To each face $\Gamma_j^{(d)}$ we put in correspondence the *truncated power–differential sum*

$$\hat{f}_j^{(d)}(X) \overset{\text{def}}{=} \sum f_Q(X) \quad \text{over} \quad Q \in \mathbf{S}(f) \cap \Gamma_j^{(d)}. \tag{4}$$

If the vector $\log |X| \overset{\text{def}}{=} (\log |x_1|, \ldots, \log |x_n|)$ tends to infinity near the normal cone $\mathbf{U}_j^{(d)}$, then the truncation $\hat{f}_j^{(d)}(X)$ is the *first approximation* of the power–differential sum $f(X)$.

Example 2 (continuation of Example 1). The convex hull $\Gamma(f)$ of the support (2) is a square with 4 vertices $\Gamma_1^{(0)} = Q_1$, $\Gamma_2^{(0)} = Q_2$, $\Gamma_3^{(0)} = Q_4$, $\Gamma_4^{(0)} = Q_6$ and 4 edges $\Gamma_1^{(1)}$, $\Gamma_2^{(1)}$, $\Gamma_3^{(1)}$, $\Gamma_4^{(1)}$ (see Fig. 1). The normal subspace $\mathbf{N}_j^{(0)}$ to each vertex $\Gamma_j^{(0)}$ is the whole plane $\mathbb{R}_*^2$. The normal subspaces $\mathbf{N}_1^{(1)}$ and $\mathbf{N}_3^{(1)}$ to the edges $\Gamma_1^{(1)}$ and $\Gamma_3^{(1)}$ respectively are the straight line $\{P = (p_1, p_2) : p_2 = p_1\}$. Normal subspaces $\mathbf{N}_2^{(1)}$ and $\mathbf{N}_4^{(1)}$ to the edges $\Gamma_2^{(1)}$ and $\Gamma_4^{(1)}$ respectively are the straight line $\{P = (p_1, p_2) : p_2 = -p_1\}$. But normal cones $\mathbf{U}_j^{(1)}$ to the edges $\Gamma_j^{(1)}$ are only halves of these straight lines:

$$\mathbf{U}_1^{(1)} = \{P = (p_1, p_2) : p_2 = p_1 < 0\},$$

$$\mathbf{U}_3^{(1)} = \{P = (p_1, p_2) : p_2 = p_1 > 0\},$$

$$\mathbf{U}_2^{(1)} = \{P = (p_1, p_2) : p_2 = -p_1 > 0\},$$

$$\mathbf{U}_4^{(1)} = \{P = (p_1, p_2) : p_2 = -p_1 < 0\}$$

(see Fig. 2). Normal cones $\mathbf{U}_j^{(0)}$ to the vertices $\Gamma_j^{(0)}$ are angles between the rays $\mathbf{U}_j^{(1)}$ and $\mathbf{U}_{j+1}^{(1)}$ for $j = 1, 2, 3, 4$ and $\mathbf{U}_5^{(1)} = \mathbf{U}_1^{(1)}$ (see Fig. 2). To the vertex $\Gamma_1^{(0)} = Q_1 = (-2, 2)$ there corresponds the truncated power–differential sum

$$\hat{f}_1^{(0)} = -2ww'' + w'^2. \tag{5}$$

To the edge $\Gamma_1^{(1)}$ there corresponds the truncated power–differential sum

$$\hat{f}_1^{(1)} = -2ww'' + w'^2 + 2b. \tag{6}$$

3. Truncated solutions and truncated equations [3]

Let Equation $f(X) = 0$ have the solution $x_n = \varphi(X')$, where $\varphi(X')$ is a *power sum*. The support $\mathbf{S}(x_n - \varphi)$ consists of the point $E_n = (0, \ldots, 0, 1)$ and of the support $\mathbf{S}(\varphi) \subset \mathbb{R}^{n-1}$. Faces of the polyhedron $\Gamma(x_n - \varphi)$ are denoted as $\gamma_l^{(e)}$. To each face $\gamma_l^{(e)}$ containing the point E_n and a point of $\mathbf{S}(\varphi)$ there correspond the *truncated solution* $x_n = \hat{\varphi}_l^{(e)}(X')$ and the normal subspace $\mathbf{n}_l^{(e)}$ and the normal cone $\mathbf{u}_l^{(e)}$. If the intersection $\mathbf{U}_j^{(d)} \cap \mathbf{u}_l^{(e)}$ is not empty then the truncated solution $x_n = \hat{\varphi}_l^{(e)}(X')$ is a solution to the *truncated equation* $\hat{f}_j^{(d)}(X) = 0$ [3, Ch. 6, Theorem 1.1].

Example 3 (continuation of Examples 1 and 2). The solution

$$x_2 = \sum_{k=1}^{\infty} c_k x_1^k \overset{\text{def}}{=} \varphi(x_1), \quad c_1 \neq 0 \tag{7}$$

has the support $\mathbf{S}(x_2 - \varphi)$ consisting of the points Q :

$$E_2 = (0,1); \quad (k,0), \quad k = 1, 2, \ldots \tag{8}$$

(see Fig. 3). The polyhedron

$$\Gamma(x_2 - \varphi) = \{Q = (q_1, q_2) : 0 \leq q_2 \leq 1, \; q_1 + q_2 \geq 1\}$$

is a part of the strip (see Fig. 3). Its only face containing the point $E_2 = (1,0)$ and a point from $\mathbf{S}(\varphi)$ is the edge $\gamma_1^{(1)}$ between points E_2 and $E_1 = (1,0)$ (see Fig. 3). The normal subspace $\mathbf{n}_1^{(1)}$ is the straight line $\{P = (p_1, p_2) : p_2 = p_1\}$, the normal cone $\mathbf{u}_1^{(1)}$ is the ray $\{P = (p_1, p_2) : p_2 = p_1 < 0\}$ (see Fig. 4). To the edge $\gamma_1^{(1)}$ there corresponds the truncated solution

$$x_2 = \hat{\varphi}_1^{(1)}(x_1) \overset{\text{def}}{=} c_1 x_1. \tag{9}$$

According to Example 2 for Equation (1) and the solution (7), the intersection $\mathbf{U}_j^{(d)} \cap \mathbf{u}_1^{(1)}$ is not empty only for $\mathbf{U}_j^{(d)} = \mathbf{U}_1^{(1)}$. So if the Equation (1) has the solution (7) then the truncated solution (9) is a solution of the corresponding truncated Equation

$$\hat{f}_1^{(1)} \overset{\text{def}}{=} -2ww'' + w'^2 + 2b = 0$$

(see (6)). Hence the constant $c_1 \neq 0$ must satisfy the algebraic equation $c_1^2 + 2b = 0$.

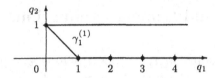

Figure 3. The support and polyhedron of the series (7)

Figure 4. The normal subspace and the normal cone to the edge $\gamma_1^{(1)}$ of Fig. 3

Note that the solution $x_2 = c_1 x_1 + \sum_{k=0}^{\infty} c_k x_1^{-k}$ has the same truncation (9) as (7) with the same normal subspace $\mathbf{n}_1^{(1)}$ but with the different normal cone $\{P = (p_1, p_2) : p_2 = p_1 > 0\} \neq \mathbf{u}_1^{(1)}$.

A priory we do not know the normal cone $\mathbf{u}_l^{(e)}$, so we come to the **problem**: find all such solutions $x_n = \hat{\varphi}(X')$ to the given truncated equation $\hat{f}_j^{(d)}(X) = 0$ that the intersection of their normal subspaces

$$\mathbf{N}(x_n - \hat{\varphi}) \cap \mathbf{N}_j^{(d)} \neq 0 \qquad (10)$$

is not zero. According to Theorem 2 [4] or 4.1 [3, Ch. 7], generically $\mathbf{N}(x_n - \hat{\varphi}) \subset \mathbf{N}_j^{(d)}$ and the property (10) means that $\mathbf{N}(x_n - \hat{\varphi}) \neq 0$. The problem is considered in the next Section, where the indices (d) and j are omitted.

4. Solution of a truncated equation [3,4]

The multiplication of the power–differential sum $f(X)$ by the monomial X^T gives the power-differential sum $X^T f(X)$ and induces the parallel translation of the support: $\mathbf{S}(X^T f) = T + \mathbf{S}(f)$.

Let $\log X = (\log x_1, \ldots, \log x_n)$. The *power transformation* $\log X = (\log Y)A$, with the property $\log X' = (\log Y')A'$, where A and A' are real nonsingular $n \times n$ and $(n-1) \times (n-1)$ matrices respectively, transforms the power–differential sum $f(X)$ into the power-differential sum $g(Y) = f(X)$. Here $\mathbf{S}(g) = \mathbf{S}(f)A^*$, where A^* is the transposed matrix A.

If the dimension $\dim \Gamma(f) \stackrel{\text{def}}{=} d = 0$, i.e. $\mathbf{S}(f) = \{Q\}$, then $g(X) = X^{-Q} f(X)$ and $g(X', X'^{R'}) = \chi(R')$, where $\chi(R')$ is a polynomial; its roots R' give all solutions $x_n = \text{const} X'^{R'}$ with the dimension $e = 1$ to Equation $f(X) = 0$.

Example 4 (continuation of Examples 1, 2, 3). According to (5), to the vertex $\Gamma_1^{(0)} = Q_1 = (-2, 2)$ there corresponds the truncated Equation

$$\hat{f}_1^{(0)}(z, w) \stackrel{\text{def}}{=} -2ww'' + w'^2 = 0. \qquad (11)$$

Here $d = 0$ and we put

$$g(z, w) = X^{-Q_1} \hat{f}_1^{(0)} = z^2 w^{-2} \hat{f}_1^{(0)} = -2w'' z^2 w^{-1} + (zw'/w)^2.$$

For $w = z^r$, we have

$$g(z, z^r) = -2r(r-1) + r^2 = -r^2 + 2r \overset{\text{def}}{=} \chi(r).$$

Roots of the polynomial $\chi(r)$ are $r = r_1 = 0$ and $r = r_2 = 2$. So Equation (11) has two types of solutions: $w = c$ and $w = cz^2$, where $c = \text{const} \neq 0$. Here $\mathbf{N}_1^{(0)} = \mathbb{R}_*^2$. The support of the first type solution is $\mathbf{S}(w - c) = \{Q : (0, 1), (0, 0)\}$ and its polyhedron γ_1 is the edge between points 0 and E_2 (see Fig. 5). Its normal subspace $\mathbf{n}_1 \overset{\text{def}}{=} \mathbf{N}(w - c) = \{P = (p_1, p_2) : p_2 = 0\}$ (see Fig. 6). The intersection $\mathbf{n}_1 \cap \mathbf{U}_1^{(0)} = \{P : p_1 < 0, p_2 = 0\}$ is a ray. Hence solutions $w_1 = c$ can be the first approximations of solutions to Equation (1).

Figure 5. The support and the polyhedron of the truncated solution $w = c$ Figure 6. The normal subspace and the normal cone of the edge γ_1 of Fig. 5

The support of the second type solutions $w = cz^2$ is $\mathbf{S}(w - cz^2) = \{Q : (0, 1), (2, 0)\}$. It and its convex hull γ_2 are shown in Fig. 7. Its normal subspace

$$\mathbf{n}_2 \overset{\text{def}}{=} \mathbf{N}(w - cz^2) = \{P : p_2 = 2p_1\} \quad \text{(see Fig. 8)}.$$

The intersection $\mathbf{n}_2 \cap \mathbf{U}_1^{(0)} = \emptyset$ is empty. Hence these solutions $w_1 = cz^2$ cannot be the first approximations for solutions to Equation (1) and must be rejected.

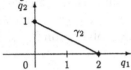

Figure 7. The support and the polyhedron of the truncated solution $w = cz^2$ Figure 8. The normal subspace of the edge γ_2 of Fig. 7

If $0 < d < n$ then by the mentioned transformations, Equation $f(X) = 0$ can be transformed into Equation $g(Y) = 0$ with $\mathbf{S}(g) \subset \{Q : q_1 = \ldots = q_{n-d} = 0\}$. Equation $g(Y) = 0$ has the *self-similar* (or quasihomogeneous) *solutions* which do not depend of several or all variables $y_1, \ldots, y_{n-d}$. Equation $g(Y) = 0$ for such solutions is simpler than for arbitrary ones and can be solved easier. If the self–similar solution $y_n = \psi(Y')$ to Equation $g(Y) = 0$ was found then the inverse power transformation $Y \to X$ gives the self–similar solution $x_n = \varphi(X')$ to Equation $f(X) = 0$.

Example 5 (continuation of Examples 1, 2, 3). According to (6), to the edge $\Gamma_1^{(1)}$ there corresponds the truncated equation

$$\hat{f}_1^{(1)}(z, w) \overset{\text{def}}{=} - 2ww'' + w'^2 + 2b = 0. \tag{12}$$

Its normal subspace $\mathbf{N}_1^{(1)} = \{P : p_2 = p_1\}$ (see Fig. 2). Hence its solution $w = \varphi(z)$ with nonempty intersection $\mathbf{N}_1^{(1)} \cap \mathbf{N}(w - \varphi(z))$ has the form $w = cz$. Substituting it into Equation (12), we obtain the algebraic equation $c^2 + 2b = 0$, which has two roots

$$c = \pm\sqrt{-2b}. \tag{13}$$

In that computation in fact, we made the power transformation $z = y_1$, $w = y_1 y_2$ with the matrix

$$A = \begin{pmatrix} 1 & 1 \\ 0 & 1 \end{pmatrix}$$

and the matrix $A' = (1)$.

5. Unique solutions [5, 9]

Let us consider Equation

$$f(X) \overset{\text{def}}{=} \mathcal{L}(X')x_n + g(X) + h(X') = 0, \tag{14}$$

where $f(X)$ is a power–differential sum, $\mathcal{L}(X')$ is a linear differential operator, $g(X)$ is such a power differential sum, that its each power-differential monomial depends on x_n, $h(X')$ is a power sum and the term $\mathcal{L}(X')x_n$ corresponds to a face Γ_0 of the polyhedron $\Gamma(f)$. Let we look for its solutions of the form

$$x_n = \sum c_{K'} X'^{K'} \quad \text{over} \quad K' \in \mathbf{K}, \tag{15}$$

where $\mathbf{K} \subset \mathbb{R}^{n-1}$ is a discrete set that can be described via $\mathbf{S}(f)$.

Condition 1 The face Γ_0 is a vertex $V = (V', 1)$ of the polyhedron $\Gamma(f)$.

On the set $\mathbf{K}$ we introduce the function

$$\nu(K') \overset{\text{def}}{=} (X')^{-K'-V'} \mathcal{L}(X') X'^{K'} \tag{16}$$

and two subsets of the set $\mathbf{K}$:

$$\mathbf{L} = \{K' : \nu(K') = 0\}, \quad \mathbf{M}' = \{K' : \nu(K') \neq 0\}. \tag{17}$$

Let $\mathbf{M}$ be a subset of the set $\mathbf{M}'$ with the following properties.

Condition 2 If $K_1', \ldots, K_l' \in \mathbf{M}$ then $K_1' + \ldots + K_l' \in \mathbf{M}$.
It means that the set $\mathbf{M}$ is closed under summation.

Condition 3 $\mathbf{S}(g(X', X'^{K'})) \subset \mathbf{M} + V'$ for any $K' \in \mathbf{K}$.
It means that after the substitution (15), the sum $g(X)$ gives only terms with vectorial power exponents laying in $\mathbf{M} + V'$.

Condition 4 $\mathbf{S}(h(X')) \subset \mathbf{M} + V'$.
If Equation (14) and the set $\mathbf{M}$ satisfy Conditions 1–4 then for any initial data

$$x_n^{**} = \sum c_{K'} X'^{K'} \quad \text{over} \ K' \in \mathbf{L}, \tag{18}$$

Equation (14) has the unique formal solution

$$x_n = x_n^* + x_n^{**}, \tag{19}$$

where

$$x_n^* = \sum c_{K'} X'^{K'} \quad \text{over} \ K' \in \mathbf{M}. \tag{20}$$

Condition 5 In each power–differential monomial of the power–differential sum $g(X)$ the sum of orders $l = \|L'\|$ of derivatives $\partial^l x_n / \partial X'^{L'}$ does not exceed the maximal order of derivations in the operator $\mathcal{L}(X')$.
Denote $\mathbf{M}_k \overset{\text{def}}{=} \mathbf{M} \cap \{K' : \|K'\| \leq 2^k\}$ and

$$\omega_k \overset{\text{def}}{=} \begin{cases} 1, & \text{if } \mathbf{M}_k \text{ is empty;} \\ \min |\nu(K')| \ \text{over} \ K' \in \mathbf{M}_k, & \text{if } \mathbf{M}_k \neq \emptyset. \end{cases}$$

Condition 6 on small divisors [2].

$$\sum_{k=0}^{\infty} \frac{\log \omega_k}{2^k} > -\infty.$$

Conditions 5 and 6 are essential for differential equations only, and Condition 6 is essential only for partial ones.

Under additional Conditions 5 and 6 on $f(X)$, the solution expansion (20) converges absolutely in a corresponding domain in $\mathbb{C}^{n-1}$ if the power-differential sum $f(X)$ and the power sum (18) are analytic. The results extend theorems of Cauchy and Kovalevskaya, but also consider small divisors.

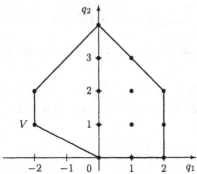

Figure 9. The support and the polyhedron of Equation (21)

Example 6 Let us consider the Equation

$$\tilde{f}(z, \tilde{w}) \stackrel{\text{def}}{=} -2(c + \tilde{w})\tilde{w}'' + \tilde{w}'^2 + 3(c + \tilde{w})^4 + 8z(c + \tilde{w})^3 +$$

$$+ 4(z^2 - a)(c + \tilde{w})^2 + 2b = 0, \tag{21}$$

where $c \neq 0$ is a constant. Its support and polyhedron are shown in Fig. 9. Equation (21) is obtained from Equation (1) after the substitution

$$w = c + \tilde{w}. \tag{22}$$

Equation (21) has the form (14), where

$$\mathcal{L}(z)\tilde{w} = -2c\tilde{w}'', \tag{23}$$

$$h(z) = 3c^4 + 8c^3 z + 4(z^2 - a)c^2 + 2b \tag{24}$$

and $g(z, w)$ contains the remaining power–differential monomials. Here $\Gamma(\mathcal{L}\tilde{w}) = (-2, 1)$ is the vertex $V = (v_1, 1)$ of $\Gamma(\tilde{f})$, where $v_1 = -2$ (see Fig. 9). For the expansion

$$\tilde{w} = \sum_{k=1}^{\infty} c_k z^k, \tag{25}$$

we have the set $\mathbf{K} = \{k : k \text{ are natural numbers}\}$ and the function

$$\nu(k) = x_1^{-v_1 - k} \mathcal{L}(x_1) x_1^k = z^{2-k} \mathcal{L}(z) z^k = -2ck(k-1).$$

Equation $\nu(k) = 0$ has two roots $k = 0$ and $k = 1$. The first root $k = 0$ is outside of the set $\mathbf{K}$ and the second one $k = 1$ is in $\mathbf{K}$. Hence the set

$\mathbf{L} = \{1\}$ and the set $\mathbf{M} = \mathbf{M}' = \{k : k \in \mathbb{Z}, \ k \geq 2\}$. It is not difficult to check that Conditions 2–4 are here satisfied. For instance, in Equation (21) according to (24) $h(z) = a_0 + a_1 z + a_2 z^2$, where $a_i = \text{const}$. Hence $\mathbf{S}(h) \subset \mathbf{M} + v_1 = \{k \in \mathbb{Z}, \ k \geq 0\}$ and Condition 4 is fulfilled. So for each constant $c \neq 0$, Equation (21) has solutions (25), where c_1 is an arbitrary constant and other c_k are uniquely determined. According to (23) and (21), Condition 5 is fulfilled. So the series (25) converges.

6. Computation of local and asymptotic expansions of solutions to Equation f(X) = 0 [5,9]

Via the support $\mathbf{S}(f)$ we compute all faces $\Gamma_j^{(d)}$ of the polyhedron $\Gamma(f)$ and their normal cones $\mathbf{U}_j^{(d)}$ [3, Ch. 1; 11]. To each face $\Gamma_j^{(d)}$ there corresponds the truncated equation $\hat{f}_j^{(d)}(X) = 0$. Its solutions $x_n = \hat{\varphi}(X')$ with the property (10) are found by methods of Section 4. Among them we select only those for which the normal subspace $\mathbf{N}(x_n - \hat{\varphi})$ intersects the normal cone $\mathbf{U}_j^{(d)}$. Let $x_n = \hat{\varphi}(X')$ be one of such solutions and after the coordinate change

$$x_n = \hat{\varphi}(X') + \tilde{x}_n \tag{26}$$

we obtain the power–differential sum

$$\tilde{f}(X', \tilde{x}_n) \overset{\text{def}}{=} f(X', \hat{\varphi}(X') + \tilde{x}_n).$$

Equation $\tilde{f}(X', \tilde{x}_n) = 0$ has form (14) where x_n means $\tilde{x}_n$ and the operator $\mathcal{L}(X')$ is the *Frechet derivative* $\delta \hat{f}_j^{(d)} / \delta x_n$ taken on the solution $x_n = \hat{\varphi}(X')$. Let $\mathcal{L}(X') \not\equiv 0$ and the face $\Gamma_0 = \Gamma(\mathcal{L}\tilde{x}_n)$ be a vertex of the polyhedron $\Gamma(\tilde{f})$. If the set $\mathbf{L}$ is empty and Conditions 2–4 are satisfied then the further expansion of the solution $\tilde{x}_n = \hat{\varphi}(X')$ exists and unique. If the set $\mathbf{L}$ is not empty, but Conditions 2–4 are satisfied, then there exists a family of solutions of the form (19), (18), (20) with parameters $c_{K'}$ for $K' \in \mathbf{L}$. If the set $\mathbf{L} \neq \emptyset$ is not empty, but Conditions 2–4 are not fulfilled, then we must compute the initial part of the expansion (15), to come to the situation, described in the previous sentence. Such initial part can exist under some restrictions on coefficients of terms in the power-differential sum $f(X)$. If $\mathcal{L} \equiv 0$ or the face $\Gamma(\mathcal{L}\tilde{x}_n)$ is not a vertex of the polyhedron $\Gamma(\tilde{f})$, then computing the support $\mathbf{S}(\tilde{f})$ and the polyhedron $\Gamma(\tilde{f})$ we can find the next approximation of the solution. We must continue such computations step by step till we come to the situation of Section 5.

62

Example 7 (continuation of Examples 1, 2, 3, 4, 6). For Equation (1), we take the truncated Equation (11). According to Example 4, it has the appropriate solution $w = c = \text{const} \neq 0$. Here the coordinate change (26) is (22) and it transforms Equation (1) into Equation (21). Here the Frechet derivative

$$\frac{\delta \hat{f}_1^{(0)}}{\delta w} = -2w'' - 2w\frac{d^2}{dz^2} + 2w'\frac{d}{dz}. \tag{27}$$

On the truncated solution $w = c$ it is

$$\mathcal{L}(z) = -2c\frac{d^2}{dz^2} \neq 0$$

in accordance with (23). Thus here results of Example 6 are applicable.

7. The fourth Painleve equation

Here we apply the exposed theory to find all power expansions of solutions to the fourth Painleve equation

$$f(z, w) \stackrel{\text{def}}{=} -2ww'' + w'^2 + 3w^4 + 8zw^3 + 4(z^2 - a)w^2 + 2b = 0, \tag{28}$$

some times repeating Examples 1–7. We assume that constants $a, b \neq 0$. Other cases can be considered by the reader as exercises. Here $n = 2$, $x_1 = z$, $x_2 = w$ and the support $\mathbf{S}(f)$ consists of 6 points 5

$$Q_1 = (-2, 2), \quad Q_2 = (0, 4), \quad Q_3 = (1, 3), \quad Q_4 = (2, 2),$$

$$Q_5 = (0, 2), \quad Q_6 = (0, 0)$$

(see Fig. 1). The convex hull $\Gamma(f)$ of the support $\mathbf{S}(f)$ is a square with 4 vertices $\Gamma_1^{(0)} = Q_1$, $\Gamma_2^{(0)} = Q_2$, $\Gamma_3^{(0)} = Q_4$, $\Gamma_4^{(0)} = Q_6$ and 4 edges $\Gamma_1^{(1)}, \Gamma_2^{(1)}, \Gamma_3^{(1)}, \Gamma_4^{(1)}$ (see Fig. 1). Their normal cones are shown in Fig. 2.

7.1 The vertex $\Gamma_1^{(0)}$ We begin from the vertex $\Gamma_1^{(0)} = Q_1$. The corresponding truncated equation is

$$\hat{f}_1^{(0)}(z, w) \stackrel{\text{def}}{=} -2ww'' + w'^2 = 0. \tag{29}$$

Here $d = 0$ and according to Section 4, we put

$$g(z, w) = X^{-Q_1}\hat{f}_1^{(0)} = z^2w^{-2}\hat{f}_1^{(0)} = -2w''z^2w^{-1} + (zw'/w)^2.$$

For $w = z^r$, we have

$$g(z, z^r) = -2r(r - 1) + r^2 = -r^2 + 2r \stackrel{\text{def}}{=} \chi(r).$$

Roots of the polynomial $\chi(r)$ are $r = r_1 = 0$ and $r = r_2 = 2$. So Equation (29) has two types of solutions: $w = c$ and $w = cz^2$, where $c = \text{const} \neq 0$. Here the normal subspace $\mathbf{N}_1^{(0)} = \mathbb{R}_*^2$. The support of the first type solution is $\mathbf{S}(w - c) = \{Q : (0, 1), (0, 0)\}$ (see Fig. 5). Its normal subspace $\mathbf{n}_1 \stackrel{\text{def}}{=} \mathbf{N}(w - c) = \{P = (p_1, p_2) : p_2 = 0\}$ (see Fig. 6). The intersection $\mathbf{n}_1 \cap \mathbf{U}_1^{(0)} = \{P : p_1 < 0, p_2 = 0\}$ is a ray. Hence solutions $w_1 = c$ can be the first approximations of solutions to Equation (28).

The support of the second type solutions $w = cz^2$ is $\mathbf{S}(w - cz^2) = \{Q : (0, 1), (2, 0)\}$ (see Fig. 7). Its normal subspace $\mathbf{n}_2 \stackrel{\text{def}}{=} \mathbf{N}(w - cz^2) = \{P : p_2 = 2p_1\}$ (see Fig. 8). The intersection $\mathbf{n}_2 \cap \mathbf{U}_1^{(0)} = \emptyset$ is empty. Hence these solutions $w_1 = cz^2$ cannot be the first approximations for solutions to Equation (28) and must be rejected.

Now for the first type solutions, we make the coordinate change (26)

$$w = c + \tilde{w},$$

then Equation (28) takes the form

$$\tilde{f}(z, \tilde{w}) \stackrel{\text{def}}{=} \mathcal{L}(z)\tilde{w} + g(z, \tilde{w}) + h(z) = 0. \tag{30}$$

The support $\mathbf{S}(\tilde{f})$ and the polyhedron $\Gamma(\tilde{f})$ are shown in Fig. 9. The Frechet derivative

$$\frac{\delta \hat{f}_1^{(0)}}{\delta w} = -2w'' - 2w\frac{d^2}{dz^2} + 2w'\frac{d}{dz}. \tag{31}$$

On the truncated solution $w = c$ it is

$$\mathcal{L}(z) = -2c\frac{d^2}{dz^2} \neq 0. \tag{32}$$

$\Gamma(\mathcal{L}\tilde{w}) = (-2, 1)$ is the vertex $V = (v_1, 1)$ of $\Gamma(\tilde{f})$, where $v_1 = -2$ (see Fig. 9). For the expansion

$$\tilde{w} = \sum_{k=1}^{\infty} c_k z^k, \tag{33}$$

we have the set $\mathbf{K} = \{k : k \text{ are natural numbers}\}$ and the function

$$\nu(k) = x_1^{-v_1 - k}\mathcal{L}(x_1)x_1^k = z^{2-k}\mathcal{L}(z)z^k = -2ck(k - 1).$$

Equation $\nu(k) = 0$ has two roots $k = 0$ and $k = 1$. The first root $k = 0$ is outside of the set $\mathbf{K}$ and the second one $k = 1$ is in $\mathbf{K}$. Hence the

64

set $L = \{1\}$ and the set $M = M' = \{k : k \in \mathbb{Z}, \ k \geq 2\}$. It is not difficult to check that Conditions 2–4 are here satisfied. For instance, in Equation (30) here $h(z) = a_0 + a_1 z + a_2 z^2$, where $a_i = \text{const}$. Hence $S(h) \subset M + v_1 = \{k \in \mathbb{Z}, \ k \geq 0\}$ and Condition 4 is fulfilled. So for each constant $c \neq 0$, Equation (30) has solutions (33), where c_1 is an arbitrary constant and other c_k are uniquely determined. Thus to the vertex $\Gamma_1^{(0)}$ there corresponds the two parameter family of solutions to Equation (28)

$$w = c + \sum_{k=1}^{\infty} c_k w^k, \tag{34}$$

where $c \neq 0$ and c_1 are parameters of the family. The series converges, because Condition 5 for Equation (30), (32) is satisfied.

7.2 The edge $\Gamma_1^{(1)}$ To it there corresponds the truncated equation

$$\hat{f}_1^{(1)}(z, w) \overset{\text{def}}{=} -2ww'' + w'^2 + 2b = 0. \tag{35}$$

Its normal subspace $N_1^{(1)} = \{P : p_2 = p_1\}$ (see Fig. 2). Hence its solution $w = \varphi(z)$ with nonempty intersection $N_1^{(1)} \cap N(w - \varphi(z))$ has the form $w = cz$, $c \neq 0$. Substituting it into Equation (35), we obtain the algebraic equation $c^2 + 2b = 0$, which has two roots $c = \pm\sqrt{-2b}$. After the substitution (26) $w = cz + \tilde{w}$, Equation (28) takes the form (30). The support $S(\tilde{f})$ and the polygon $\Gamma(\tilde{f})$ are shown in Fig. 10. The Frechet derivative $\delta \hat{f}_1^{(1)} / \delta w$ equals to (31). On the truncated solution $w = cz$ it gives

$$\mathcal{L}(z) = -2cz \frac{d^2}{dz^2} + 2c \frac{d}{dz} \neq 0.$$

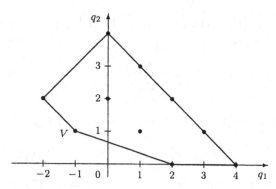

Figure 10. The support and its polyhedron of the sum $\tilde{f}(z, \tilde{w}) \overset{\text{def}}{=} f(z, cz + \tilde{w})$

Here $\mathbf{S}(\mathcal{L}\tilde{w}) = V = (v_1, 1) = (-1, 1)$ (see Fig. 10). For the expansion

$$\tilde{w} = \sum_{k=2}^{\infty} c_k z^k, \tag{36}$$

we have the set $\mathbf{K} = \{k : k \in \mathbb{Z}, \ k \geq 2\}$ and the function

$$\nu(k) = z^{-v_1-k}\mathcal{L}(z)z^k = -2ck(k-1) + 2ck = -2ck(k-2).$$

Hence the set $\mathbf{L} = \{2\}$ and $\mathbf{M} = \mathbf{M}' = \mathbf{K} \cap \{k \geq 3\}$. It is easy to check that Conditions 2–5 are fulfilled. So in the solution expansion (36) the coefficient c_2 is arbitrary and all other coefficients c_k are uniquely determined and the series converges. Thus to the edge $\Gamma_1^{(1)}$ there correspond two one-parameter (c_2) families of solutions

$$w = cz + \sum_{k=2}^{\infty} c_k z^k, \tag{37}$$

where $c = \pm\sqrt{-2b}$, $c_2 \in \mathbb{C}$ is arbitrary and all other c_k are uniquely determined and the series converges.

7.3 The edge $\Gamma_2^{(1)}$ To it there correspond the truncated equation

$$\hat{f}_2^{(1)} \stackrel{\text{def}}{=} -2ww'' + w'^2 + 3w^4 = 0 \tag{38}$$

and the normal cone $\mathbf{U}_2^{(1)} = \{P : p_2 = -p_1 > 0\}$. Hence we must look for the solutions $w = cz^{-1}$ with $c \neq 0$ to the truncated Equation (38). After such substitution in Equation (38), we obtain the algebraic Equation $-4c^2 + c^2 + 3c^4 = 0$. Its nonzero solutions are $c = \pm 1$. The Frechet derivative

$$\frac{\delta \hat{f}_2^{(1)}}{\delta w} = -2w'' - 2w\frac{d^2}{dz^2} + 2w'\frac{d}{dz} + 12w^3.$$

On the truncated solution $w = cz^{-1}$ it is

$$\mathcal{L}(z) = -\frac{2c}{z}\frac{d^2}{dz^2} - \frac{2c}{z^2}\frac{d}{dz} + \frac{8c}{z^3} \neq 0. \tag{39}$$

Here $\Gamma(\mathcal{L}\tilde{w}) = (-3, 1)$ is the vertex $V = (v_1, 1)$ of $\Gamma(\tilde{f})$, where $v_1 = -3$. For the expansion

$$\tilde{w} = \sum_{k=0}^{\infty} c_k z^k, \tag{40}$$

we have the set $\mathbf{K} = \{k : k \in \mathbb{Z}, \; k \geq 0\}$ and the function

$$\nu(k) = x_1^{-\nu_1 - k} \mathcal{L}(x_1) x_1^k = z^{3-k} \mathcal{L}(z) z^k = -2ck(k-1) - 2ck + 8c =$$

$$= -2c(k^2 - 4) = -2c(k-2)(k+2).$$

It has two roots $k = -2$ and $k = 2$. Only one root $k = 2$ of $\nu(k)$ is in the set $\mathbf{K}$. So the set $\mathbf{L} = \{k = 2\}$, and the set $\mathbf{M}' = \mathbf{K} \setminus \mathbf{L}$ does not satisfy Condition 2, because $k = 1 \in \mathbf{M}'$ and $1 + 1 = 2 \notin \mathbf{M}'$. If we take the set $\mathbf{M} = \{k : k = 0; \; k \in \mathbb{Z}, \; k \geq 3\}$, then Condition 2 is fulfilled but Condition 4 is violated. Indeed, after the substitution (26) $w = cz^{-1} + \tilde{w}$, Equation (28) has the form (30) where

$$h(z) = 8c^3 z^{-2} + 4c^2(z^2 - a)z^{-2} + 2b = 8cz^{-2} + 4(z^2 - a)z^{-2} + 2b,$$

because $c^2 = 1$. $\mathbf{S}(h) = \{k = -2, \; k = 0\}$ and $\mathbf{S}(h) \not\subset \mathbf{M} + v_1 = \{k : k = -3; \; k \in \mathbb{Z}, \; k \geq 0\}$. So in the solution expansion (40) we must find coefficients c_0 and c_1. After the substitution $w = cz^{-1} + c_0 + c_1 z$ with $c^2 = 1$ in Equation (28), and the comparison of coefficients for z^{-3} and z^{-2}, we obtain equations $c_0 = 0$ and $\mathcal{L}(z)c_1 z + 8cz^{-2} - 4az^{-2} = 0$. According to (39), the last equation gives $-2cc_1 + 8cc_1 + 8c - 4a = 0$. Hence $c_1 = (2ac - 4)/3$. Now after the substitution

$$w = cz^{-1} + c_1 z + \tilde{\tilde{w}},$$

Equation (28) has the form (30) with $\tilde{\tilde{w}}$ instead of $\tilde{w}$ and with $\mathbf{S}(h) \subset \{k : k \in \mathbb{Z}, \; k \geq 0, \; k \text{ is even}\}$. Now for the sets $\mathbf{L} = \{k = 2\}$ and $\mathbf{M} = \{k : k \geq 3, \; k \in \mathbb{Z}\}$, all Conditions 2–4 are fulfilled. Hence solutions to the mentioned equation for $\tilde{\tilde{w}}$ have the form

$$\tilde{\tilde{w}} = \sum_{k=2}^{\infty} c_k z^k, \tag{41}$$

where c_2 is arbitrary and other c_k are uniquely determined. If $c_2 = 0$ then according to [5, Theorem 3.2], the expansion (41) contains the odd powers k only. The series (41) converge, because here according to (39), Condition 5 is fulfilled. Thus to the edge $\Gamma_2^{(1)}$ there correspond two one-parameter families of solutions

$$w = cz^{-1} + \sum_{k=1}^{\infty} c_k z^k, \tag{42}$$

where $c = \pm 1$, $c_1 = -4/3 + 2ac/3$, c_2 is arbitrary and other c_k are uniquely determined; if $c_2 = 0$, then the expansion (42) consists of terms of odd powers k only. The series converges.

7.4 The edge $\Gamma_3^{(1)}$ To it there correspond the algebraic truncated equation

$$\hat{f}_3^{(1)} \stackrel{\text{def}}{=} 3w^4 + 8zw^3 + 4z^2w^2 = 0 \tag{43}$$

and the normal cone $\mathbf{U}_2^{(1)} = \{P : p_2 = p_1 > 0\}$. So the corresponding truncated solution has the form $w = cz$, $c \neq 0$. Substituting it into the truncated Equation (43), we obtain the algebraic equation $\xi(c) \stackrel{\text{def}}{=} 3c^4 + 8c^3 + 4c^2 = 0$. It has two nonzero roots

$$c = -2 \quad \text{and} \quad c = -2/3. \tag{44}$$

After the substitution (26) $w = cz + \tilde{w}$, Equation (28) takes the form (30). The Frechet derivative

$$\frac{\delta \hat{f}_3^{(1)}}{\delta w} = \frac{\partial \hat{f}_3^{(1)}}{\partial w} = 12w^3 + 24zw^2 + 8z^2w.$$

On the truncated solution $w = cz$ it gives

$$\mathcal{L}(z) = (12c^3 + 24c^2 + 8c)z^3 = \xi'(c)z^3.$$

As $\xi'(c) \neq 0$ for values (44), then $\mathcal{L}(z) \not\equiv 0$. Here $\mathbf{S}(\mathcal{L}\tilde{w}) = V = (v_1, 1) = (3, 1)$. For the expansion

$$\tilde{w} = \sum_{l=0}^{\infty} c_l z^{-l}, \tag{45}$$

we have the set $\mathbf{K} = \{k : k \in \mathbb{Z}, \ k \leq 0\}$ and the function

$$\nu(k) = z^{-v_1 - k}\mathcal{L}(z)z^k = \xi'(c) = \text{const}.$$

Hence the set $\mathbf{L}$ is empty and $\mathbf{M} = \mathbf{M}' = \mathbf{K}$. It is easy to check that Conditions 2–4 are fulfilled. So in the solution expansion (45) all coefficients c_l are uniquely determined. According to Theorem 3.2 from [5], the expansion (45) contains odd powers l only. But the series (45) diverges, because Condition 5 is violated. Thus, to the edge $\Gamma_3^{(1)}$ there correspond two unique formal solutions to Equation (28)

$$w = cz + \sum_{l=1}^{\infty} c_l z^{-l}, \tag{46}$$

containing odd powers l only, where c has two values -2, $-2/3$. The series diverges.

7.5 The edge $\Gamma_4^{(1)}$ To it there correspond the algebraic truncated equation

$$\hat{f}_4^{(1)} \stackrel{\text{def}}{=} 4z^2w^2 + 2b = 0$$

and two formal solutions

$$w = cz^{-1} + \sum_{l=3}^{\infty} c_l z^{-l}, \tag{47}$$

containing the odd powers l only, where $c = \pm\sqrt{-b/2}$ and all constants c_l are uniquely determined. The series diverges. All that is found similarly to the edge $\Gamma_3^{(1)}$.

7.6 The vertices $\Gamma_2^{(0)}$, $\Gamma_3^{(0)}$, $\Gamma_4^{(0)}$ To them there correspond the algebraic monomial truncated equations $\text{const}\, z^\alpha w^\beta = 0$, which have no appropriate solutions.

7.7. Summary The fourth Painleve Equation (28) has only the following power expansions of its solutions.

(i) The two-parameter ($c \neq 0$ and c_1) family (34) with converging series.

(ii) The two one-parameter (c_2) families (37) with $c = \pm\sqrt{-2b}$ and converging series.

(iii) The two one-parameter (c_2) families (42) with $c = \pm 1$, $c_1 = -4/3 + 2ac/3$ and converging series.

(iv) The two formal solutions (46) with $c = -2$ and $c = -2/3$ and diverging series.

(v) The two formal solutions (47) with $c = \pm\sqrt{-b/2}$ and diverging series.

If $a = 0$ and $b \neq 0$ in Equation (28) then the support $\mathbf{S}(f)$ consists of points $Q_1 - Q_4$, Q_6 and the polyhedron $\Gamma(f)$ is the same square again and all results listed in Subsection 7.7 are valid. If $b = 0$ in Equation (28) then the support $\mathbf{S}(f)$ does not contain the point $Q_6 = 0$ and the polygon $\Gamma(f)$ is a triangle with vertices Q_1, Q_2, Q_4. In this case solutions, corresponding to the edges $\Gamma_2^{(1)}$ and $\Gamma_3^{(1)}$, are the same, but to the vertex $\Gamma_1^{(0)} = Q_1$ there correspond other solutions. Besides, solutions, corresponded to edges $\Gamma_1^{(1)}$ and $\Gamma_4^{(2)}$, are absent.

All power expansions of solutions to the first Painleve equation are given in [9].

8. The system of equations [3,6,10]

Let $W = (w_1, \ldots, w_m) \in \mathbb{C}^m$ be dependent variables, and $X' = (x_1, \ldots, x_{n-1})$ be independent variables as before. The *power–differential monomial* $a(X', W)$ is a product of the usual monomial const $X'^{R'} W^T$ and of a finite number of derivatives $\partial^k w_j / \partial X'^{K'}$, where $1 \leq j \leq m$ and $k = \|K'\|$. To each power–differential monomial $a(X', W)$ we put in correspondence its *vectorial power exponent* $Q(a) \in \mathbb{R}^{m+n-1}$ by the following rules:

$$Q(\text{const } X'^{R'} W^T) = (R', T),$$

$$Q(\partial^k w_j / \partial X'^{K'}) = (-K', E_j), \quad Q(ab) = Q(a) + Q(b),$$

where $b = b(X', W)$ is a power–differential monomial and E_j denotes the j-th unit vector. The sum $f(X', W) \overset{\text{def}}{=} \sum a_i(X', W)$, where $a_i(X', W)$ are power–differential monomials, is called the *power–differential sum*. Its *support* $\mathbf{S}(f) \overset{\text{def}}{=} \{Q(a_i)\} \subset \mathbb{R}^{m+n-1}$ is formed by power exponents of its monomials.

If we have the system

$$f_i(X', W) = 0, \quad i = 1, \ldots, m, \tag{48}$$

where all f_i are power–differential sums, then to each f_i there correspond its support $\mathbf{S}(f_i)$, its polyhedron $\Gamma(f_i)$ with faces $\Gamma_{ij}^{(d)}$, normal subspaces $\mathbf{N}_{ij}^{(d)}$, normal cones $\mathbf{U}_{ij}^{(d)}$ and truncated equations $\hat{f}_{ij}^{(d)}(X', W) = 0$. The system of truncated equations

$$\hat{f}_{ij_i}^{(d_i)}(X', W) = 0, \quad i = 1, \ldots, m \tag{49}$$

is the *truncated system*, if the intersection of corresponding normal cones

$$\mathbf{U}_{1j_1}^{(d_1)} \cap \ldots \cap \mathbf{U}_{mj_m}^{(d_m)} \overset{\text{def}}{=} \mathbf{U}_J^D \neq \emptyset \tag{50}$$

is not empty, where $D = (d_1, \ldots, d_m)$, $J = (j_1, \ldots, j_m)$.

The cone (50) is the *normal cone of the truncated system* (49). For the systems (48), there were developed a theory and algorithms which allow to find all power expansions

$$w_i = X'^{R'_i} \sum_{K'} c_{iK'} X'^{K'}, \quad i = 1, \ldots, m$$

of solutions to the system (48) and are similar to the described theory and algorithms for one equation. The theory for the system takes the especially simple form for the dynamical systems (48) when the number of independent variables $n - 1 = 1$ and all power–differential sums f_i have the form

$$f_i \overset{\text{def}}{=} - dw_i/dx_1 + \psi_i(x_1, W), \ i = 1, ..., m,$$

where ψ_i are polynomials or power series.

In [10] the approach was applied to the Lorenz system

$$\dot{w}_1 = s(w_2 - w_1), \quad \cdot \overset{\text{def}}{=} d/dt,$$
$$\dot{w}_2 = -w_1 w_3 + r w_1 - w_2, \ \dot{w}_3 = w_1 w_2 - b w_3. \tag{51}$$

It was shown, that for

$$r = 1 \ \text{and} \ b = -2s(s + 1)/(4s + 1),$$

it has the formal power expansions of solutions

$$x_1 = t^{-1/2}\left(c_1 + \sum_{k=1}^{\infty} c_{1k} t^{-k}\right),$$
$$x_2 = x_1 + t^{-3/2}\left(c_2 + \sum_{k=1}^{\infty} c_{2k} t^{-k}\right), \ x_3 = t^{-1}\left(c_3 + \sum_{k=1}^{\infty} c_{3k} t^{-k}\right), \tag{52}$$

where

$$c_1 = \pm\sqrt{b(s + 1)/(2s)}, \ c_2 = -c_1/(2s), \ c_3 = (s + 1)/(2s);$$

c_{11} is arbitrary constant and other constants c_{ik} are uniquely determined. The series diverge. If constants s and c_{11} are real and $4s + 1 < 0$ then all coefficients in the expansions (52) are real. The expansions (52) represent solutions with all $x_i \to 0$ when $t \to \infty$.

In [10] there are also all power expansions of solutions to the Lorenz system (51) and to the Hill problem.

Remark If some coordinate presents in Equation $f(X) = 0$ or in the system (48) but it is absent in all derivatives presented in them, then it may be chosen as an independent variable or as an dependent variable. The only rule here is that the number of dependent variables must be equal to the number of independent equations. Hence, equations and systems with small and big parameters can be included in the described approach.

Acknowledgments

The work was supported by RFBR, Grant 02-01-01067.

References

[1] Bruno, A.D.: The asymptotic behavior of solutions to nonlinear systems of differential equations. DAN SSSR **143**:4 (1962), 763–766 (Russian) = Soviet Math. Dokl. **3** (1962), 464-467.

[2] Bruno, A.D.: Local Methods in Nonlinear Differential Equations. Nauka, Moscow, 1979 (Russian) = Springer, Berlin, 1989.

[3] Bruno, A.D.: Power Geometry in Algebraic and Differential Equations. Fizmatlit, Moscow, 1998 (Russian) = Elsevier Science, Amsterdam, 2000.

[4] Bruno, A.D.: Self-similar solutions and power geometry. Uspekhi Mat. Nauk **55**:1 (2000) 3–44 (Russian) = Russian Math. Surveys **55**:1 (2000), 1–42.

[5] Bruno, A.D.: Power expansions of solutions to a single algebraic or differential equation. DAN **380**:2 (2001), 155–159 (Russian) = Doklady Mathematics **64**:2 (2001), 160–164.

[6] Bruno, A.D.: Power expansions of solutions to a system of algebraic and differential equations. DAN **380**:3 (2001) (Russian) 298–304 = Doklady Mathematics **64**:2 (2001), 180–186.

[7] Bruno, A.D., Varin, V.P.: The limit problems for the equation of oscillations of a satellite. Celestial Mechanics and Dynamical Astronomy **61**:1 (1997), 1–40.

[8] Denk, R., Menniken, R., Volevich, L.: The Newton polygon and elliptic problems with parameter. Mathematische Nachrichten **192** (1998), 125–157.

[9] Bruno, A.D.: Power expansions of solutions to one algebraic or differential equation. Preprint No. 63 of Inst. Appl. Math., Moscow, 2000 (Russian).

[10] Bruno, A.D.: Power expansions of solutions to a system of algebraic and differential equations. Preprint No. 68 of Inst. Appl. Math., Moscow, 2000 (Russian).

[11] Aranson, A.B.: Computation and applications of the Newton polyhedrons. Mathematics and Computers in Simulation **57**:3 (2001), 155–160.

[12] Vasiliev, M.M.: Obtaining the self-similar asymptotics of solutions to the Navier-Stokes equations by Power Geometry. Proceedings of the 3rd ISAAC Congress, Subsection I-2. World Scientific, Singapore, to appear.

[13] Bruno, A.D., Lunev, V.V.: The modified system of equations, describing motions of a rigid body. Preprint No. 49 of Inst. Appl. Math., Moscow, 2001 (Russian).

[14] Bruno, A.D., Lunev, V.V.: Local expansions of modified motions of a rigid body. Preprint No. 73 of Inst. Appl. Math., Moscow, 2001 (Russian).

[15] Bruno, A.D., Lunev, V.V.: Asymptotical expansions of modified motions of a rigid body. Preprint No. 90 of Inst. Appl. Math., Moscow, 2001 (Russian).

[16] Bruno, A.D., Lunev, V.V.: Properties of expansions of modified motions of a rigid body. Preprint No. 23 of Inst. Appl. Math., Moscow, 2002 (Russian).

A SURVEY OF Q–SPACES AND $Q^\#$-CLASSES

Matts Essén

Department of Mathematics

Uppsala University

P.O.Box 480, S-75106, Uppsala, Sweden

matts.essen@math.uu.se

Abstract We study Q-spaces and $Q^\#$-classes which are Möbius-invariant, weighted Dirichlet spaces or classes of analytic or meromorphic functions in the unit disc in the plane. Generalizations to half-spaces in higher dimensions are also possible. In the analytic case, they are subspaces of BMOA or of the Bloch space $\mathcal{B}$. In the meromorphic case, they are subclasses of the class of normal functions $\mathcal{N}$ or of the class of spherical Bloch functions $\mathcal{B}^\#$. In the first part of the survey, we discuss concrete examples where different kinds of p-Carleson measures ($0 < p < 1$) are important. In the last section, we discuss a more general theory which gives both new results and new proofs of several results from the first part.

Mathematics Subject Classification 30-02, 30D45, 30D50, 42B35

1.   Introduction

Let Δ be the unit disk in the complex plane, let $\partial\Delta$ be the unit circle and $dA(z)$ be the Euclidean area element on Δ. $H(\Delta)$ and $M(\Delta)$ denote the class of functions analytic and meromorphic in Δ, respectively. Green's function in Δ with pole at $a \in \Delta$ is given by $g(z,a) = \log \frac{1}{|\varphi_a(z)|}$, where $\varphi_a(z) = \frac{a-z}{1-\bar{a}z}$ is a Möbius transformation of Δ. For $0 < r < 1$, let $\Delta(a,r) = \{z \in \Delta : |\varphi_a(z)| < r\}$ be the pseudohyperbolic disk with center $a \in \Delta$ and radius r. In the meromorphic case, we shall also need the spherical derivative $f^\#(z) = \frac{|f'(z)|}{1+|f(z)|^2}$.

If $K : [0,\infty) \to [0,\infty)$ is a right-continuous and nondecreasing function, we define the spaces Q_K and $Q_K^\#$ as follows

H.G.W. Begehr et al. (eds.), Analysis and Applications - ISAAC 2001, 73–87.

Definition 1 $f \in H(\Delta)$ belongs to the space Q_K if

$$\|f\|_{Q_K}^2 = \sup_{a \in \Delta} \iint_\Delta |f'(z)|^2 K(g(z,a)) \, dA(z) < \infty. \tag{1}$$

Definition 2 $f \in M(\Delta)$ belongs to the class $Q_K^\#$ if

$$\sup_{a \in \Delta} \iint_\Delta f^{\#^2}(z) K(g(z,a)) \, dA(z) < \infty. \tag{2}$$

Modulo constants, Q_K is a Banach space under the norm defined in (1.1). $Q_K^\#$ is not a linear space.

A space X or a class X of functions in Δ is *Möbius invariant* if $f \in X$ implies that $f \circ \varphi_a \in X$ for all $a \in \Delta$. It is clear that Q_K and $Q_K^\#$ are Möbius invariant.

For $0 < p < \infty$, we define spaces Q_p and M_p by

$$Q_p = \{ f \in H(\Delta) : \sup_{a \in \Delta} \iint_\Delta |f'(z)|^2 (g(z,a))^p \, dA(z) < \infty \},$$

$$M_p = \{ f \in H(\Delta) : \sup_{a \in \Delta} \iint_\Delta |f'(z)|^2 (1 - |\varphi_a(z)|^2)^p \, dA(z) < \infty \}.$$

If we replace the derivative $|f'|$ in these definitions by the spherical derivative $f^\#$ and the analytic functions $H(\Delta)$ by the meromorphic functions $M(\Delta)$, we obtain the classes $Q_p^\#$ and $M_p^\#$.

We see that the choice $K(t) = t^p$ in (1) and (2) gives the space Q_p and the class $Q_p^\#$ while the choice $K(t) = (1 - e^{-2t})^p$ gives us M_p and $M_p^\#$. Let $\|f\|_{Q_p}^2$ denote the supremum in (1.1) when we have chosen $K(t)$ to be t^p.

It is also possible to take $p = 0$ in the definition of Q_p. This gives us the Dirichlet space $\mathcal{D}$.

Finally, we introduce the *Bloch space* (cf. [20])

$$\mathcal{B} = \{ f \in H(\Delta) : \|f\|_\mathcal{B} = \sup_{z \in \Delta} (1 - |z|^2) |f'(z)| < \infty \},$$

the class of *normal functions* defined by (cf. [20])

$$\mathcal{N} = \{ f \in M(\Delta) : \|f\|_N = \sup_{z \in \Delta} (1 - |z|^2) f^\#(z) < \infty \},$$

and the class of *spherical Bloch functions* (cf. [28])

$$\mathcal{B}^{\#} = \{\, f \in M(\Delta) : \sup_{a \in \Delta} \iint\limits_{\Delta(a,r)} f^{\#^2}(z)\, dA(z) < \infty \text{ for some } r \in (0,1) \,\}.$$

It is known that $Q_1 = \text{BMOA}$ (cf. [15] and Theorem 5 in [6]), that $Q_p = \mathcal{B}$ for $p > 1$ (cf. Aulaskari and Lappan [2]), that $Q_1^{\#} = \text{UBC}$, the class of meromorphic functions of uniformly bounded characteristic discovered by Yamashita in 1982 (cf. [28]), and that $Q_p^{\#} = \mathcal{N}$ for $p > 1$ (cf. [2]). We note that UBC is the meromorphic counterpart of BMOA.

In Sections 2-5, we discuss p-Carleson measures and the spaces Q_p, M_p, $Q_p^{\#}$ and $M_p^{\#}$. The emphasis will be on the case $0 < p < 1$. In Section 6, we discuss the more general classes Q_K and $Q_K^{\#}$ and present a theory due to Essén and Wulan [12]. Preliminary versions can be found in Essén [9] and Wu and Wulan [23].

For a more complete list of references, we refer to the surveys of Q_p and $Q_p^{\#}$-theory which can be found in Essén [9] and Essén and Xiao [14]. We would also like to mention Xiao's book [27].

Let us start by quoting some basic results. We have always $\mathcal{D} \subset Q_p \subset Q_q$ and $Q_p^{\#} \subset Q_q^{\#}$, $0 < p < q < \infty$ (cf. Wulan [22] for the second result).

Theorem 1 *(cf. Yamashita [28] and Aulaskari, Xiao and Zhao [5]).*

$$Q_p \subsetneq Q_q \subsetneq Q_1 = \text{BMOA}, \quad 0 < p < q < 1,$$
$$Q_p^{\#} \subsetneq Q_q^{\#} \subsetneq Q_1^{\#} = \text{UBC}, \quad 0 < p < q < 1.$$

Remark 1 Aulaskari, Xiao and Zhao [5] have also proved that the gap series

$$f(z) = \sum_{k=0}^{\infty} a_k z^{n_k}, \qquad n_{k+1}/n_k \geq \lambda > 1, \quad k = 0, 1, \ldots,$$

is in Q_p, $0 < p \leq 1$, if and only if

$$\sum_{k=1}^{\infty} 2^{k(1-p)} \sum_{n_j \in I(k)} |a_j|^2 < \infty,$$

where $I(k) = \{n : 2^k \leq n < 2^{k+1}, n \in \mathbb{N}\}$. This result is used in their proof of the strict inclusions in Theorem 1.

Theorem 2 *(cf. Aulaskari, Stegenga and Xiao [3], Aulaskari, Wulan and Zhao [4] and Wulan [22])*

$$Q_p = M_p, \quad 0 < p < \infty,$$
$$Q_p^\# \subsetneqq M_p^\#, \quad 0 < p \le 1,$$
$$M_p^\# = B^\#, \quad 1 < p < \infty.$$

For more information on results of this type, we refer to Theorems 12 and 22 and Corollary 7.

2. p-Carleson measures

For $p \in (0, \infty)$, we say that a positive Borel measure μ on Δ is a p-Carleson measure provided that

$$\sup_{I \subset \partial\Delta} \frac{\mu(S(I))}{|I|^p} < \infty,$$

where $S(I)$ is the Carleson square based on I:

$$S(I) = \{z \in \Delta : 1 - \frac{|I|}{2\pi} \le |z| < 1, \ \frac{z}{|z|} \in I\}.$$

Here $|I|$ denotes normalized arc-length on the unit circle $\partial\Delta$. We note that $p = 1$ gives us the classical Carleson measure.

Remark 2 *There are also results on p-Carleson measures in the case $1 < p < \infty$ in [1].*

At a conference in Hongkong in 1993, D.F.Shea suggested that there was a connection between p-Carleson measures and Q_p-spaces. This led Aulaskari, Stegenga and Xiao to the following important results (which are wellknown in the case $p = 1$: then we have $Q_1 = $ BMOA (cf. e.g. Lemma 3.3 in [16])).

Lemma 1 *(cf. [3]) Let $0 < p < \infty$. Then μ is a bounded p-Carleson measure if and only if*

$$\sup_{a \in \Delta} \iint_\Delta \left(\frac{1 - |a|^2}{|1 - \bar{a}z|^2} \right)^p d\mu(z) < \infty.$$

Theorem 3 *(cf. [3]) Let $p \in (0, \infty)$ and $f \in H(\Delta)$. Then $f \in Q_p$ if and only if $|f'(z)|^2(1 - |z|)^p dA(z)$ is a p-Carleson measure.*

3. The boundary value characterization

In the proof of the following results, we use p-Carleson measures. Let

$$K_p(f, I) = \left\{ \int_I \int_I \frac{|f(e^{i\theta}) - f(e^{i\varphi})|^2}{|e^{i\theta} - e^{i\varphi}|^{2-p}} d\theta d\varphi \right\}^{1/2}. \tag{3}$$

Theorem 4 *(cf. Essén and Xiao [13]) Let $p \in (0,1)$. Then $f \in H^2(\Delta)$ is in Q_p if and only if*

$$\sup_{I \in \partial\Delta} |I|^{-p} K_p(f, I)^2 < \infty, \tag{4}$$

where the supremum is taken over all intervals I on the unit circle.

An inner function is a bounded analytic function in Δ which has boundary values of modulus 1 a.e. on $\partial\Delta$. Every inner function can be written as the product of a Blaschke product and a singular inner function. It is also known that every function in $H^2(\Delta)$ is the product of an inner and an outer factor (cf. e.g. Section 2.4 in [8]). What is surprising in the following result is that it tells us that inner functions in Q_p have no singular inner factor.

Theorem 5 *(cf. Essén and Xiao [13]) Let $p \in (0,1)$ If I is an inner function, the following are equivalent:*
(i) $I \in Q_p$.
(ii) I is a multiplier of $H^\infty \cap Q_p$, i.e.

$$I \cdot (H^\infty \cap Q_p) \subset H^\infty \cap Q_p.$$

(iii) I is a Blaschke product with zeros $\{z_n\}_{n=0}^\infty$ which are such that

$$d\mu(z) = \sum_{n=0}^\infty (1 - |z_n|^2)^p \delta_{z_n}$$

is a p-Carleson measure, where δ_{z_n} is a Dirac measure at z_n.

We have also

Theorem 6 *(cf. Xiao [26])*
(i) Let $p \in (0,1)$ and let $f \in H^2(\Delta)$. Then $f \in Q_p$ if and only if

$$\sup_{a \in \Delta} \int\int_\Delta \left\{ \int_{\partial\Delta} |f|^2 d\mu_z - |f(z)|^2 \right\} (1 - |\varphi_a(z)|^2)^p \, d\lambda(z) < \infty,$$

where $d\lambda(z) = \frac{dA(z)}{(1-|z|^2)^2}$ and $d\mu_z$ is the Poisson measure.

(ii) Let $p \in (0,1)$ and let $f \in H^2(\Delta)$ with $f \not\equiv 0$. Then $f \in Q_p$ if and only if $f = IO$ where I is an inner function and O is an outer function in Q_p for which

$$\sup_{a \in \Delta} \iint_\Delta |O(z)|^2 (1 - |I(z)|^2)(1 - |\varphi_a(z)|^2)^p \, d\lambda(z) < \infty.$$

A famous result of Nevanlinna says that every function in the Nevanlinna class can be written as the quotient of two bounded analytic functions (cf. e.g. Section 2.1 in [8]). Xiao has proved that if $p \in (0,1)$, every function in Q_p can be written as the quotient of two functions in $H^\infty \cap Q_p$ (cf. [26]).

Let $Q_p(\partial\Delta)$ be the class of all measurable functions on $\partial\Delta$ which are such that (4) holds. The Poisson extension of a function in $Q_p(\partial\Delta)$ to Δ is harmonic in Δ. Let for us while replace the unit disc by a halfplane or halfspace. We can take functions defined on the boundary of the halfspace which satisfy an analogue of (4) and study their Poisson extensions to the halfspace. In this way, we can generalize our theory to higher dimensions: these results are due to Essén, Janson, Peng and Xiao.

Let $\mathbb{R}^n$ be n-dimensional Euclidean space, where $n \geq 1$, and let $\mathbb{R}^{n+1}_+$ be the upper halfspace based on $\mathbb{R}^n$. We consider "cubes" I in $\mathbb{R}^n$ with edges parallell to the coordinate axis, with sidelength $l(I)$ and volume $|I|$ (it is clear that $l(I) = |I|^{1/n}$). For $\alpha \in (-\infty, \infty)$, we define $Q_\alpha(\mathbb{R}^n)$ to be the space of all measurable functions on $\mathbb{R}^n$ which satisfy

$$\|f\|^2_{Q_\alpha(\mathbb{R}^n)} = \sup_I [l(I)]^{2\alpha-n} \int_I \int_I \frac{|f(x) - f(y)|^2}{|x-y|^{n+2\alpha}} dx dy < \infty,$$

where the supremum is taken over all cubes I described above. $Q_\alpha(\mathbb{R}^n)$ is a Banach space of functions modulo constants. It is known that $\mathrm{BMO}(\mathbb{R}^n)$ is related to Poisson integrals and Carleson measures and can be studied via wavelet coefficients and dyadic cubes. We give analogues for $Q_\alpha(\mathbb{R}^n)$.

Theorem 7 *(cf. [10]) $Q_\alpha(\mathbb{R}^n)$ is invariant under translations, rotations and dilations. Furthermore,*

(i) $Q_\alpha(\mathbb{R}^n) \supseteq Q_\beta(\mathbb{R}^n)$, $\alpha \leq \beta$.

(ii) If $n \geq 2$ and $\alpha \geq 1$, $Q_\alpha(\mathbb{R}^n)$ contains only functions that are constant a.e.

(iii) If $\alpha < 0$, $Q_\alpha(\mathbb{R}^n) = \mathrm{BMO}(\mathbb{R}^n)$.

It remains to study the case $0 < \alpha < 1$. As an example of our results, we mention

Theorem 8 *(cf. [10]). Let* $\alpha \in (0,1)$, *let* $f \in L^2_{loc}(\mathbb{R}^n)$ *be such that*

$$\int_{\mathbb{R}^n} |f(x)|(1 + |x|^{n+1})^{-1}dx < \infty,$$

and let $f(x,t)$ *be the Poisson extension of* f *to* $\mathbb{R}^{n+1}$. *Then* $f \in Q_\alpha(\mathbb{R}^n)$ *if and only if*

$$\int_{S(I)} |\nabla f(x,t)|^2 t^{1-2\alpha}dxdt < M[l(I)]^{n-2\alpha}, \tag{5}$$

for some $M < \infty$ *and all cubes* $I \subset \mathbb{R}^n$. *Here* $S(I)$ *is the Carleson box built on* I, *i.e.* $S(I) = \{(x,t) : x \in I, 0 < t < l(I)\}$.

We note that (5) is a generalization of our p-Carleson condition. For more about harmonic analysis on Q-spaces in higher dimensions, we refer to [10].

Let us now return to the unit disc in the plane. Theorem 4 tells us that $H(\Delta) \cap Q_p(\partial\Delta) = Q_p$, $0 < p < 1$. Xiao has proved that when $p = 1$, this relation is no longer true. Then we have $H(\Delta) \cap Q_1(\partial\Delta) \subsetneq Q_1 = \text{BMOA}$ (cf. [25]).

We now know quite a lot about what kinds of functions we can find in Q_p-spaces, $0 < p < 1$. In addition to the information in Theorems 1, 2, 4, 5 and 6, we have also the gap series described in Remark 1 which can be used to show that (cf. Section 6 in Essén and Xiao [13])

$$H^\infty \setminus \left(\bigcup_{0<p<1} Q_p \right) \neq \phi.$$

We have $H^\infty \subset \text{BMOA} = Q_1$. Since $\log(1 - z) \in Q_p$, $0 < p \leq 1$, we have also

$$\left(\bigcap_{0<p<1} Q_p \right) \setminus H^\infty \neq \phi.$$

4. The Q_p-corona theorem

Assume that $0 < p < \infty$. To solve the Q_p-corona problem means to find necessary and sufficient conditions on n given functions $\{f_k\}^n_1$ in Q_p ensuring that for any $g \in Q_p$, there exist functions $\{g_k\}^n_1$ in Q_p such that $\sum^n_{k=1} f_k g_k = g$. The terminology comes from the corresponding H^∞-

corona problem which was solved by Carleson in 1962 (cf. [7]). In the case $p > 1$, the Q_p-problem was solved by Ortega and Fabrèga in 1996 (cf. [19]). The cases $p = 1$ and $p = 0$ are due to Tolokonnikov in 1991 (cf. [21]) and Nicolau in 1990 (cf. [18]).

Theorem 9 *(cf. Xiao [24]) Assume that $0 < p < 1$. The corona problem can be solved in Q_p if and only if $f_1 \ldots f_n$ are multipliers of Q_p, i.e. $f_k h \in Q_p$ for all $h \in Q_p$, $k = 1, \ldots n$, and*

$$\inf_{z \in \Delta} \sum_1^n |f_k(z)| > 0.$$

Let $\mathcal{M}(Q_p)$ denote the multipliers of Q_p. A partial characterization of $\mathcal{M}(Q_p)$ is given by

Theorem 10 *(cf. Xiao [24]). Let $0 < p < 1$.*
(i) $f \in \mathcal{M}(Q_p)$ only if $f \in H^\infty$ and

$$\iint_{S(I)} |f'(z)|^2 (1 - |z|^2)^p dA(z) \leq \frac{Const.|I|^p}{(\log \frac{2}{|I|})^2},$$

for all intervals $I \subset \partial \Delta$.
(ii) $f \in \mathcal{M}(Q_p)$ if $f \in H^\infty$ and

$$\iint_{S(I)} |f'(z)|^2 (1 - |z^2|)^p \log^2(1 - |z|) dA(z) \leq Const.|I|^p,$$

for all intervals $I \subset \partial \Delta$, i.e., $|f'(z)|^2 (1 - |z|^2)^p \log^2(1 - |z|) dA(z)$ is a p-Carleson measure.

5. The meromorphic case

To start our discussion, we introduce

$$\mathcal{B}_1 = \{f \in H(\Delta) : \sup_{a \in \Delta} \iint_{\Delta(a,r)} |f'(z)|^2 dA(z) < \infty \text{ for some } r \in (0,1)\}.$$

Since $|f'(z)|^2$ is subharmonic in Δ when $f \in H(\Delta)$, we can prove that $\mathcal{B}_1 = \mathcal{B}$. We lose this subharmonicity when working with the spherical derivative $f^\#(z)$ The meromorphic analogues of the classes $\mathcal{B}$ and $\mathcal{B}_1$ are the normal functions $\mathcal{N}$ and the spherical Bloch functions $\mathcal{B}^\#$. It turns out that we have $\mathcal{N} \subsetneq \mathcal{B}^\#$ (cf. [17]). We have also

Theorem 11 *(cf. Yamashita [28]) A meromorphic function f is in $\mathcal{N}$ if and only if*

$$\sup_{a \in \Delta} \iint_{\Delta(a,r)} f^{\#}(z)^2 dA(z) < \pi$$

for some $r \in (0, 1)$.

To Theorems 1 and 2, we now add

Theorem 12 *(cf. Wulan [22]).*

(i) $Q_p^{\#} = \mathcal{N} \cap M_p^{\#}, \qquad 0 < p < \infty.$

(ii) $M_p^{\#} \subsetneqq M_q^{\#} \subsetneqq \mathcal{B}^{\#}, \quad 0 < p < q \leq 1.$

Theorem 13 *(cf. Wu-Wulan [23])*
    There exists a non-normal function in $M_p^{\#} \setminus Q_p^{\#}, \quad 0 < p < \infty.$

We also recall that $Q_p^{\#} = \mathcal{N}$, $p > 1$ (cf. [2]), and that $M_p^{\#} = \mathcal{B}^{\#}$, $p > 1$ (cf. Theorem 2). To characterize the classes $\mathcal{N}$ and $\mathcal{B}^{\#}$, we can also use a new type of Carleson measure introduced in Wulan [22].

For $w \in \Delta$, $w \neq 0$, and $r \in (0, 1]$ given, we define

$$S(w,r) = \{z \in \Delta : |w| \leq |z| < |w| + r(1 - |w|), |arg(z/w)| < r(1 - |w|)\},$$

and $S(0, r) = \{z : |z| < r\}$. For $r = 1$, $S(w, 1) = S(w)$ is a classical Carleson box.

Definition 3 *Let $0 < p < \infty$. A positive measure μ defined on Δ is called a p-Carleson measure of type I if there exists a positive constant C such that*

$$\mu(S(w,r)) \leq C(r(1 - |w|))^p, \tag{6}$$

for all $w \in \Delta$ and some r, $0 < r \leq 1$. For $r = 1$, we get the p-Carleson measure introduced in Section 2.

Definition 4 *A positive measure μ defined on Δ is called a p-Carleson measure of type II if there exist constants $C > 0$ and $R \in (0, 1]$ such that (6) holds for all $w \in \Delta$ and all $r \in (0, R]$.*

Theorem 14 *(cf. Essén and Wulan [11]). Let $f \in M(\Delta)$.*
(i) $f \in \mathcal{B}^{\#}$ if and only if $f^{\#}(z)^2(1 - |z|^2)^p dA(z)$ is a p-Carleson measure of type I for all $p \in (0, \infty)$.

(ii) $f \in \mathcal{N}$ if and only if $f^{\#}(z)^2(1-|z|^2)^p dA(z)$ is a p-Carleson measure of type II for all $p \in (0,2]$.

In both cases, "for all" can be replaced by "for some".

In [11], there are further applications of these ideas to classes such as

$$B^{\#}(p) = \{f \in M(\Delta) : \sup_{a \in \Delta} \iint_{\Delta(a,r)} f^{\#}(z)^p (1-|z|^2)^{p-2} dA(z) < \infty\},$$

where $0 < p < \infty$.

In Section 6, we shall explain why $Q_p^{\#} \subsetneq M_p^{\#}$ for $0 < p < \infty$.

6.   The general case. Q_K-spaces and $Q_K^{\#}$-classes

We recall that for $0 < p < \infty$, $K(t) = t^p$ gives the space Q_p and the class $Q_p^{\#}$. Choosing $K(t) = (1 - e^{-2t})^p$, we obtain M_p and $M_p^{\#}$. Is there a more general structure behind the results in Theorem 2? What properties of K_1 and K_2 imply that

$$Q_{K_1} = Q_{K_2},$$

or

$$Q_{K_1}^{\#} = Q_{K_2}^{\#}?$$

In [12], Essén and Wulan give a general theory for Q_K-spaces and $Q_K^{\#}$-classes which answers these questions and which also gives new theorems which as corollaries have several results given above. The present section contains a survey of our theory.

Let us first note that if for a fixed r, $0 < r < 1$, we choose

$$K_0(t) = \begin{cases} 1, & t \geq \log(1/r), \\ 0, & 0 < t < \log(1/r), \end{cases}$$

we obtain $\mathcal{B} = \mathcal{B}_1$ (cf. Section 5) and $\mathcal{B}^{\#}$. To avoid trivialities, we assume without loss of generality that $K(1) > 0$. It is now easy to see that $Q_K \subset \mathcal{B}$ and that $Q_K^{\#} \subset \mathcal{B}^{\#}$. In the analytic case, this is clear:

$$\sup_{a \in \Delta} \iint_{\Delta(a,1/e)} |f'(z)|^2 \, dA(z) \leq \sup_{a \in \Delta} \frac{1}{K(1)} \iint_{\Delta} |f'(z)|^2 K(g(z,a)) \, dA(z),$$

which immediately gives that $Q_K \subset \mathcal{B}$. The proof is the same in the meromorphic case.

For a nondecreasing function $K : [0,\infty) \to [0,\infty)$, we say that Q_K or $Q_K^{\#}$ is trivial if Q_K contains only constant functions.

Proposition 1 *Assume that the integral*

$$\int_0^{1/e} K(\log(1/\rho))\rho \, d\rho = \int_1^\infty K(t)e^{-2t} \, dt \qquad (7)$$

is divergent. Then the space Q_K and the class $Q_K^{\#}$ will be trivial.

From now on, we assume that the integral in (7) is convergent.

Theorem 15 *We assume that $K(1) > 0$ and define $K_1(r) = \inf(K(r), K(1))$. Then $Q_K = Q_{K_1}$.*

As we shall see below, Theorem 15 is not correct in the meromorphic case (cf. Theorem 20).

Corollary 1 *Let $0 < p < \infty$. Assume that $K(r) \approx r^p$, $r \to 0$. Then $Q_K = Q_p = M_p$.*

This is clear since the choices of the kernel given above for these two classes behave in this way near the origin. We have a new proof of the first part of Theorem 2.

Theorem 16
(i) $Q_K = \mathcal{B}$ if and only if

$$\int_0^1 K(\log(1/r))(1 - r^2)^{-2} r \, dr < \infty. \qquad (8)$$

(ii) $Q_K = \mathcal{D}$ if and only if $K(0) > 0$.

Choosing $K(r) = r^p$, we obtain that (8) holds for $p > 1$ and that (8) fails for $0 < p \le 1$, that is

Corollary 2 *(cf. Aulaskari and Lappan [2] and Aulaskari, Xiao and Zhao [5].) $Q_p = \mathcal{B}$ when $p > 1$ and $Q_p \subsetneq \mathcal{B}$ when $0 < p \le 1$.*

Applying the open mapping theorem, we deduce

Theorem 17 *Let $K_1 \le K_2$ in $(0,1)$ and assume furthermore that $K_1(r)/K_2(r) \to 0$ as $r \to 0$. If (6.2) doesn't hold when we choose*

$K = K_2$, then $Q_{K_2} \subsetneqq Q_{K_1}$.

Theorem 17 gives a new proof of the analytic part of Theorem 1: choosing $K_1(r) = r^q$ and $K_2(r) = r^p$, we obtain

Corollary 3 $Q_p \subsetneqq Q_q \subsetneqq Q_1 = \text{BMOA}, \ 0 < p < q < 1.$

In the meromorphic case, we begin with (cf. Theorem 13)

Theorem 18 *(cf. Wu and Wulan [23])*
(i) Let $0 < p < \infty$. Assume that K is bounded and that $K(r) = O(r^p)$ as $r \to 0$. Then there exists a non-normal function $f_0 \in Q_K^\#$.
(ii) Assume that $K(r) \to \infty$ as $r \to \infty$. Then $Q_K^\# \subset \mathcal{N}$.

The proof of (i) uses an example of Aulaskari, Wulan and Zhao (cf. [4]) while the proof of (ii) is an application of Theorem 11.

Thus a necessary and sufficient condition to have $Q_K^\# \subset \mathcal{N}$ is that the function K is unbounded. For K bounded, we have (cf. Theorem 11)

Theorem 19 *Assume that $K(\infty) = 1$. Then $f \in N$ if and only if*

$$\sup_{a \in \Delta} \iint_{\Delta(a,r)} f^{\#2}(z) K(g(z,a)) \, dA(z) < \pi,$$

for some $r \in (0,1)$.

Corollary 4 *Assume that $K(\infty) = 1$. If $f \in Q_K^\#$ and*

$$\sup_{a \in \Delta} \iint_{\Delta} f^{\#}(z)^2 K(g(z,a)) \, dA(z) < \pi,$$

then $f \in N$.

The meromorphic analogue of Theorem 15 is given by (cf. Theorem 12(i))

Theorem 20 *We assume that $K(1) > 0$, that K is unbounded and define $K_1(r) = \inf(K(r), K(1))$. Then $Q_K^\# = \mathcal{N} \cap Q_{K_1}^\#$.*

Proof It is easy to see that $Q_K^\# \subset \mathcal{N} \cap Q_{K_1}^\#$ (cf. Theorem 18(ii)). Now assume that $f \in \mathcal{N} \cap Q_{K_1}^\#$. We note that $K(g(z,a)) = K_1(g(z,a))$ in

$\Delta \setminus \Delta(a, 1/e)$ (in this domain, we have $g(z, a) \leq 1$). To compare the two suprema in the integrals defining $Q_K^\#$ and $Q_{K_1}^\#$, it suffices to deal with integrals over $\Delta(a, 1/e)$. Using our assumption that $f \in \mathcal{N}$, we see that

$$
\iint\limits_{\Delta(a,1/e)} f^\#(z)^2 \; K(g(z,a)) \, dA(z)
$$

$$
\leq \|f\|_{\mathcal{N}}^2 \iint\limits_{\Delta(a,1/e)} (1 - |z|^2)^{-2} K(g(z,a)) \, dA(z)
$$

$$
= \|f\|_{\mathcal{N}}^2 \iint\limits_{\Delta(0,1/e)} (1 - |w|^2)^{-2} K(\log|1/w|) \, dA(w)
$$

$$
= 2\pi \|f\|_{\mathcal{N}}^2 \int_0^{1/e} r(1 - r^2)^{-2} K(\log(1/r)) \, dr.
$$

The right hand member gives a bound for the supremum over $a \in \Delta$ of the first term in this chain of inequalities. Hence $f \in Q_K^\#$ and Theorem 20 is proved.

Next, we state conditions on K_1 and K_2 which imply that $Q_{K_1}^\# = Q_{K_2}^\#$.

Theorem 21 *Assume that K_1 and K_2 are both either bounded or unbounded and that $K_1(r) \approx K_2(r)$ as $r \to 0$. Then $Q_{K_1}^\# = Q_{K_2}^\#$.*

Corollary 5 *Let $0 < p < \infty$ and assume furthermore that K is bounded and that $K(r) \approx r^p$ as $r \to 0$. Then $Q_K^\# = M_p^\#$.*

Combining Theorems 20 and 21 and Corollary 5, we deduce (cf. Theorem 12(i))

Corollary 6 *Let $0 < p < \infty$ and assume furthermore that K is unbounded and that $K(r) \approx r^p$ as $r \to 0$. Then $Q_K^\# = Q_p^\# = \mathcal{N} \cap M_p^\#$.*

The analogue of Theorem 16 holds in the meromorphic case in the following sense.

Theorem 22
i) If K is unbounded and (8) holds, then $Q_K^\# = N$.
ii) If K is bounded and (8) holds, then $Q_K^\# = \mathcal{B}^\#$.
iii) In (i) and (ii), (8) is a necessary condition that we have $Q_K^\# = N$

or $Q_K^\# = B^\#$, *respectively.*

Corollary 7 *(Aulaskari and Lappan [2] and Wulan [22])*
i) $Q_p^\# = N$ *when* $p > 1$ *and* $Q_p^\# \subsetneq N$ *when* $0 < p \le 1$.
ii) $M_p^\# = B^\#$ *when* $p > 1$ *and* $M_p^\# \subsetneq B^\#$ *when* $0 < p \le 1$.

References

[1] Arazy, J., Fisher, S., Peetre, J.: Möbius invariant function spaces. J. Reine Angew. Math. 363 (1985), 110–145.

[2] Aulaskari, R., Lappan, P.: Criteria for an analytic function to be Bloch and a harmonic or meromorphic function to be normal. Complex analysis and its applications Pitman Research Notes in Mathematics 305. Longman Scientific & Technical Harlow 1994, 136–146.

[3] Aulaskari, R., Stegenga, D., Xiao, J.: Some subclasses of BMOA and their characterization in terms of Carleson measures. Rocky Mountain J. Math. 26 (1996), 485–506.

[4] Aulaskari, R., Wulan, H., Zhao, R.: Carleson measure and some classes of meromorphic functions. Proc. Amer. Math. Soc. 128 (2000), 2329–2335.

[5] Aulaskari, R., Xiao, J., Zhao,R.: On subspaces and subsets of $BMOA$ and UBC. Analysis 15 (1995), 101–121.

[6] Baernstein, A. II: Analytic functions of bounded mean oscillation. Aspects of contemporary complex analysis. Academic Press, London, 1980, 3–36.

[7] Carleson, L.: Interpolations by bounded analytic functions and the corona problem. Ann. Math. 76 (1962), 547–559.

[8] Duren, P.: Theory of H^p spaces. Academic Press, New York and London, 1970.

[9] Essén, M.: Q_p-spaces. "Complex Function Spaces", Mekrijärvi 1999 (ed.Aulaskari), Department of Mathematics, University of Joensuu, Finland, Report no 4, 2001, 9–40.

[10] Essén, M., Janson, S., Peng, L., Xiao, J.: Q-spaces of several real variables. Ind. Univ. Math. J. 49 (2000), 575–615.

[11] Essén, M., Wulan, H.: Carleson type measures and their applications. Complex Variables, Theory Appl. 42 (2000), 67–88.

[12] Essén, M., Wulan, H.: On analytic and meromorphic functions and spaces of Q_K-type. Department of Mathematics, Uppsala University, Report 2000:32.

[13] Essén, M., Xiao, J.: Some results on Q_p-spaces, $0 < p < 1$. J. Reine Angew. Math. 485 (1997), 173–195.

[14] Essén, M., Xiao, J.: Q_p-spaces - a survey. Complex Function Spaces, Mekrijärvi 1999, (ed. Aulaskari), Department of Mathematics, University of Joensuu, Finland, Report no 4, 2001, 41–60.

[15] Fefferman, C.: Characterizations of bounded mean oscillation. Bull. Amer. Math. Soc. 77 (1971), 587–588.

[16] Garnett , J.: Bounded analytic functions. Academic Press, New York 1981.

[17] Lappan, P.: A non-normal locally uniformly univalent function. Bull. London Math. Soc. 5 (1973), 291–294 .

[18] Nicolau, A.: The corona property for bounded analytic functions in some Besov spaces. Proc. Amer. Math. Soc. 110 (1990), 135–140.

[19] Ortega, J.M., Fàbrega, J.: The corona type decomposition in some Besov spaces. Math. Scand. 78 (1996), 93–111.

[20] Pommerenke, Ch.: Boundary behaviour of conformal maps. Springer, Berlin 1992.

[21] Tolokonnikov, V.A.: The corona theorem in algebras of bounded analytic functions. Amer. Math. Soc. Trans. 149 (1991), 61–93.

[22] Wulan, H.: On some classes of meromorphic functions. Ann. Acad. Sci. Fenn. Math. Diss. 116 (1998), 1–57.

[23] Wu, P., Wulan, H.: Characterizations of Q_T spaces. J. Math. Anal. Appl. 254 (2001), 484–497.

[24] Xiao, J.: The Q_p corona theorem. Pac. J. of Math. 194 (2000), 491–509.

[25] Xiao, J.: Some essential properties of $Q_p(\partial\Delta)$-spaces. J. of Fourier Analysis and Applications 6 (2000), 311–323.

[26] Xiao, J.: Outer functions in Q_p and $Q_{p,0}$. Preprint 1999.

[27] Xiao, J.: Holomorphic Q classes. Lecture Notes in Mathematics 1767. Springer, Berlin 2001.

[28] Yamashita, S.: Functions of uniformly bounded characteristic. Ann. Acad. Sci. Fenn. Ser. A I Math. 7 (1982), 349–367.

HYPERBOLICALLY CONVEX FUNCTIONS

Diego Mejía

Departamento de Matemáticas, Universidad Nacional, A.A. 3840
Medellín, Colombia
dmejia@perseus.unalmed.edu.co

Christian Pommerenke

Fachbereich Mathematik, 8-2, Technische Universität, D-10623 Berlin, Germany
pommeren@math.tu-berlin.de

Abstract   A conformal map f of the unit disk D of the complex plane into itself
is called hyperbolically convex if the hyperbolic segment between any
two points of $f(D)$ also lies in $f(D)$. These functions form a non-linear
space invariant under Moebius transformations of D onto itself. The
fact that this space is non-linear makes it impossible to use many of the
standard methods.
This survey talk will concentrate on
– Analytic characterizations of h-convex functions
– Inequalities for h-convex functions
– Hausdorff dimension of image sets
A few proofs will be sketched.

1.   Let $\mathbb{D}$ denote the unit disk and $\mathbb{T} = \partial\mathbb{D}$ and let f be a conformal
map of $\mathbb{D}$ onto G. One of the principal aims of geometric function theory
is to establish correspondences of the form

$$\text{geometric property of } G \iff \text{analytic property of } f.$$

Here are three examples:

(i) A theorem of Carathéodory [9, p. 24] states that

$$G \text{ is a Jordan domain} \iff f \text{ is continuous and injective in } \overline{\mathbb{D}}.$$

(ii) A theorem of Riesz and Privalov [9, p. 134] states that

H.G.W. Begehr et al. (eds.), Analysis and Applications - ISAAC 2001, 89–95.
© 2003 *Kluwer Academic Publishers.*

length $\partial G < \infty \iff f' \in H^1$ (the Hardy space).

(iii) G is convex $\iff$ Re $\left[1 + z \frac{f''(z)}{f'(z)}\right] > 0 \quad (z \in \mathbb{D})$.

The map f is called convex if G is convex.

2. The domain $G \subset \mathbb{D}$ is called *hyperbolically convex* (h-convex) if

$$a, b \in G \Rightarrow (\text{hyperbolic segment from } a \text{ to } b) \subset G.$$

Now let f be h-convex, that is, $G = f(\mathbb{D})$ is h-convex. It follows from (i) that f has a continuous injective extension to $\overline{\mathbb{D}}$ and from (ii) that $f' \in H^1$. The space of h-convex functions was first intensively studied by Ma and Minda [1] who showed that

(iv) G is h-convex $\iff$ Re $\left[1 + z \frac{f''(z)}{f'(z)} + \frac{2z\, f'(z)\, \overline{f(z)}}{1 - |f(z)|^2}\right] > 0.$

The characterization (iii) of convex functions can be used to derive many facts, whereas this is not the case for the characterization (iv) of h-convex functions. The reason is that the right-hand side is harmonic in (iii) but not in (iv).

This difference reflects an important analytic difference between the euclidean and hyperbolic geometries: The euclidean automorphism group Möb($\mathbb{C}$) consists of the transformations

$$\tau(z) = az + b, \quad a, b \in \mathbb{C}, \ a \neq 0,$$

which depend complex-analytically on the parameters, whereas the hyperbolic automorphism group Möb($\mathbb{D}$) consists of the transformations

$$\tau(z) = \frac{az + b}{\bar{b}z + \bar{a}}, \quad a, b \in \mathbb{C}, \quad |a| > |b|,$$

which only depend real-analytically on the parameters.

The two function spaces have the invariance properties

$$f \text{ is convex}, \ \sigma \in \text{Möb}(\mathbb{C}), \tau \in \text{Möb}(\mathbb{D}) \Rightarrow \sigma \circ f \circ \tau \text{ is convex}, \qquad (1)$$

$$f \text{ is h-convex, } \sigma, \tau \in \text{Möb}(\mathbb{D}) \Rightarrow \sigma \circ f \circ \tau \text{ is h-convex.} \qquad (2)$$

Choosing σ suitably we can achieve the normalizations

convex: $\qquad f(z) = z + a_2 z^2 + \dots$,

h-convex: $\qquad f(z) = \alpha z + a_2 z^2 + \dots$, $0 < \alpha \le 1$.

3. A standard example is the h-convex function

$$k_\alpha(z) = \frac{2\alpha z}{1 - z + \sqrt{(1 - z)^2 + 4\alpha^2 z}} = \alpha z + \alpha(1 - \alpha^2)z^2 + \dots \qquad (3)$$

which maps $\mathbb{D}$ onto a hyperbolic halfplane. It is extremal in the following basic estimate:

Theorem 1 [4] *If* $f(z) = \alpha z + a_2 z^2 + \dots$ *is h-convex then*

$$\text{Re}\left[z\,\frac{f'z}{f(z)}\right] > \frac{1}{2} \quad \text{for } z \in \mathbb{D}.$$

With 0 instead of $\frac{1}{2}$ this would be trivial, it would just say that G is starlike with respect to 0. The proof is based on the following fact. If $\varphi : \mathbb{D} \to \mathbb{D}$ is analytic and $\varphi(0) = 0$ then the radial limits satisfy

$$\zeta \in \mathbb{T}, \ \varphi(\zeta) \in \mathbb{T} \Rightarrow 1 \le \zeta\varphi'(\zeta)/\varphi(\zeta) \le +\infty. \qquad (4)$$

Proof Since $f' \in H^1$ is suffices to show that

$$\text{Re}\left[\zeta\,\frac{f'(\zeta)}{f(\zeta)}\right] \ge \frac{1}{2} \text{ for almost all } z \in \mathbb{T}, \qquad (5)$$

namely when $f'(\zeta)$ exists. If $f'(\zeta) \in \mathbb{T}$ then (5) follows at once from (4).

Now let $f(\zeta) \in \mathbb{D}$. Since G is h-convex there is an h-line through $f(\zeta)$ such that G lies in the corresponding h-halfplane H containing 0. There are ϑ and β such that $e^{i\vartheta}k_\beta$ maps $\mathbb{D}$ onto H. Since $G \subset H$ we have

$$f = e^{i\vartheta}k_\beta \circ \varphi, \qquad (6)$$

where $\varphi : \mathbb{D} \to \mathbb{D}$ is analytic and $\varphi(0) = 0$, $\varphi(\zeta) \in \mathbb{T}$. Since $\zeta\varphi'(\zeta)/\varphi(\zeta)$ is real and ≥ 1 by (4), we obtain from (6) that

$$\mathrm{Re}\left[\zeta\,\frac{f'(\zeta)}{f(\zeta)}\right] = \zeta\,\frac{\varphi'(\zeta)}{\varphi(\zeta)}\,\mathrm{Re}\left[\varphi(\zeta)\,\frac{k_\beta'(\varphi(\zeta))}{k_\beta(\varphi(\zeta))}\right] \geq \frac{1}{2}$$

because the last factor is $= \frac{1}{2}$ as we easily see from (3).

4. A consequence of Theorem 1 is that

$$\mathrm{Re}\left[\frac{\zeta}{\zeta - z}\,\frac{f(z) - f(\zeta)}{f(\zeta)}\right] > \frac{1}{2} \text{ for } z, \zeta \in \mathbb{D}. \tag{7}$$

Using the invariance property (2), we can deduce inequalities with three variables that generalize the inequalities for convex functions of Ruscheweyh and Sheil-Small [10], which they used to prove the Pólya-Schoenberg conjecture.

Theorem 1 can be expressed as follows: The function

$$h(z) = \frac{f(z)^2}{z} = \alpha^2 z + 2\alpha a_2 z^2 + \ldots$$

maps $\mathbb{D}$ conformally onto a starlike domain in $\mathbb{D}$. Using a classical inequality for bounded univalent functions, we deduce the estimate

$$|a_2| \leq \alpha(1 - \alpha^2) \leq 2\sqrt{3}/9$$

of Ma and Minda [1]. The sharp bounds for the higher coefficients a_n are not known.

Conjecture 1 [3] $a_n = O\left(\dfrac{1}{n(\log n)^2}\right)$ *as* $n \to \infty$.

It is only known that $a_n = o(1/n)$ which follows at once from the fact that $f' \in H^1$.

5. Since Theorem 1 is connected with the example k_α it does not give the exact growth of the derivative for $|z|$ near to 1.

Theorem 2 [3] *If* f *is h-convex then*

$$f'(z) = O\left(\left(1 - |z|\right)^{-1}\left(\log\frac{1}{1 - |z|}\right)^{-2}\right) \text{ as } |z| \to 1.$$

The growth rate is attained if G has a cusp. A more precise and invariant form is

$$\frac{\left(1 - |z|^2\right) |f'(z)| \left(1 - |\zeta|^2\right) |f'(\zeta)|}{\left|f(z) - f(\zeta)\right|^2} \lambda_{\mathbb{D}} (z, \zeta)^2 \leq c \quad \text{for } z, \zeta \in \mathbb{D},$$

where $\lambda_{\mathbb{D}}$ is the hyperbolic distance and where $c > 1$ is an unknown absolute constant. See [8] for further results about the growth of the derivative.

The most important differential invariant under Möbius transformations is the Schwarzian derivative

$$S_f = \left(\frac{f''}{f'}\right)' - \frac{1}{2} \left(\frac{f''}{f'}\right)^2.$$

Theorem 3 [6] *If f is h-convex then*

$$\sup_{z \in \mathbb{D}} \left(1 - |z|^2\right)^2 |S_f(z)| < 2.54. \tag{8}$$

The left-hand side remains invariant if f is replaced by $\sigma \circ f \circ \tau$ where $\sigma \in \text{Möb}$ and $\tau \in \text{Möb}(\mathbb{D})$. Ma and Minda [2] had previously established the bound 3.

Conjecture 2 [4] *The sharp bound in (8) is*

$$\sup_{\vartheta} \left(\frac{\pi^2}{2K(\cos \vartheta)^2} + 2\cos 2\vartheta\right) \approx 2.384.$$

Here K is the complete elliptic integral of the first kind. The conjectured extremal function maps $\mathbb{D}$ onto the intersection of two h-halfplanes.

6.   Another important difference between the euclidean geometry in $\mathbb{C}$ and the hyperbolic geometry in $\mathbb{D}$ is that $\mathbb{D}$ has so many "points at infinity", namely the points on $\mathbb{T}$.

A fundamental domain G of a Fuchsian group is an interesting example of an h-convex domain. If f maps $\mathbb{D}$ conformally on G, then the points $\zeta \in \mathbb{T}$ with $f(\zeta) \in \mathbb{T}$ are of particular interest.

For $0 < p \leq 2$, let $\Lambda_p(E)$ denote the Hausdorff measure of E. In particular Λ_1 is the linear measure, the length. The Hausdorff dimension of E is defined by

$$\dim E = \inf\{p : \Lambda_p(E) = 0\}.$$

Thus $\dim \mathbb{T} = 1$, and we have

$$\Lambda_1(E) > 0 \Rightarrow \dim E \geq 1 \tag{9}$$

but not vice versa.

Let f be h-convex. It easily follows from (7) that

$$\left|\frac{f(z) - f(\zeta)}{z - \zeta}\right| \geq \frac{1}{2} \inf_{s \in \mathbb{D}} \left|\frac{f(s)}{s}\right| > 0 \quad \text{for } z, \zeta \in \mathbb{D}$$

which implies that $\dim f(A) \geq \dim A$ for $A \subset \mathbb{T}$.

The other direction is much more difficult. Since $f' \in H^1$ the Riesz-Privalov theorem [9, p. 134] shows that

$$\Lambda_1(A) = 0 \Rightarrow \Lambda_1(f(A)) = 0.$$

In view of (9) this suggests:

Conjecture 3 *If $A \subset \mathbb{T}$ and $f(A) \subset \mathbb{T}$ then*

$$\dim A < 1 \Rightarrow \dim f(A) < 1. \tag{10}$$

This conjecture is known only in a very special case. The compact set E is called *uniformly perfect* if the doubly connected subdomains of $\hat{\mathbb{C}} \setminus E$ that separate E have bounded moduli. The classical Cantor set is a typical example.

Theorem 4 [7] *Let f be h-convex. If $A \subset \mathbb{T}$ and if $f(A)$ is a uniformly perfect subset of $\mathbb{T}$, then A is also uniformly perfect and (10) holds.*

Added in proof-sheets: Steffen Rohde (personal communication) has now given a counter-example to Conjecture 3.

References

[1] Ma, W., Minda, D.: Hyperbolically convex functions. Ann. Math. Polon. 60 (1994), 81–100.

[2] Ma, W., Minda, D.: Hyperbolically convex functions II. Ann. Math. Polon. 71 (1999), 273–285.

[3] Mejía, D., Pommerenke, Ch.: Sobre aplicaciones conformes hiperbólicamente convexas. Rev. Colombiana Mat. 32 (1998), 29–43.

[4] Mejía, D., Pommerenke, Ch.: On hyperbolically convex functions. J. Geom. Analysis 10 (2000), 365–378.

[5] Mejía, D., Pommerenke, Ch.: On the derivative of hyperbolically convex functions. Ann. Acad. Sci. Fennicae Ser I A Math, 27 (2002), 47–56.

[6] Mejía, D., Pommerenke, Ch.: Sobre la derivada schwarziana de aplicaciones conformes hiperbólicamente convexas. Rev. Colombiana Mat., to appear.

[7] Mejía, D., Pommerenke, Ch.: Hyperbolically convex functions, dimension and capacity. Complex Variables, Theory Appl., to appear.

[8] Mejía, D., Pommerenke, Ch., Vasil'ev, A.: Distortion theorems for hyperbolically convex functions. Complex Variables, Theory Appl. 44 (2001), 117–130.

[9] Pommerenke, Ch.: Boundary behaviour of conformal maps. Springer-Verlag, Berlin 1992.

[10] Ruscheweyh, S., Sheil-Small, T.: Hadamard products of schlicht functions and the Pólya-Schoenberg conjecture. Comment. Math. Helv. 48 (1973), 119–135.

NEVANLINNA THEORY IN CHARACTERISTIC P AND APPLICATIONS

Abdelbaki Boutabaa, Alain Escassut
Laboratoire de Mathématiques Pures
Université Blaise Pascal
(Clermont-Ferrand) Les Cézeaux
63177 AUBIERE CEDEX, FRANCE
boutabaa@math.univ-bpclermont.fr , escassut@math.univ-bpclermont.fr

Abstract   Let K be a complete ultrametric algebraically closed field of characteristic p. We show that Nevanlinna's main Theorem holds, with however some corrections. Then, many results obtained in characteristic zero have generalization. When $p \neq 0$, we have to make new proofs in most of the cases. Many algebraic curves admit no parametrizations by meromorphic functions in K, or by unbounded meromorphic functions inside a disk, like in zero characteristic, provided we assume one of the function to have a non zero derivative. More generally, certain functional equations have no solution. In zero characteristic, results previously obtained are somewhat generalised, and then are extended to any characteristic. In functional equations $f^m + g^n = 1$, conclusions also are similar to those obtained in zero characteristic, provided we replace m, n by $\widetilde{m} = m|m|_p$, $\widetilde{n} = n|n|_p$. We consider the Yoshida Equation in charactersitic $p \geq 0$ and characterise all solutions when it has constant coefficients: this generalizes previous results in characteristic zero but with a more general form involving polynomials with a zero derivative. Proofs are given in a preprint where applications to the *abc*-problem are also obtained.

1.   Nevanlinna Theory

Definitions and notation   We denote by K an algebraically closed field complete for an ultrametric absolute value. We denote by p the

[0]2000 *Mathematics Subject Classification:* Primary 12E05 Secondary 11C08 11S80 30D25

H.G.W. Begehr et al. (eds.), Analysis and Applications - ISAAC 2001, 97–107.
© *2003 Kluwer Academic Publishers.*

characteristic of K and by q its characteristic exponent, i.e. $q = p$ if $p \neq 0$, and $q = 1$ if $p = 0$.

When $p \neq 0$, we denote by θ the $\mathbb{F}_p$-automorphism of K defined by $\theta(x) = \sqrt[p]{x}$. More generally this mapping has continuation to a K-algebra automorphism of $K[X]$ as $\theta(A \prod_{j=1}^{n}(X - a_j)) = \theta(A) \prod_{j=1}^{n}(X - \theta(a_j))$.

We denote by $\mathcal{A}(K)$ the set of entire functions in K, and by $\mathcal{M}(K)$ the set of meromorphic functions in K, i.e. the field of fractions of $\mathcal{A}(K)$. Given $a \in K$ and $r > 0$, $d(a, r)$ is the disk $\{x \in K | \, |x - a| \leq r\}$ and $d(a, r^-)$ is the disk $\{x \in K | \, |x - a| < r\}$. In the same way, we denote by $\mathcal{A}(d(a, r^-))$ the set of analytic functions in $d(a, r^-)$, i.e. the K-algebra of power series $\sum_{n=0}^{\infty} a_n(x - a)^n$ converging in $d(a, r^-)$, and by $\mathcal{M}(d(a, r^-))$ the set of meromorphic functions inside $d(a, r^-)$, i.e. the field of fractions of $\mathcal{A}(d(a, r^-))$.

We easily verify that

$$\mathcal{A}(K) = \bigcap_{r>0} \mathcal{A}(d(0, r^-)) \text{ and } \mathcal{M}(K) = \bigcap_{r>0} \mathcal{M}(d(0, r^-)).$$

We denote by $\mathcal{A}_b(d(a, R^-))$ the K-subalgebra of $\mathcal{A}(d(a, R^-))$ consisting of the bounded analytic functions $f \in \mathcal{A}(d(a, R^-))$, and by $\mathcal{M}_b(d(a, R^-))$ the field of fractions of $\mathcal{A}_b(d(a, R^-))$. Then, we denote by $\mathcal{A}_u(d(a, R^-))$ the set $\mathcal{A}(d(a, R^-)) \backslash \mathcal{A}_b(d(a, R^-))$, and similarly, we put $\mathcal{M}_u(d(a, R^-)) = \mathcal{M}(d(a, R^-)) \backslash \mathcal{M}_b(d(a, R^-))$.

Proposition 1 *We assume K to have characteristic $p \neq 0$. Let $r > 0$ and let $f \in \mathcal{M}(d(a, r^-))$. Then $\sqrt[p]{f}$ belongs to $\mathcal{M}(d(a, r^-))$ if and only if $f' = 0$. Moreover, there exists a unique $t \in \mathbb{N}$ such that $\sqrt[p^t]{f} \in \mathcal{M}(d(a, r^-))$ and $(\sqrt[p^t]{f})' \neq 0$.*

Definition and notation Let K has characteristic $p \neq 0$, given, $f \in \mathcal{M}(d(a, r^-))$, we will call *ramification index of f*, the integer t such that $\sqrt[p^t]{f} \in \mathcal{M}(d(a, r^-))$ and $(\sqrt[p^t]{f})' \neq 0$.

In the same way, given an algebraically closed field B of characteristic p and $P(x) \in B[x]$, we call *ramification index of P* the unique integer t such that $\sqrt[p^t]{P} \in B[x]$ and $(\sqrt[p^t]{P})' \neq 0$. This ramification index will be denoted by $u(f)$ for any $f \in \mathcal{M}(d(a, r^-))$ or $f \in \mathcal{M}(K)$ and similarly it will be denoted by $u(P)$ for any $P \in B[x]$.

Remark Let K have characteristic $p \neq 0$ and let $f \in \mathcal{M}(d(a, r^-))$ have ramification index t. For every $r' \in]0, r[$ f has the same ramification index as an element of $\mathcal{M}(d(a, r'^-))$ because of course, on one hand, $\sqrt[p^t]{f} \in \mathcal{M}(d(a, r'^-))$ and on the other hand (by classical elementary properties of analytic functions [6]), $(\sqrt[p^t]{f})'$ is not identically zero inside $d(a, r')$.

Let $\alpha \in d(0, R^-)$ and $h \in \mathcal{M}(d(0, R^-))$. If h has a zero (resp. a pole) of order n at α, we put $\omega_\alpha(h) = n$ (resp. $\omega_\alpha(h) = -n$). If $h(\alpha) \neq 0$ and ∞, we put $\omega_\alpha(h) = 0$.

For every $n \in \mathbb{Z}$, $|n|_\infty$ will denote the archimedean absolute value of n, and $|n|_p$ will be the p-adic absolute value defined on $\mathbb{Z}$.

Given $(n, m) \in \mathbb{Z}^* \times \mathbb{Z}$, $n | m$ means that n divides m and $n \not| m$ means that n does not divide m.

Finally, we denote by log be the real logarithm function of base $a > 1$.

Corollary 1.a *Suppose $p \neq 0$ and let $f, g \in \mathcal{M}(d(a, R^-)) \setminus K$ satisfy an equation of the form $g(t) = G(f(t))$ and $G \in K(Y)$. Then $u(g) \geq u(f)$ and, putting $s = u(f)$, $g_s = \sqrt[p^s]{g}$, $f_s = \sqrt[p^s]{f}$, $G_s = \Theta^s(G)$, they satisfy $g_s = G_s \circ f_s$.*

Proposition 2 *Let $f(x) \in \mathcal{M}(K)$, let S be the set of zeros and poles of f in K and let d be the g.c.d. of $\{\omega_\alpha(f) \mid \alpha \in S \cup T\}$. Then d is the greatest of the integers n such that there exists $g \in \mathcal{M}(K)$ satisfying $g^n = f$.*

Remark This property does not hold for meromorphic functions inside a disk $d(a, r^-)$.

Corollary 1.b *Let $p \neq 0$. Let $f(x) \in \mathcal{M}(K)$, let S be the set of zeros and poles of f in K and let d be the g.c.d. of $\{\omega_\alpha(f) \mid \alpha \in S \cup T\}$. Then $u(f) = -\log(|d|_p)$.*

Notation In the sequel, I will denote an interval of the form $[\rho, R[$, with $\rho > 0$, and J will denote an interval of the form $[\rho, +\infty[$.

Let $R \in]0, +\infty[$ and let $f \in \mathcal{M}(d(0, R^-))$ such that 0 is neither a zero nor a pole of f. Let $r \in]0, R[$. We denote by $Z(r, f)$ the counting function of zeros of f in $d(0, r)$:

$$Z(r, f) = \sum_{\omega_\alpha(f) > 0, \, |\alpha| \leq r} \omega_\alpha(f) \log \frac{r}{|\alpha|}.$$

Next, denoting

$$\Delta(r,f) = \{a \in d(0,r) \mid \omega_a(f) > 0, \ p^{u(f)+1}/|\omega_a(f)|\},$$

we put

$$\overline{Z}(r,f) = \sum_{\alpha \in \Delta(r,f)} \log \frac{r}{|\alpha|}.$$

We shall also consider the counting functions of poles of f in $d(0,r)$:

$$N(r,f) = Z(r,\frac{1}{f}) \text{ and } \overline{N}(r,f) = \overline{Z}(r,\frac{1}{f}).$$

The Nevanlinna function $T(r,f)$ is defined by $(\mathcal{T})$ $T(r,f) = \max[Z(r,f) + \log|f(0)|, \ N(r,f)]$. Now, assume that f' is not identically 0 and has neither a zero nor a pole at 0. Let $\Xi(r,f) = \{a \in d(0,r) \mid \omega_a(f) < 0, \ p^{u(f)+1}|\omega_a(f)|\}$. We put

$$N_0(r,f') = \sum_{\alpha \in \Xi(r,f)} [\omega_\alpha(f') - \omega_\alpha(f)] \log \frac{r}{|\alpha|}.$$

Now, given a finite subset S of K, we put $\Lambda'(r,f,S) = \{a \in d(0,r) \mid f'(a) = 0, \ f(a) \notin S\}$ and $\Lambda''(r,f,S) = \{a \in d(0,r) \mid p^{u(f)+1}|\omega_a(f-f(a)), \ f(a) \in S\}$. Then we can define

$$Z_0^S(r,f') = \sum_{\alpha \in \Lambda'(r,f,S)} \omega_\alpha(f') \log \frac{r}{|\alpha|} + \sum_{\alpha \in \Lambda''(r,f,S)} [\omega_\alpha(f') - \omega_\alpha(f - f(\alpha))] \log \frac{r}{|\alpha|}.$$

Remarks 1) It is easily verified that all the above functions are positive.
2) If $p = 0$, we have:

(i) $\overline{Z}(r,f) = \displaystyle\sum_{\omega_a(f)>0, \ |\alpha|\leq r} \log \frac{r}{|\alpha|}$; hence this function counts each

zero of f in $d(0,r)$ only one time (i.e., without multiplicities), and similarly, $\overline{N}(r,f)$ counts each pole of f in $d(0,r)$ without multiplicities.

(ii) $N_0(r,f') = 0$.

(iii) $Z_0^S(r,f') = \displaystyle\sum_{s \in S, \ \omega_a(f-s)=0, \ |\alpha|\leq r} \omega_\alpha(f') \log \frac{r}{|\alpha|}$; hence this func-

tion counts the zeros of f' in $d(0,r)$ which are not zeros of any $f - s$ for $s \in S$.

Now, here is the ultrametric Nevanlinna main Theorem in a basic form :

Theorem 1 *Let $\alpha_1, ..., \alpha_n \in K$, with $n \geq 2$, and let $f \in \mathcal{M}(d(0,R^-))$ (resp. $f \in \mathcal{M}(K)$) of ramification index s, have no zero and no pole at*

0. *Let* $S = \{\sqrt[q^s]{\alpha_1}, ..., \sqrt[q^s]{\alpha_n}\}$. *Assume that* $f, , (\sqrt[q^s]{f})')$, $f - \alpha_j$ *have no zero and no pole at 0* $(1 \leq j \leq n)$. *Then we have:*

$$\frac{(n-1)T(r,f)}{q^s} \leq \sum_{i=1}^{n} \overline{Z}(r, f - \alpha_i) + Z(r, (\sqrt[q^s]{f})') - Z_0^S(r, (\sqrt[q^s]{f})')$$
$$+ \quad O(1) \forall r \in I \; (resp. \; \forall \, r \in J).$$

Moreover, if f *belongs to* $\mathcal{A}(d(0, R^-))$ *(resp.* $f \in \mathcal{A}(K)$*), then*

$$\frac{nT(r,f)}{q^s} \leq \sum_{i=1}^{n} \overline{Z}(r, f - \alpha_i) + Z(r, (\sqrt[q^s]{f})') - Z_0^S(r, (\sqrt[q^s]{f})')$$
$$+ \quad O(1) \; \forall \, r \in I \; (resp. \forall \, r \in J).$$

A second form of the ultrametric Nevanlinna main Theorem is the following:

Theorem 2 *Let* $\alpha_1, ..., \alpha_n \in K$, *with* $n \geq 2$, *and let* $f \in \mathcal{M}(d(0, R^-))$ *(resp.* $f \in \mathcal{M}(K)$*) of ramification index* s, *have no zero and no pole at 0. Let* $S = \{\sqrt[q^s]{\alpha_1}, ..., \sqrt[q^s]{\alpha_n}\}$. *Assume that* f, $(\sqrt[q^s]{f})')$, $f - \alpha_j$ *have no zero and no pole at 0* $(1 \leq j \leq n)$. *Then we have:*

$$\frac{(n-1)T(r,f)}{q^s} \leq \sum_{i=1}^{n} \overline{Z}(r, f - \alpha_i) + \overline{N}(r, f)$$
$$-Z_0^S(r, (\sqrt[q^s]{f})') - N_0(r, (\sqrt[q^s]{f})') - \log r + O(1) \; \forall \, r \in I \; (resp. \forall \, r \in J).$$

Corollary 1.d *Let* $f \in \mathcal{M}(d(a, R^-))$, *(resp.* $f \in \mathcal{M}(K)$*), have no zero and no pole at 0 and be such that* $f' \neq 0$ *and let* $\alpha_1, ..., \alpha_n \in K$, *with* $n \geq 2$. *Assume that* f, f', $f - \alpha_j$ *have no zero and no pole at 0* $(1 \leq j \leq n)$. *Then we have:*

$$\sum_{i=1}^{n} Z(r, f - \alpha_i) - \overline{Z}(r, f - \alpha_i) \leq T(r, f) + \overline{N}(r, f)$$
$$-Z_0^S(r, f') - N_0(r, f') - \log r + O(1) \; \forall \, r \in I \; (resp. \forall \, r \in J).$$

2.   Functional equations

Notation We denote by D an infinite set included in $d(a, R^-)$, of diameter $r < R..$ Given $(m, n) \in \mathbb{N} \times \mathbb{N} \setminus \{(0, 0)\}$ we denote by (m, n) the greatest common divisor of m and n.

In Theorems 3 and 4, functional equalities true whenever $x \in D$ obviously hold for all $x \in d(a, R^-)$ (resp. for all $x \in K$ when $f, g \in \mathcal{M}(K)$)

provided x is not a pole for f or g. Theorems 3 and 4 may apply to curves of genus 0 as well as curves of genus ≥ 1: they generalize Theorems 1 and 2 obtained in [4].

Theorem 3 *Let P, $Q \in K[X]$ be two relatively prime polynomials of degrees s and t respectively and assume that Q has no factor whose power is multiple of p. Let n be the number of distinct zeros of Q, let $m \in \mathbb{N}^*$ and let $g \in \mathcal{M}(d(a, R^-))$ be such that all poles of g have an order either multiple of p, or $\geq m$, except maybe a finite number l of them. Suppose that there exists a function $f \in \mathcal{M}(d(a, R^-))$ such that $f' \neq 0$, satisfying $g(x)Q(f(x)) = P(f(x))$ for every $x \in D$ which is not a pole of f or g.*

   i)   Assume that $f \notin \mathcal{M}_b(d(a, R^-))$. Then $mn \leq t + 2m$. Moreover, if $s > t$, then $mn \leq \min(t + 2m, s + m)$.

   ii)   Assume that $f \in \mathcal{A}(d(a, R^-)) \notin \mathcal{A}_b(d(a, R^-))$. Then $mn \leq t + m$.

   iii) Assume $f \in \mathcal{A}(d(a, R^-)) \setminus \mathcal{A}_b(d(a, R^-))$. Then $mn \leq t + 2m$. Moreover, if $s > t$, then $mn \leq \min(t + 2m, s + m)$.

   iv) Assume $f \in \mathcal{A}(K) \setminus K[x]$. Then $mn \leq t + m$.

   v) Assume $f \in \mathcal{M}(K)$ and that $l = 0$ or 1. Then $mn < t + 2m$. Moreover, if $s > t$, then $mn < \min(t + 2m, s + m)$.

   vi) Assume $f \in \mathcal{A}(K)$ and that $l = 0$ or 1. Then $mn < t + m$.

Examples 1) Suppose $p \neq 2$. Let Γ be the curve of equation $y^4(x - b')(x-b'') = (x-c)^3$ (with b', b'', c all distinct) and let f, $g \in \mathcal{M}(d(a, R^-))$ be such that $(f(u), g(u)) \in \Gamma$ for all $u \in D$. Let $s = u(f)$. Since $p \neq 2$, we can check that $u(f) = u(g)$. Consequently, putting $s = u(f)$ and $f_s = \sqrt[p^s]{f}$, $g_s = \sqrt[p^s]{g}$, we can see that f_s, g_s satisfy the same hypothesis as f, g respectively, with $f'_s \neq 0$. Then by Theorem 3 f_s, g_s belong to $\mathcal{M}_b(d(a, R^-))$ and so do f and g.

2) Suppose $p \neq 3$. Let Γ be the curve of equation $y^3(x - b')(x - b'') = (x - c)^3$ (with b', b'', c all distinct) and let f, $g \in \mathcal{M}(K)$ be such that $(f(u), g(u)) \in \Gamma$ for all $u \in D$. Since $p \neq 3$, have $u(f) = u(g)$. Consequently, putting $s = u(f)$ and $f_s = \sqrt[p^s]{f}$, $g_s = \sqrt[p^s]{g}$, we see that f_s, g_s satisfy the same hypothesis as f, g respectively, with $f'_s \neq 0$. Then by Theorem 3 f_s, g_s are constant and so are f and g.

3) Consider the functional equation $\phi(t)^2 h(t) Q(f(t)) = P(f(t))$, where $f \in \mathcal{A}(K)$, $f' \neq 0$, $\phi \in \mathcal{M}(K)$, $h \in K(t)$ and $Q \in K[X]$ has four distinct zeros and is of degree 4 or 5. Then f belongs to $K[t]$, and g belongs to $K(t)$.

Notation In Theorems 4, 5, 6, m is a positive integer relatively prime to p, h belongs to $K(x)$ and l is the total number of distinct poles and zeros of h, $P(X) = A \prod\limits_{i=1}^{k}(X - a_i)^{s_i}$, $Q(X) = B \prod\limits_{j=1}^{n}(X - b_j)^{t_j}$ are polynomials of degrees s and t respectively and f, g belong to $\mathcal{M}(d(a, R^-))$ and satisfy $h(x)(g(x))^m Q(f(x)) = P(f(x))$ for all $x \in D$. Moreover, we assume that either $h' = 0$, or $f' \neq 0$.

Theorem 4

i) If $k + n > 1 + \dfrac{1}{m}\left((m, |s - t|_\infty) + \sum\limits_{i=1}^{k}(m, s_i) + \sum\limits_{j=1}^{n}(m, t_j)\right)$ then both f and g lie in $\mathcal{M}_b(d(a, R^-))$.

ii) Moreover, if f lies in $\mathcal{A}(d(a, R^-))$, and if

$$k + n > 1 + \frac{1}{m}\left(\sum_{i=1}^{k}(m, s_i) + \sum_{j=1}^{n}(m, t_j)\right)$$

then $f \in \mathcal{A}_b(d(a, R^-))$ and $g \in \mathcal{M}_b(d(a, R^-))$.

Theorem 5

i) If both f, g lie in $\mathcal{M}(K) \setminus K$, and if $k + n \geq 1 + \dfrac{1}{m}\Big((m, |s - t|_\infty) + \sum\limits_{i=1}^{k}(m, s_i) + \sum\limits_{j=1}^{n}(m, t_j)\Big)$, then $l > 1$.

ii) Moreover, if $k + n > 1 + \dfrac{1}{m}\Big((m, |s - t|_\infty) + \sum\limits_{i=1}^{k}(m, s_i) + \sum\limits_{j=1}^{n}(m, t_j)\Big)$, then $f, g \in K(x)$.

iii) Further, if f, $g \in \mathcal{M}(K) \setminus K$ and if

$$k + n > 1 + \frac{1}{m}\left(1 + \sum_{i=1}^{k}(m, s_i) + \sum_{j=1}^{n}(m, t_j)\right),$$

then f admits at least one pole of order $< m$. In the same way, if

$$f, g \in \mathcal{M}(K) \setminus K, \ if k + n \geq 1 + \frac{1}{m}\left(1 + \sum_{i=1}^{k}(m, s_i) + \sum_{j=1}^{n}(m, t_j)\right)$$

and if $l \leq 1$, then f admits at least one pole of order $< m$.

Theorem 6

i) *If both f, g lie in $\mathcal{A}(K)$ and if $k+n > 1+\dfrac{1}{m}\left(\displaystyle\sum_{i=1}^{k}(m,s_i)+\sum_{j=1}^{n}(m,t_j)\right)$,*

then $f \in K[x]$ and $g \in K(x)$.

  ii) *Moreover, if both f, g lie in $\mathcal{A}(K) \setminus K$ and if $k + n \geq 1 +$*

$\dfrac{1}{m}\left(\displaystyle\sum_{i=1}^{k}(m,s_i) + \sum_{j=1}^{n}(m,t_j)\right)$ *then $l > 1$.*

Examples 3) We assume $p \neq 3$. Let c', $c'' \in K$ (with $c' \neq c''$) and let Γ be the curve of equation $y^3 = (x - c')^2(x - c'')$. Let f, $g \in \mathcal{M}(K)$ be such that $(f(t), g(t)) \in \Gamma$ for all $t \in D$. If f and g lie in $\mathcal{A}(K)$, by Theorem 5 they are constant. If f and g are not constant, then f admits at least one pole of order 1 or 2. Here the genus is clearly 0, therefore there exist f, $g \in K(t)$ satisfying $g^3 = (f - c')^2(f - c'')$.

4) We assume $p \neq 3$. Let Γ be the curve of equation $y^3(x-b)^2 = (x-c)$ (with $b \neq c$) and let f, $g \in \mathcal{M}(K)$ be such that $(f(t), g(t)) \in \Gamma$ for all $t \in D$. If f and g lie in $\mathcal{A}(K)$, by Theorem 6 they are constant. If f and g are not constant, then f admits at least one pole of order 1 or 2.

5) We assume $p \neq 3$. Let Γ be the curve of equation $y^3(x - b)^2 = (x - c')^2(x - c'')$ (with b, c', c'' all distinct) and let f, $g \in \mathcal{M}(d(a, R^-))$ be such that $(f(t), g(t)) \in \Gamma$ for all $t \in D$. Then by Theorem 4 f, $g \in \mathcal{M}_b(d(a, R^-))$.

6) We assume $p \neq 2$. Let Γ be the curve of equation $y^2(x-b')(x-b'') = (x - c)$ (with b', b'', c all distinct) and let $f \in \mathcal{A}(d(a, R^-))$ and let $g \in \mathcal{M}(d(a, R^-))$ be such that $(f(t), g(t)) \in \Gamma$ for all $t \in D$. Then by Theorem 4 $f \in \mathcal{A}_b(d(a, R^-))$ and $g \in \mathcal{M}_b(d(a, R^-))$.

7) We assume $p \neq 2$. Consider the equation $g(t)^2 h(t)(f(t) - b_1)(f(t) - b_2) = (f(t) - a_1)(f(t) - a_2)(f(t) - a_3)$ with f, $g \in \mathcal{M}(K)$, $f' \neq 0$ and $h \in K(t)$. By Theorem 4 β) both f, g belong to $K(t)$. And we have the same conclusion with the equation $g(t)^2 h(t)(f(t) - b_1)(f(t) - b_2) = (f(t) - a_1)(f(t) - a_2)(f(t) - a_3)^2$.

8) Similarly, we assume $p = 0$. Consider the equation $g(t)^2(t-\alpha)(f(t)-b_1) = (f(t) - a_1)(f(t) - a_2)$ with f, $g \in \mathcal{M}(K)$, $f' \neq 0$, $\alpha \in K$. By Theorem 5 both f, g are constant, and we have the same conclusion with the equation $g(t)^2(t - \alpha)(f(t) - b_1)(f(t) - b_2) = (f(t) - a_1)(f(t) - a_2)$.

Corollary 2.a *Let Γ be an algebraic curve on K of genus 1 or 2 and let f, $g \in \mathcal{M}(K)$ be such that $(f(u), g(u)) \in \Gamma \; \forall u \in D$. Then f and g are constant.*

Corollary 2.b *Let Γ be a non degenerate elliptic curve on K and let $f,\ g \in \mathcal{A}(d(a, R^-))$ be such that $(f(u), g(u)) \in \Gamma \ \forall u \in D$. Then f and g are bounded.*

Corollary 2.c *Let Γ be an algebraic curve on K of genus 2 and let $f,\ g \in \mathcal{M}(d(a, R^-))$ be such that $(f(u), g(u)) \in \Gamma \ \forall u \in D$. Then both f and g lie in $\mathcal{M}_b(d(a, R^-))$.*

3. Equality $f^m + g^n = 1$

In [3] it was proven that the equation $f^m + g^n = 1$ in $\mathcal{M}(K)$ leads to $f,\ g \in K$ as soon as the least common multiple q of m and n satisfies: $\dfrac{1}{m} + \dfrac{1}{n} + \dfrac{1}{q} \geq 1$ and that in $\mathcal{A}(K)$ it leads to $f,\ g \in K$ as soon as $\min(m, n) \geq 2$. These results were improved in [4] in the case when the field K has characteristic 0. Here we are now able to widen conclusions obtained in [4].

Notation Given $n \in \mathbb{N}^*$, we put $n = \tilde{n}q^l$ with $(\tilde{n}, q) = 1$. In other words, if $p = 0$, then $\tilde{n} = n$, and if $p > 0$, then we put $\tilde{n} = n|n|_p$.

Theorem 7 *Let $f,\ g \in \mathcal{M}(d(a, r))$ satisfy $f^m + g^n = 1$. If $\min(\tilde{m}, \tilde{n}) \geq 3$ and $\max(\tilde{m}, \tilde{n}) \geq 4$, then both f and g lie in $\mathcal{M}_b(d(a, r))$. Moreover, if $f,\ g \in \mathcal{M}(K)$, if $\min(\tilde{m}, \tilde{n}) \geq 2$ and $\max(\tilde{m}, \tilde{n}) \geq 3$ then $f,\ g$ are constant.*

Theorem 8 *Let $f,\ g \in \mathcal{A}(d(a, r^-))$ satisfy $f^m + g^n = 1$, with $\min(\tilde{m}, \tilde{n}) \geq 2$. Then f and g are bounded in $d(a, r^-)$. Moreover, if $f,\ g$ belong to $\mathcal{A}(K)$ then $f,\ g$ are constant.*

Finally, Proposition 3 lets us complete this topic:

Proposition 3 *Let $n \in \mathbb{N}$ be such that $\tilde{n} \geq 2$. Let $f,\ g \in \mathcal{A}(d(a, r^-))$ and let $h \in \mathcal{A}_b(d(a, r^-)), h \neq 0$, satisfy $f^n + g^n = h$. Then f and g are bounded inside $d(a, r^-)$.*

4. Application to Yoshida's equation

We call Yoshida's Equation a differential equation of the form $(\mathcal{E})\ (y')^m = F(x, y)$ (with $F(x, y) \in K(x, y) \setminus K$). Several results were obtained in characteristic 0. In [2], it was shown that if $(\mathcal{E})$ admits solutions in $\mathcal{M}(K) \setminus K(x)$, then $F \in K(x)[y]$, and $\deg_y(F) \leq 2m$. Moreover, in [4], it was shown that if $F \in K(y)$, then any solution of the equation lying in $\mathcal{M}(K)$ is a rational function with a very specific form. Here we can generalize these results in characteristic $p > 0$.

First, we notice that Theorems 4.1 and 4.2, proven in [2] when the ground field is $\mathbb{C}_p$, have an immediate generalization in any algebraically closed field complete for an ultrametric absolute value, such as K.

Theorem 9 *Let $F(x,y)$, $G(x,y) \in K(x,y)$ and suppose that there exists a non constant solution $f \in \mathcal{M}(K) \setminus K(x)$ of the differential equation $G(x,y^{(n)}) = F(x,y)$. Let $\dfrac{A(x,y)}{B(x,y)}$ (resp. $\dfrac{P(x,y)}{Q(x,y)}$) be an irreducible form of F (resp. G), let $d = \max(\deg_y(A), \deg_y(B))$ and let $t = \max(\deg_y(P), \deg_y(Q))$. Then $(n+1)t \geq d$.*

Thanks to Theorem 9 it is easy to show Theorem 10.

Theorem 10 *Let $F(x,y) \in K(x,y) \setminus \{0\}$ and suppose that there exists a non constant solution $f \in \mathcal{M}(K) \setminus K(x)$ of the differential equation $(\mathcal{E})$ $(y')^m = F(x,y)$.*
Then F belongs to $K(x)[y]$, and $\deg_y(F) \leq 2m$.

We now generalize Theorem 6 in [4]:

Theorem 11 *Let $F(Y) \in K(Y) \setminus K$ and suppose that there exists a non constant solution $f \in \mathcal{M}(K)$ of the differential equation $(\mathcal{E})$ $(y')^m = F(y)$.*
Then F is a polynomial $A(Y-b)^d$ $(0 \leq d \leq 2m)$, with $A \in K$, such that $m - d$ divides m. In that case, the solutions $f \in \mathcal{M}(K) \setminus K$ of $(\mathcal{E})$ are the functions of the form $f(x) = b + \lambda(x + Q(x^p))^{\frac{m}{m-d}}$, where λ satisfies $\lambda^{m-d}\left(\dfrac{m}{m-d}\right)^m = A$ and $Q \in K[x]$.

In particular, when $m = 1$, we obtain the solutions of Malmquist's equation when $F \in K(y)$:

Corollary 3.a *Let $F \in K(Y) \setminus K$ be such that the equation $y' = F(y)$ admits a non constant solution in $\mathcal{M}(K)$. Then $F(Y)$ is of the form $A(Y-b)^2$ and the non constant solutions in $\mathcal{M}(K)$ are the functions of the form $b + \dfrac{1}{A(x + Q(x^p))}$ with $Q \in K[x]$.*

Remark Since any field can be provided with the trivial absolute value for which it is obviously complete, we can easily apply Theorem 11 to rational functions in an algebraically closed field.

Corollary 3.b *Let L be an algebraically closed field of characteristic p and let $F \in L(Y)$. If the equation $(y')^m = F(y)$ admits a non constant solution $f \in L(x)$, then F is a polynomial of the form $A(Y - b)^d$ ($0 \leq d \leq 2m$), with $A \in L$, such that $m - d$ divides m. In that case, the solutions $f \in L(x) \setminus L$ of $(\mathcal{E})$ are the functions of the form $f(x) = b + \lambda(x + Q(x^p))^{\frac{m}{m-d}}$, where λ satisfies $\lambda^{m-d}\left(\dfrac{m}{m-d}\right)^m = A$ and $Q \in L[x]$.*

References

[1] Berkovich, V.: Spectral Theory and Analytic Geometry over Non-archimedean Fields. AMS Surveys and Monographs 33, 1990.

[2] Boutabaa, A.: Théorie de Nevanlinna p-adique. Manuscripta Mathematica 67 (1990), 251–269.

[3] Boutabaa, A.: On some p-adic functional equations, p-adic functional analysis. Lecture Notes in Pure and Applied Mathematics 192, Marcel Dekker, 1997.

[4] Boutabaa, A., Escassut, A.: Applications of the p-adic Nevanlinna theory to functional equations. Annales de l'Institut Fourier, 50 (3) (2000), 751–766.

[5] Boutabaa, A., Escassut, A.: Nevanlinna Theory in characteristic p and applications. Preprint.

[6] Escassut, A.: Analytic Elements in p-adic Analysis. World Scientific Publishing Co. Pte. Ltd., Singapore, 1995.

[7] Hu, P.C., Yang, C.C.: Meromorphic functions over non-Archimedean fields. Kluwer Academic Publishers, Dordrecht, 2000.

A NEW PROPERTY OF MEROMORPHIC FUNCTIONS AND ITS APPLICATIONS

G.A.Barsegian

Institute of Mathematics
National Acad. of Sci. of Armenia
Marshal Bagramian ave. 24-b
Yerevan 375019, Republic of Armenia
barseg@instmath.sci.am

C.C. Yang*

Department of Mathematics
The Hong Kong University of Science and Technology
Kowloon, Hong Kong
mayang@ust.hk

Abstract   In the present paper we start consideration of a new type problem in the theory of meromorphic functions w. It turns out that any set of finite polygons which are far from each other, has many w^{-1}-preimages whose geometric shapes are quite similar to the shapes of initial polynomials. The obtained results have generalized one of versions of so called proximity property which describes geometric locations of simple a-points and, in turn, implies the main conclusions of classical value distribution theory describing these points only quantitatively. The newly obtained properties can be used to study meromorphic functions w whose a-points lie on finite non-parallel lines for a belonging to a given set.

Keywords: Nevanlinna theory, proximity property of a-points, functions with a-points lying on lines
AMS Classification: 30D30, 30D35

1.   Introduction

Let w be a meromorphic function in the complex plane. In the present paper we establish a new regularity, which is the so call *similarity property* that qualitatively speaking shows: any set of finite polygons P_ν,

---

*The work was supported by the UGC grant of Hong Kong , Project No. 6134/00P HKUST

H.G.W. Begehr et al. (eds.), Analysis and Applications - ISAAC 2001, 109–120.

which are far from each other, has many w^{-1}-preimages $z_i(P_\nu)$ whose geometric shapes are quite similar to the shapes of initial polynomials. On the other hand these similar preimages $z_i(P_\nu)$ are small and close to each other. Similarity of a polygon P_ν and and its curvilinear polygon preimage curvilinear preimage $z_i(P_\nu)$ is described in terms of closeness of some metric and angular proportions taken to P_ν and $z_i(P_\nu)$. The obtained similarity property generalizes the proximity property of a-points of meromorphic functions [1, 2, 3, 4] which, in turn, gives an addition to the main conclusions of value distribution theory [5]: the proximity property describes locations of a-points, in addition to numbers of a-points that are studied in value distribution. The following idea clearly shows a way for possible applications of these results: since vertices of convex polygons can not lie on a straight line, we can conclude that vertices of the curvilinear polygons $z_i(P_\nu)$ can not lie on straight lines. Thus one can expect that the above properties will be useful in studying of different classes of functions w with a-points lying on straight lines, that were carried out since three decades ago by Ozawa [6], Baker[7] and etc. We are able to show that there are only 15 values (with a certain geometry) such that any functions w which is meromorphic in the complex plane and has preimages of all these values lying on a finite collection of non parallel straight lines must be a rational function. Earlier similar problem conjectured by Ozawa [6] required infinite many values for deriving the same conclusion.

2. Main result and discussions

We study shapes of figures under mapping by meromorphic functions w. Obtained results can be utlized to generalize the so called Proximity Property of a-points of w [1, 2] that describes geometric locations of these points and also gives a generalization of Nevanlinna Value Distribution Theory as well (see, e.g., [5]).

We come to a situation when some main conclusions of proximity property (and consequently Nevanlinna theory) can be derived from results describing shapes of figures under mappings by w.

In order to present the progress and make some further applications on concepts that describing shapes of domains and geometric locations of a-points mentioned we need to give some new definitions.

Definition 1 Let $w(z)$ be a meromorphic function in C and let a_1, $a_2, \ldots, a_q, q > 4$, be distinct complex values. Let $\Omega(r, w)$ be a subset of sets $\{z_i(a_\nu, w)\}$, $i = 1, 2, \ldots$, $\nu = 1, 2, \ldots, q$, where $z_i(a_\nu, w)$ are simple a_ν-points in the disks $D := \{z : |z| \leq r\}$; $n_0(\Omega(r, w), a_\nu)$ is the number of simple a_ν-points belonging to $\Omega(r, w)$. We call sets $\Omega(r, w)$ *Ahlfors' sets of simple a-points* if for any $r \notin E$, where E is a set of finite logarithmic measure,

$$\sum_{\nu=1}^{q} n_0(\Omega(r, w), a_\nu) \geq (q - 4)A(r, w) - o(A(r, w)), \ r \to \infty, \ r \notin E. \quad (1)$$

Remark 1 For entire functions $(q - 4)$ is replaced by $(q - 2)$.

The main conclusion of value distribution theory (in version of Ahlfors theory of covering surfaces, see [5], ch.13), related to simple a-points is as follows.

Theorem A *Given a meromorphic function $w(z)$ in C and distinct complex values $a_1, a_2, \ldots, a_q$ there exists Ahlfors' sets $\Omega(r, w)$ of simple a-points.*

Definition 2 Let [x] denote the greatest integer not exceeding $[x]$. We say that an Ahlfors sets $\Omega(r, w)$ of simple a-points is *right distributed* if $\Omega(r, w)$ is a totality of $[A(r, w)]$ sets (*accumulations*) $\Omega_i(r, w)$, $i = 1, 2, \ldots, [A(r, w)]$, such that each accumulation consists of some simple a_ν-points, $\nu = 1, 2, \ldots, q$, or empty and in every $\Omega_i(r, w)$ there is no more than one simple a_ν-point (which we denote $z_i(a_\nu, w)$).

Remark 2 The concept of right distributed Ahlfors sets touches upon some intristic properties of meromorphic functions related to an interdependence of a_ν-points for different values ν. The right distribution means that $A(r, w)$ accumulations $\Omega_i(r, w)$ implice Ahlfors set of simple $a_1, a_2, \ldots, a_q$-points so that we conclude that in average any of the accumulations $\Omega_i(r, w)$ contains one simple a_ν-point for almost any value a_ν but 4. Thus we conclude that $a_1, a_2, \ldots, a_q$-points occurring in accumulations $\Omega_i(r, w)$ are interdependent; in fact they correspond those values on Riemann surface of w^{-1} which lie on the same sheet.

Definition 3 Let $\varphi(r)$, $r \in (0, \infty)$, be an arbitrary monotone increasing function tending to infinity as $r \to \infty$. We say that an right distributed Ahlfors' sets $\Omega(r, w)$ of simple a-points is *closely located (for given function $\varphi(r) > 0$)* if for any its accumulation $\Omega_i(r, w)$, $i = 1, 2, \ldots, [A(r, w)]$, and for any $z_i(a_\nu, w), z_i(a_\mu, w) \in \Omega_i(r, w)$, $\nu, \mu \in \{1, 2, \ldots, q\}$

$$|z_i(a_\nu, w) - z_i(a_\mu, w)| \leq \frac{\varphi(r)r}{A^{1/2}(r, w)} \quad (2)$$

Respectively points $z_i(a_\nu, w), z_i(a_\mu, w) \in \Omega_i(r, w)$ satisfying (2) will be called *close a_ν, a_μ-points*.

In a number of papers [1, 2, 3, 4] so called proximity property of a-points was studied describing geometric locations of a-points and involving main conclusions of value distribution theory which relates to numbers of a-points. Below we need some proximity property that relates to the case of simple a-points, see for instance [3, 4]. We give here a new aspect of the property, which follows easily from our Theorem 1 below.

Theorem B *Given a meromorphic function $w(z)$ in C and distinct complex values $a_1, a_2, \ldots, a_q$ there exists right distributed and closely located Ahlfors' sets $\Omega(r, w)$ of simple $a_1, a_2, \ldots, a_q$-points.*

Thus proximity property not only enables us to obtain the main conclusion related to the numbers of a-points (since theorem B implies theorem A) but also information of locations of a-points. Since $[A(r, w)]$ accumulations $\Omega_i(r, w)$ implies Ahlfors set of simple $a_1, a_2, \ldots, a_q$-points and due to right distribution of the set we conclude that in average any of the accumulations $\Omega_i(r, w)$ implies one simple a_ν-point for almost any value a_ν but 4. But then all a_ν-points occurring in any accumulation $\Omega_i(r, w)$ must be close to each other due to close distribution.

Moreover, if instead of a_ν-points, certain polygons P_ν are considered, then w^{-1}-images $z_i(P_\nu)$ of these polygons P_ν will be also close to each other and they must be small in general. It is not surprising since we need only to adjust proximity property to the considered case. On the other hand we study a new type problem in value distribution: shapes of preimages $z_i(P_\nu)$ of polygons P_ν. It appears that there must be many preimages $z_i(P_\nu)$ whose shapes are similar to that of P_ν.

In order to describe about quantitative and geometric properties of $w^{-1}(P_\nu)$ more clearly, we need to introduce the follwoing

Definition 4 Let $P_1, P_2, \ldots, P_q$ be a totality of some non overlapping polygons without common boundary points each involving respectively values $a_1, a_2, \ldots, a_q$. We call sets $\Omega^P(r, w)$ *Ahlfors' sets of right distributed preimages* $z_i(P_\nu)$ *of polygons* P_ν if $\Omega^P(r, w)$ is a totality of $A(r, w)$ sets (accumulations) $\Omega_i^P(r, w)$, $i = 1, 2, \ldots, [A(r, w)]$, such that 1) each accumulation consists of some preimages (curvilinear polygons) $z_i(P_\nu)$, $\nu = 1, 2, \ldots, q$, or empty, 2) for every $z_i(P_\nu) \subset \Omega_i^P(r, w)$ function w accomplishes one to one mapping on the closure of $z_i(P_\nu)$, 3) each accumulation implies no more than one curvilinear polygon for given index ν, 4) for any $r \notin E$, where E is a set of finite logarithmic measure, is valid

$$\sum_{\nu=1}^{q} n_0(\Omega^P(r,w), P_\nu) \geq (q-4)A(r,w) - o\left(A(r,w)\right), \ r \to \infty, \ r \notin E.$$

(3)

where $n_0(\Omega^P(r,w), P_\nu)$ is number of all $z_i(P_\nu)$ belonging to $\Omega^P(r,w)$.

Let's denote by $d(X)$ diameter of the set X, that is $\sup_{xy\in X} |x-y|$.

Definition 5 We call sets $\Omega^P(r,w)$ *Ahlfors' sets of right distributed, closely located, small preimages* $z_i(P_\nu)$ of polygons P_ν if in addition to previous definition each $z_i(P_\nu) \subset \Omega_i^P(r,w)$ is small for any i and ν so that

$$d\left(z_i(P_\nu)\right) \leq \frac{\varphi(r)r}{A^{1/2}(r,w)}$$

(4)

and any totality of preimages of curvilinear polygons $z_i(P_1), z_i(P_2), ..., z_i(P_q)$ belonging to the same $\Omega_i(r,w)$ is close to each other for any i so that

$$d\left(\Omega_i(r,w)\right) \leq \frac{\varphi(r)r}{A^{1/2}(r,w)},$$

(5)

In terms of definitions 4 and 5 we can talk about numbers, smallness and closeness of preimages $z_i(P_\nu)$.

Now we pass to definition relating to shapes of the sets $z_i(P_\nu)$. Let P be a polygon with vertices $t_k(P)$ and sides $s_k(P)$ connecting $t_k(P)$ and $t_{k+1}(P)$. Consider a curvilinear polygon $\mathcal{P}$ that is a simply connected domain which has the same number of vertices $t_k(\mathcal{P})$ and side $s_k(\mathcal{P})$ that are smooth curves. Denote by $m_k(P)$, (respectively by $m_k(\mathcal{P})$), the ratio of lengths two sides of P emanating from a vertex $t_k(P)$, (respectively emanating from the vertex $t_k(\mathcal{P})$). And denote by $\alpha_k(P)$, (respectively by $\alpha_k(\mathcal{P})$), angle formed by points $t_{k-1}(P), t_k(P), t_{k+1}(P)$, (respectively by points $t_{k-1}(\mathcal{P}), t_k(\mathcal{P}), t_{k+1}(\mathcal{P})$).

Definition 6 We say that a curvilinear polygon $\mathcal{P}$ is ε-similar to P if for any k

$$\left(\frac{1-\varepsilon}{1+\varepsilon}\right)^4 m_k(P) \leq m_k(\mathcal{P}) \leq \left(\frac{1+\varepsilon}{1-\varepsilon}\right)^4 m_k(P)$$

(6)

and

$$|\alpha_k(P) - \alpha_k(\mathcal{P})| \leq \ln\left(\frac{1}{1-2\varepsilon}\right)^4.$$

(7)

Geometrically it is clear, that if ε is small then the above definition reflect a similarity between the shapes of P and $\mathcal{P}$.

Finally we need to define distances for a collection of q polygons P_ν, $\nu = 1, 2, ..., q$ each containing a point a_ν, $\nu = 1, 2, ..., q$.

Definition 7 A *a collection of ε-distant polygons P_ν means all the diameters $d(P_\nu)$ of these polygons are less than or equal to $\varepsilon\rho^*$, where $\varepsilon = const < 1/4$, $\rho^* := \min_{\nu \neq \mu} \frac{|a_\nu - a_\mu|}{4}$, $\nu, \mu \in \{1.2...., q\}$.*

With the above definitions and notations we prove the following results of similarity property, in terms of the shapes, closeness , smallness of the preimages of polygons, which also implies Theorems B and generalizes Theorem A.

Theorem 1 *Suppose $w(z)$ is a meromorphic function in C, a_ν, $\nu = 1, 2, ..., q$, are distinct complex values and $P_\nu \ni a_\nu$, $\nu = 1, 2, ..., q$, is a collection of ε- distant polygons $\varepsilon = const < 1/4$. Then there exists Ahlfors' sets $\Omega^P(r, w)$ of right distributed, closely located, small preimages $z_i(P_\nu)$ of polygons P_ν each ε-similar to P_ν.*

Geometrically it is clear, that the smaller ε is the more similar are shapes of $z_i(P_\nu)$ to that of P_ν in Theorem 1.

3. Proof of Theorem 1

In order to prove the theorem, we need to recall some notations and results. Next lemma is a part of Theorem 2 in [4] that describes particularly locations of a_ν-points of meromorphic functions for different values a_ν and the geometry of neighborhoods of these a_ν-points.

Lemma 1 *Suppose conditions of Theorem 1 are valid and suppose ρ' is a constant, $0 < \rho' < 1$, $\rho^{(w)} = \min_{\nu \neq \mu} \frac{|a_\nu - a_\mu|}{3}$, $\nu, \mu \in \{1.2...., q\}$, $Y(a_\nu, x)$ is the disk $\{w : |w - a_\nu| < x\}$. Then there exist $\Phi(r, w)$ having no common points simply connected domains $E_i(r, w)$, $i = 1, 2, ...\Phi(r, w)$ in the disks $D(r) := \{z : |z| < r\}$, $r \notin E$, where E is a set of finite logarithmic measure, such that the following assertions are valid.*

1) Function w is univalent in any of the domains $E_i(r, w)$, $i = 1, 2, ...\Phi(r, w)$, consequently the inverse function $F(w)$ accomplishes one-to-one mapping in any of the domains $w(E_i(r, w))$ considered on the covering surface $\{w(z) : |z| < r\}$ and thus determines that of branches $F_i(w)$ of the inverse function $F(w)$ for which $F_i(w(E_i(r, w))) = E_i(r, w)$.

2) Suppose for given ν the domain $E_i(r, w)$ implies an a_ν-point $z_i(a_\nu, w)$. Then $F_i(Y(a_\nu, \rho'\rho^{(w)})) \subset E_i(r, w)$.

3) Totality of all simple $a_1, a_2, \cdots, a_q$ -points lying in $\cup_{i=1}^{\Phi(r,w)} E_i(r, w)$ compose an Ahlfors' sets $\Omega(r, w)$ of simple a-points in the disks $D(r)$, $r \notin E$ and the set $\Omega(r, w)$ is a totality of accumulations $\Omega_i(a_1, a_2, \ldots, a_q, r, w)$, $i = 1, 2, \ldots, \Phi(r, w)$ which, in turn, are totalities of close and simple $a_1, a_2, \ldots, a_q$-points of w belonging to $E_i(r, w)$. In other words

$$\sum_{\nu=1}^{q} n_0^*(r, a_\nu, w) \geq (q - 4)A(r, w) - o(A(r)), \ r \to \infty, \ r \notin E \qquad (8)$$

where $n_0^*(r, a_\nu, w)$ is the number of all simple a-points of w in $\cup_{k=1}^{\Phi(n,w)} E_i(r, w)$.

4) For diameters $d(E_i(r, w))$

$$d(E_i(r, w)) \leq \frac{\varphi(r)r}{A^{1/2}(r, w)}, i = 1, 2, \ldots \Phi(r, w), \qquad (9)$$

hold, consequently all $a_1, a_2, \ldots, a_q$-points of w belonging to $E_i(r, w)$ are close.

5) For the number of the domains $E_i(r, w)$ take place

$$\frac{\Phi(r, w)}{A(r, w)} \to 1, \ r \to \infty, \ r \notin E. \qquad (10)$$

The items 1,3,and 4 coincide and are proved as items 1,2,and 3 of Theorem 1 in [4], and item [2] is a part of item 4 of Theorem 1 in [4].

Now we want to deal with exactly $[A(r, w)]$ domain $E_i(r, w)$ instead of $\Phi(r, w)$ such a domains in Lemma 1. Suppose first that for given value r we have $\Phi(r, w) > [A(r, w)]$. Then we remove from the set of all domains $E_i(r, w)$, $i = 1, 2, \ldots, \Phi(r, w)$, arbitrary $\Phi(r, w) - [A(r, w)]$ domains $E_i(r, w)$. If $\Phi(r, w) < [A(r, w)]$ then we add to the domains $E_i(r, w)$, $i = 1, 2, \ldots, \Phi(r, w)$, some $[A(r, w)] > \Phi(r, w)$ empty domains as domains $E_i(r, w)$. Retaining all previous notations for these new totality of domains $E_i(r, w)$, $i = 1, 2, \ldots, [A(r, w)]$ we notice that due to (10) difference between the number of all a_ν-points in old and new totalities of domains $E_i(r, w)$ is $O\left(\Phi(r, w) - [A(r, w)]\right) = o\left(A(r, w)\right)$. Thus we obtain

Lemma 2 *Lemma 1 is valid if $\Phi(r, w)$ is substituted by $[A(r, w)]$.*

Clearly then assertion 5) of Lemma 1 becomes trivial.

Also we need the following classical result (see [8] ch. 4 or [10] item 6.10) in the theory of functions of known class S : regular, univalent functions $f = z + a_2 z^2 + ...$ in the unit disk.

Lemma 3 *For any $f \in S$ and z in the unit disk*

$$\frac{1 - |z|}{(1 + |z|)^3} \leq |f'(z)| \leq \frac{1 + |z|}{(1 - |z|)^3}, \tag{11}$$

$$-\ln \frac{1 + |z|}{1 - |z|} \leq \arg \frac{f(z)}{z} \leq \ln \frac{1 + |z|}{1 - |z|}. \tag{12}$$

Now we proceed to the proof of Theorem 1. By choosing $\rho' = \rho^{(w)}/\rho^*$ in item 2 of Lemma 2, and if the domain $E_i(r, w)$ contains an a_ν-point then $w(E_i(r, w))$ contains $Y(a_\nu, \rho^*)$. Therefore each a_ν-point $z_i(a_\nu, w)$ belonging to a domain $E_i(r, w)$ belongs also to a domain $z_i(P_\nu, w)$ which is F_i-image of polygon P_ν since $F_i(Y(a_\nu, \rho^*)) \subset E_i(r, w)$ and $P_\nu \subset Y(a_\nu, \rho^*)$.

The domain $z_i(P_\nu, w)$ is a simply connected since $F_i(w)$ accomplishes one-to-one mapping in $Y(a_\nu, \rho^*)$, see item 1 in Lemma 2.

Clearly we have one to one correspondence of defined above a_ν-points and curvilinear polygons. Consequently number $n_0(\Omega^P(r, w), P_\nu)$ of defined above curvilinear polygons $z_i(P_\nu)$ coincides with the number $n_0(\Omega(r, w), a_\nu)$ for any ν and therefore, due to item 3, all these curvilinear polygons taken for all values ν compose an Ahlfors sets $\Omega^P(r, w)$ of curvilinear polygons which is right distributed if as accumulations $\Omega_i^P(r, w)$, $i = 1, 2, \ldots, [A(r, w)]$ we consider totalities of $z_i(P_\nu)$ belonging to a given $E_i(r, w)$.

From estimates for diameters of $E_i(r, w)$ by item 4 of Lemma 2 we immediately conclude that each of these curvilinear polygons $z_i(P_\nu)$, (implied in mentioned totalities of right distributed Ahlfors sets), is small and each totality of curvilinear polygons $z_i(P_\nu)$ in each accumulation $\Omega_i^P(r, w)$ is close to each other.

Thus to complete proof of Theorem 1 we need only to show that the above curvilinear polygons $z_i(P_\nu)$ are ε-similar to polygons P_ν. Again due to item 1 $F_i(Y(a_\nu, \rho^*)) \subset E_i(r, w)$ and $P_\nu \subset Y(a_\nu, \rho^*)$ and therefore we have that function

$$\mathcal{F}_i(W) = \frac{F_i(\rho^* W + a_\nu) - F_i(a_\nu)}{(F_i)'_w(a_\nu)}, \quad w = \rho^* W + a_\nu$$

is defined in $|W| < 1$ and belongs to the class S.

Applying inequality (12) to this function and taking $(F_i)'_w = (F_i)'_W$ $W'_w = (F_i)'_W/\rho^*$ into account and hence

$$(\mathcal{F}_i)'_W = \frac{(F_i)'_w \rho^*}{(F_i)'_w(a_\nu)}$$

we have

$$\frac{\rho^* \left(\rho^* - |w - a_\nu|\right) |(F_i)'_w(a_\nu)|}{(\rho^* + |w - a_\nu|)^3} \leq |(F_i)'_w(w)|$$

$$\leq \frac{\rho^* \left(\rho^* + |w - a_\nu|\right) |(F_i)'_w(a_\nu)|}{(\rho^* - |w - a_\nu|)^3}.$$

Since any two sides $s_k(P_\nu)$ and $s_{k-1}(P_\nu)$ of polynomial P_ν lie in $Y(a_\nu, \varepsilon \rho^*)$, so that $|w - a_\nu| < \varepsilon \rho^*$ when $w \in s_k(P_\nu)$ or $w \in s_{k-1}(P_\nu)$ we conclude that for any such a w take place inequality

$$\frac{(1 - \varepsilon) |(F_i)'_w(a_\nu)|}{\rho^* (1 + \varepsilon)^3} \leq |(F_i)'_w(w)| \leq \frac{(1 + \varepsilon) |(F_i)'_w(a_\nu)|}{\rho^* (1 - \varepsilon)^3}.$$

Consequently maximal and minimal distortions under mapping by F_i in $Y(a_\nu, \varepsilon \rho^*)$ are determined respectively by magnitudes

$$\left(\frac{1 + \varepsilon}{1 - \varepsilon}\right)^4, \quad \left(\frac{1 - \varepsilon}{1 + \varepsilon}\right)^4$$

So we conclude that for for P_ν and $z_i(P_\nu)$ inequality (6) holds.

Finally we prove that the curvilinear polygons satisfy (7). First we note that since diameters of polygons $z_i(P_\nu)$ are less or equal to $\varepsilon \rho^*$, where $\varepsilon = const < 1/4$, then for vertices $t_k(P_\nu)$ of the polygon P_ν we have $P_\nu \subset Y(t_k(P_\nu), \rho^* - \varepsilon \rho^*) \subset Y(a_\nu, \varepsilon \rho^*)$. Therefore if an accumulation $\Omega_i^P(r, w)$ implies a curvilinear polygon $z_i(P_\nu)$ then function F_i is a one to one mapping on $Y(t_k(P_\nu), \rho^* - \varepsilon \rho^*)$ and consequently function

$$\mathcal{F}_i^*(W) := \frac{F_i \left((\rho^* - \varepsilon \rho^*)W + t_k(P_\nu)\right) - F_i(t_k(P_\nu))}{(F_i)'_w(t_k(P_\nu))},$$

$$w = (\rho^* - \varepsilon \rho^*)W + t_k(P_\nu)$$

is defined in $|W| < 1$ and belongs to the class S. Applying double inequality (12) to this function and taking into account that

$$\frac{\mathcal{F}_i^*(W)}{W} = \frac{F_i(w) - F_i(t_k(P_\nu))}{w - t_k(P_\nu)} \frac{\rho^* - \varepsilon \rho^*}{F_i'(t_k(P_\nu))}$$

we obtain for $w \in Y(t_k(P_\nu), \rho^* - \varepsilon\rho^*)$

$$- \ln \frac{\rho^* - \varepsilon\rho^* + |w - t_k(P_\nu)|}{\rho^* - \varepsilon\rho^* - |w - t_k(P_\nu)|} \leq \arg \frac{F_i(w) - F_i(t_k(P_\nu))}{w - t_k(P_\nu)}$$

$$- \arg \frac{F_i'(t_k(P_\nu))}{\rho^* - \varepsilon\rho^*} \leq \ln \frac{\rho^* - \varepsilon\rho^* + |w - t_k(P_\nu)|}{\rho^* - \varepsilon\rho^* - |w - t_k(P_\nu)|}$$

and, consequently, we have

$$\arg \frac{F_i'(t_k(P_\nu))}{\rho^* - \varepsilon\rho^*} - \ln \frac{\rho^* - \varepsilon\rho^* + |w - t_k(P_\nu)|}{\rho^* - \varepsilon\rho^* - |w - t_k(P_\nu)|} \leq$$

$$\leq \arg(F_i(w) - F_i(t_k(P_\nu))) - \arg(w - t_k(P_\nu)) \leq$$

$$\leq \arg \frac{F_i'(t_k(P_\nu))}{\rho^* - \varepsilon\rho^*} + \ln \frac{\rho^* - \varepsilon\rho^* + |w - t_k(P_\nu)|}{\rho^* - \varepsilon\rho^* - |w - t_k(P_\nu)|}.$$

Applying the above inequalities to points w coinciding with vertices $t_{k-1}(P_\nu)$ and $t_{k+1}(P_\nu)$ and taking into account that for $w = t_{k-1}(P_\nu)$ or $w = t_{k+1}(P_\nu)$ take place $|w - t_k(P_\nu)| < \varepsilon\rho^*$ we obtain

$$|\arg(F_i(t_{k-1}(P_\nu)) - F_i(t_k(P_\nu))) - \arg(t_{k-1}(P_\nu) - t_k(P_\nu))| \leq$$

$$\leq 2\ln \frac{\rho^* - \varepsilon\rho^* + |w - t_k(P_\nu)|}{\rho^* - \varepsilon\rho^* - |w - t_k(P_\nu)|} < 2\ln \frac{1}{1 - 2\varepsilon} \qquad (13)$$

and similarly

$$|\arg(F_i(t_{k-1}(P_\nu)) - F_i(t_k(P_\nu))) - \arg(t_{k-1}(P_\nu) - t_k(P_\nu))| < 2\ln \frac{1}{1 - 2\varepsilon}. \qquad (14)$$

And (7) follows from these estimates.

This also completes the proof of Theorem 1.

4. An application of similarity property: theorem about 15 points.

Consider meromorphic functions w for which there a set is $L^{(w)}$ of values a such that all a-points of w, $a \in L^{(w)}$, lie on a finite collection of non-parallel straight lines. The following known conjectuire in this aspect have been posed by Ozawa [6]: if $L^{(w)} = \{a_n\}_{n=1}^\infty$, $|w| \to \infty$, for an entire function w then w must be a polynomial. This conjecture was confirmed by Qiao [10]. Here we show that in a similar problem the same conclusion holds for some sets $L^{(w)}$ consisting only on 9 points.

Theorem 2 *There are 15 values a_ν, $\nu = 1, 2, ..., 15$, (with a certain geometry) such that any meromorphic function w in C with $L^{(w)} = \{a_\nu\}_{\nu=1}^{15}$ must be a rational function.*

As such a set of 15 values a_ν, $\nu = 1, 2, ..., 15$, one can choose vertices of arbitrary five ε-distant acute triangles P_ν, $\nu = 1, 2, ..., 5$, with $\varepsilon < \left(e^{\pi/8} - 1\right)/2e^{\pi/8}$.

In case of entire function we have

Theorem 3 *There are 9 values a_τ, $\nu = 1, 2, ..., 9$, (with a certain geometry) such that any entire function w with $L^{(w)} = \{a_\tau\}_{\tau=1}^{9}$ must be a polynomial.*

As such a set of 9 values a_ν, $\nu = 1, 2, ..., 9$, one can take vertices of arbitrary 3 ε-distant acute triangles, with $\varepsilon < \left(e^{\pi/8} - 1\right)/2e^{\pi/8}$.

Proof of Theorem 2 We apply Theorem 1 with above defined triangles P_ν, $\nu = 1, 2, ..., 5$, value ε and function $\varphi(r) = A^{1/4}(r, w)$. Since number of these triangles is equal to 5 we obtain that there is a ceratin number $\mathcal{N}(r) \geq A(r, w) + O\left(A(r, w)\right)$, $r \to \infty$, $r \notin E$, of small preimages $z_i(P_\nu)$ of triangles P_ν each is ε-similar to P_ν (here we consider all values $\nu = 1, 2, ..., 5$). Consequently for at least one value ν, (say value $\nu = 1$) we have a certain number $\mathcal{N}^{(1)}(r) \geq \frac{1}{5}A(r, w) + O\left(A(r, w)\right)$, $r \to \infty$, $r \notin E$, of small preimages $z_j^*(P_1)$ $(\subset \{z_i(P_1)\})$ of polygons P_1, $j = 1, 2, ..., \mathcal{N}^{(1)}(r)$, each ε-similar to P_1. Denoting by a_1, a_2, a_3 vertices of triangle P_1 we consider vertices $z_i^*(a_1)$, $z_i^*(a_2)$, $z_i^*(a_3)$ of these preimages which are, in fact, a_1, a_2, a_3-points of w belonging to an accumulation with index i. Due to inequalities (13)-(14) we have with the above ε

$$|\arg(z_i^*(a_1) - z_i^*(a_2)) - \arg(a_1 - a_2)| < 2\ln\frac{1}{1 - 2\varepsilon} \leq \frac{\pi}{4}$$

and

$$|\arg(z_i^*(a_3) - z_i^*(a_2)) - \arg(a_3 - a_2)| < 2\ln\frac{1}{1 - 2\varepsilon} \leq \frac{\pi}{4}.$$

Taking into account that points a_1, a_2, a_3 form an acute triangle we conclude that angle with vertex $z_i(a_2)$ formed by points $z_i^*(a_1)$, $z_i^*(a_2)$, $z_i^*(a_3)$ is less than π. Clearly the same is true for angle in vertices $z_i^*(a_1)$ and $z_i^* z_i(a_3)$. So that points $z_i^*(a_1)$, $z_i^*(a_2)$, $z_i^*(a_3)$ also form a triangle (they do not lie on a line). On the other hand due to smallness of $z_i^*(P_1)$ we have

$$|z_i^*(a_1) - z_i^*(a_2)|, |z_i^*(a_2) - z_i^*(a_3)|, |z_i^*(a_1) - z_i^*(a_3)| \leq$$

$$\leq \frac{\varphi(r)r}{A^{1/2}(r, w)} = \frac{r}{A^{1/8}(r, w)}. \tag{15}$$

According to condition of Theorem 2, these points $z_i^*(a_1)$, $z_i^*(a_2)$, $z_i^*(a_3)$ must be on finite non-parallel lines (say $L_1, L_2, ..., L_c$). It is easy to see that excepting for a finite number of indexes i all triples $z_i^*(a_1)$, $z_i^*(a_2)$, $z_i^*(a_3)$ with the same index must lie on the same line (one belonging to $L_1, L_2, ..., L_c$). Indeed, since $L_1, L_2, ..., L_c$ are non-parallel we conclude that there is a value $r_0 > 0$ such that for $r > r_0$ distances between sets $L_{c_1} \setminus D(r_0)$ and $L_{c_2} \setminus D(r_0)$, $c_1 \neq c_2$, must be greater than $r_0/A^{1/8}(r_0, w)$. Therefore if we consider values $r > r_0 + r_0/A^{1/8}(r_0, w)$ then the above defined triples can not lie on different lines since for them (15) holds. However, these triples can not lie on any of the straight lines $L_1, L_2, ..., L_c$ since each of these triples composes an triangle. This contradiction shows that function w can't be transcendental and hence Theorem 2 is proved.

Proof of Theorem 3 follows similarly by taking Remark 1 into account.

References

[1] Barsegian, G.: On the geometry of meromorphic functions, Math. Sbornic, 114(156), no. 2 (1981), 179–226. (in Russian); Math. USSR Sbornic 42, no. 2 (1982), 155–196.

[2] Barsegian, G.: A proximity property of the a-points of meromorphic functions. Math. Sbornic, 120(162) (1983), 42–63 (in Russian); Math. USSR Sbornik 48 (1984), 41–63.

[3] Barsegian, G.A., Yang C.C.: Some new generalizations in the theory of meromorphic functions and their applications. Part 1. On the magnitudes of functions on the sets of a-points of derivatives, Complex Variables, Theory Appl. 41 (2000), 293–313.

[4] Barsegian, G.A., Yang C.C.: Some new generalizations in the theory of meromorphic functions and their applications. Part 2. On the derivatives of meromorphic and inverse functions. Complex Variables, Theory Appl. 44 (2001), 13–28.

[5] Nevanlinna, R.: Eindeutige analytische Funktionen. Springer, Berlin, 1936.

[6] Ozawa, M.: On the solution of the functional equation $f \circ g(z) = F(z)$. Y. Kodai Math. Sem. Rep. 20 (1968), 305–313.

[7] Baker, I.N.: Entire functions whose A-points lie on a system of lines. Lectutre Notes in Pure and Applied Mathematics 78, Marcel Dekker 1982, 1–186.

[8] Golusin, G.M.: Geometric function theory of complex variables. Nauka, Moscow 1966 (in Russian).

[9] Hayman, W.K.: Multivalent functions. Cambridge University Press 1958.

[10] Qiao, J.: Factorizations of meromorphic functions. Chinese Annals of Math. 10A (1989) 692–698 (in Chinese).

COMPLEX ZERO DECREASING SEQUENCES AND THE RIEMANN HYPOTHESIS II

George Csordas

Department of Mathematics
University of Hawaii
Honolulu, HI 96822
george@math.hawaii.edu

Abstract   A long-standing open problem (the Karlin–Laguerre problem) in the theory of distribution of zeros of real entire functions requires the characterization of all real sequences $T = \{\gamma\}_{k=0}^{\infty}$ such that for any real polynomial $p(x) := \sum_0^n a_k x^k$, the polynomial $\sum_0^n \gamma_k a_k x^k$ has no more nonreal zeros than $p(x)$ has. The sequences T which satisfy the above property are called complex zero decreasing sequences. While the Karlin–Laguerre problem has remained open, recently there has been significant progress made in a series of papers by A. Bakan, T. Craven, A. Golub and G. Csordas. In particular, it follows that under a mild growth restriction, an entire function, $f(z)$, of exponential type has only real zeros, if the sequence $T = \{f(k)\}_{k=0}^{\infty}$ is a complex zero decreasing sequence. These results yield new necessary and sufficient conditions for the validity of the Riemann Hypothesis. Applying these conditions to the Riemann ξ-function, some numerical results will highlight a quantitative version of the dictum that "the Riemann Hypothesis, if true, is only barely so".

Keywords: Laguerre-Pólya class, multiplier sequences, complex zero decreasing sequences, interpolation, Riemann Hypothesis
Subject Classification: Primary 26C10, 30D15; Secondary 30D10

1.   Introduction

The principal goal of this paper is to investigate several properties enjoyed by functions in the Laguerre-Pólya class and to relate these to the Riemann ξ-function. Among these properties, certain necessary conditions are expressed in terms of the classical Laguerre inequalities

H.G.W. Begehr et al. (eds.), Analysis and Applications - ISAAC 2001, 121–134.
© *2003 Kluwer Academic Publishers.*

as well as in terms of some new iterated Turán inequalities for multiplier sequences (Section 1). Section 2 focuses on a particular class of multiplier sequences, called complex zero decreasing sequences, and discusses the recent progress that has been made (Theorem 4 and Theorem 5) for solving the Karlin-Laguerre problem (Problem 5). Section 3 establishes a connection between the Mellin transform of a special theta function and the Riemann ξ-function (Proposition 1). This, together with the results of Section 2, yield necessary and sufficient conditions, expressed in terms of multiplier sequences and complex zero decreasing sequences, for the validity of the Riemann Hypothesis (Remark 3). The paper includes ten problems (or conjectures); most of these are relevant to the general theory, while others may be of special interest in the theory of the Riemann ζ-function. The concrete numerical examples (Example 1) which conclude the paper, further underscore the delicate behavior of the ξ-function on the set of nonnegative integers.

2. The Laguerre-Pólya class and the Riemann Hypothesis

We commence here with some pertinent definitions involving the Laguerre–Pólya class, multiplier sequences and the Riemann Hypothesis. Also we shall state the Laguerre and Turán inequalities. The recent extensions of these inequalities provide new necessary conditions for an entire function to belong to the Laguerre–Pólya class. We discuss these inequalities in terms of the Riemann ξ-function and formulate several questions, conjectures and problems.

Definition 1 *A real entire function $\varphi(x)$ is said to be in the Laguerre-Pólya class, written $\varphi \in \mathcal{L}\text{-}\mathcal{P}$, if $\varphi(x)$ can be represented in the form*

$$\varphi(x) := cx^m e^{-\alpha x^2 + \beta x} \prod_{k=1}^{\omega} (1 + x/x_k) e^{-x/x_k}, \qquad (0 \le \omega \le \infty), \qquad (1)$$

where c, β, x_k are real, $\alpha \ge 0, m$ is a nonnegative integer, $\sum x_k^{-2} < \infty$ Pólya and Schur ([18]) termed a real entire function $\varphi(x) := \sum_{k=0}^{\infty} \frac{\gamma_k}{k!} x^k$ as a function of type I in the Laguerre-Pólya class, written $\varphi \in \mathcal{L}\text{-}\mathcal{P}I$, if $\varphi(x)$ or $\varphi(-x)$ can be represented in the form

$$\varphi(x) := cx^m e^{\sigma x} \prod_{k=1}^{\omega} (1 + x/x_k), \qquad (0 \le \omega \le \infty), \qquad (2)$$

where c is real, $\sigma \ge 0$, m is a nonnegative integer, $x_k > 0$, and $\sum 1/x_k < \infty$. We will write $\varphi(x) = \sum_{k=0}^{\infty} \frac{\gamma_k}{k!} x^k \in \mathcal{L}\text{-}\mathcal{P}^+$, if $\varphi \in \mathcal{L}\text{-}\mathcal{P}I$ and if $\gamma_k \ge 0$

for all $k = 0, 1, 2 \ldots$. In this case the sequence $\{\gamma_k\}_0^\infty$ is called a multi-plier sequence.

Remark 1 (i) The significance of the Laguerre-Pólya class in the theory of entire functions is natural, since functions in this class, *and only these* are the uniform limits, on compact subsets of $\mathbb{C}$, of polynomials with only real zeros ([17], p. 54). (ii) Thus, it follows from this result that the class $\mathcal{L}$-$\mathcal{P}$ is closed under differentiation; that is, if $\varphi \in \mathcal{L}$-$\mathcal{P}$, then $\varphi^{(n)} \in \mathcal{L}$-$\mathcal{P}$ for $n \geq 0$. (iii) We remark that a sequence, $\{\gamma_k\}_0^\infty$, of non-negative real numbers is a multiplier sequence if and only if for any function $\varphi(x) = \sum_{k=0}^\infty \frac{\alpha_k}{k!} x^k \in \mathcal{L}$-$\mathcal{P}$, the entire function $\sum_{k=0}^\infty \frac{\gamma_k \alpha_k}{k!} x^k \in \mathcal{L}$-$\mathcal{P}$ (cf. [18]). For the various properties and characterizations of multiplier sequences see, for example, [18], ([17], 100–124) or ([14], pp 29–47).

In order to expedite our presentation, we proceed to review briefly some of the nomenclature pertaining to the Riemann Hypothesis. To begin with, the Riemann ξ-function can be defined (cf. [15] or ([17], p. 285)) by

$$\xi(iz) := \frac{1}{2}\left(z^2 - \frac{1}{4}\right)\pi^{-\frac{z}{2}-\frac{1}{4}}\Gamma\left(\frac{1}{4} + \frac{z}{2}\right)\zeta\left(\frac{1}{2} + z\right), \qquad (3)$$

where ζ is the Riemann ζ-function. It is known that $\xi(z)$ is an entire function of order one (see [19], p. 29) and note that Titchmarsh uses the symbol Ξ for ξ) and of maximal type ([7], p. 493). Moreover, the Riemann ξ-function admits the Fourier integral representation [15] or ([19], p. 255)

$$\xi_1(x) := \frac{1}{8}\xi\left(\frac{x}{2}\right) = \int_0^\infty \Phi(t)\cos(xt)\,dt = \frac{1}{2}\int_{-\infty}^\infty \Phi(t)e^{ixt}\,dt, \qquad (4)$$

where

$$\Phi(t) := \sum_{n=1}^\infty \left(2\pi^2 n^4 e^{9t} - 3\pi n^2 e^{5t}\right)\exp\left(-\pi n^2 e^{4t}\right). \qquad (5)$$

The Riemann Hypothesis is the statement that all the zeros of $\xi_1(x)$ are real and so $\xi_1 \in \mathcal{L}$-$\mathcal{P}$. Since

$$\xi_1(x) = \sum_{m=0}^\infty \frac{b_k}{(2k)!}(-x^2)^k, \qquad (6)$$

where the b_k's are the moments defined by

$$b_k = \int_0^\infty t^{2k}\,\Phi(t)\,dt, \quad k = 0, 1, 2, \ldots, \qquad (7)$$

the change of variables $u = -x^2$ in (1.4) yields

$$\xi_2(u) := \sum_{k=0}^{\infty} \frac{b_k}{(2k)!} u^k = \int_0^{\infty} \Phi(t) \cosh(t\sqrt{u}) \, dt. \tag{8}$$

Thus, ξ_2 is an entire function of order $\frac{1}{2}$ and the Riemann Hypothesis is equivalent to the statement that ξ_2 has only real negative zeros (so that $\xi_2 \in \mathcal{L}\text{-}\mathcal{P}^+$). Since $\xi \in \mathcal{L}\text{-}\mathcal{P} \Leftrightarrow \xi_1 \in \mathcal{L}\text{-}\mathcal{P} \Leftrightarrow \xi_2 \in \mathcal{L}\text{-}\mathcal{P}$, in the sequel it will be convenient to refer to ξ_1 (as well as to ξ_2) as the Riemann ξ-function. For the reader's convenience, we summarize in the next two theorems some of the basic properties of the kernel, Φ.

Theorem 1 ([6], Theorem A) *Consider the function Φ defined by (1.5). Then, the following are valid:*

 (i) $\Phi(t) > 0$ *for all* $t \geq 0$;

 (ii) $\Phi(z)$ *is analytic in the strip* $-\pi/8 < \operatorname{Im} z < \pi/8$;

 (iii) $\Phi(t)$ *is an even function, so that* $\Phi^{(2m+1)}(0) = 0$ $(m = 0, 1, \dots)$;

 (iv) *for any* $\varepsilon > 0$ *and for each* $n = 0, 1, \dots$

$$\lim_{t \to \infty} \Phi^{(n)}(t) \exp\left((\pi - \varepsilon)e^{4t}\right) = 0;$$

 (v) $\Phi'(t) < 0$ *for all* $t > 0$.

Theorem 2 ([9], Theorem 2.1) *The function $\Phi(\sqrt{t})$, where Φ is defined by (1.5), is logarithmically concave for $t > 0$; that is,*

$$\left(\log \Phi(\sqrt{t})\right)'' < 0 \qquad \text{for } t > 0. \tag{9}$$

Preliminaries aside, we will now describe a few properties that are enjoyed by all functions in the Laguerre-Pólya class and examine these necessary conditions in relation to the function ξ_1 (or ξ_2). One of the simplest properties, shared by all functions in $\mathcal{L}\text{-}\mathcal{P}$, is logarithmic concavity; that is, if $\varphi(x) \in \mathcal{L}\text{-}\mathcal{P}$, then $(\log \varphi(x))'' < 0$ for all real numbers x which are not zeros of φ. This condition implies, in light of the closure property of $\mathcal{L}\text{-}\mathcal{P}$ under differentiation (see Remark 1 (ii)), that if $\varphi \in \mathcal{L}\text{-}\mathcal{P}$, then φ satisfies the following inequalities, known as the *Laguerre inequalities*,

$$L_k(\varphi(x)) := (\varphi^{(k)}(x))^2 - \varphi^{(k-1)}(x)\varphi^{(k+1)}(x) \geq 0,$$
$$k = 1, 2, 3 \dots, \quad \text{for all} \quad x \in \mathbb{R}, \tag{10}$$

see, for example, [3] or [8]. Given the extensive and profound research dealing with the theory of zeta functions, it is remarkable that, today, we still do *not* know whether or not the Riemann ξ-function satisfies the Laguerre inequalities for *all* $x \in \mathbb{R}$.

Problem 1 (The simplest Laguerre inequality for the Riemann ξ-function.)
With ξ_2 defined by (8), prove that

$$\left(\xi_2'(x)\right)^2 - \xi_2(x)\,\xi_2''(x) \geq 0 \quad \text{for all} \quad x \in \mathbb{R}.$$

We hasten to remark that there is a complex analog of the Laguerre inequalities which is *both* necessary and sufficient for a real entire function, $f(z)$, of order less than 2, to have only real zeros. Indeed, it is known that $f(z) \in \mathcal{L}\text{-}\mathcal{P}$ if and only if $|f'(z)|^2 \geq \text{Re}\{f(z)\overline{f''(z)}\}$ for all $z \in \mathbb{C}$ ([10], Theorem 2.10). The concavity condition becomes clear if we observe that the foregoing inequality can also expressed in the form

$$\frac{1}{2}\frac{\partial^2}{\partial y^2}|f(x+iy)|^2 = |f'(z)|^2 - \text{Re}\{f(z)\overline{f''(z)}\} \geq 0 \quad \text{for all} \quad x,y \in \mathbb{R}.$$

$$(11)$$

But the verification of (11) in case of the Riemann ξ-function, is ostensibly very difficult and so, striving for simplicity, we return to the Laguerre inequalities (10) in the special case when $x = 0$. To this end, let $\varphi(x) := \sum_{k=0}^{\infty} \gamma_k \frac{x^k}{k!}$ be an entire function of order, say, less than 2. Then, with $x = 0$ in (10)), we find that a *necessary condition* that $\varphi(x)$ have only real zeros is that the following *Turán inequalities* hold:

$$T_1(k) := \gamma_k^2 - \gamma_{k-1}\gamma_{k+1} \geq 0, \qquad k = 1,2,3,\ldots. \qquad (12)$$

For various extensions and inequalities related to (12), we refer to [3] and [11]. In [3], Craven and Csordas investigated certain polynomial invariants and used them to prove the following theorem.

Theorem 3 ([3], Theorem 2.13) *If $\varphi(x) := \sum_{k=0}^{\infty} \gamma_k \frac{x^k}{k!} \in \mathcal{L}\text{-}\mathcal{P}^+$, then (with $T_1(k)$ defined by (12)) the following double Turán inequalities hold*

$$T_2(k) := T_1(k)^2 - T_1(k-1)T_1(k+1) \geq 0, \qquad k = 2,3,4,\ldots. \qquad (13)$$

The question whether or not the higher iterated Turán inequalities hold for functions in the Laguerre-Pólya class remains open and we formulate this as our next problem.

Problem 2 *Let $\varphi(x) := \sum_{k=0}^{\infty} \gamma_k \frac{x^k}{k!} \in \mathcal{L}\text{-}\mathcal{P}^+$. Let $T_1(k) := \gamma_k^2 - \gamma_{k-1}\gamma_{k+1} \geq 0$, $k = 1, 2, 3, \ldots$. Then is it true that the following iterated Turán inequalities*

$$T_n(k) := (T_{n-1}(k))^2 - T_{n-1}(k-1)T_{n-1}(k+1) \geq 0 \quad (k \geq n \geq 2) \quad (14)$$

hold?

Note that the iterated Turán inequalities (14) hold if γ_k's tend to 0, sufficiently fast, as $k \to \infty$. We next examine the foregoing results, problems and conjectures in relation to the Riemann ξ_2-function. To this end set $\gamma_k := k!b_k/(2k)!$, where the b_k's are the moments defined by (7). Then it is known that the logarithmic concavity $\Phi(\sqrt{t})$ for $t > 0$ (cf. Theorem 2) implies that γ_k's satisfy the Turán inequalities $T_1(k) := \gamma_k^2 - \gamma_{k-1}\gamma_{k+1} \geq 0$, $k = 1, 2, 3, \ldots$ ([9], Theorem 2.4) (for a different proof, see [6], Theorem 2.5). On the other hand, today we do not know if the double Turán inequalities are satisfied when $\varphi(x) := \xi_2(x) = \sum_{k=0}^{\infty} \frac{\gamma_k}{k!} x^k$. Now, Dr. H.J.J. te Riele has furnished the author with the numerical values of the b_k's, $k = 0, 1 \ldots, 500$, accurate to 200 places. Using these calculated moments, the author has verified that $T_2(k) \geq 0$ for $k = 2, 3 \ldots, 499$. In light of this and other evidence, we conjecture that $T_2(k) \geq 0$, for all $k \geq 2$, and we state this as the next problem.

Problem 3 (The double Turán inequalities for the Riemann ξ-function.)
Let $\gamma_k := k!b_k/(2k)!$, where the b_k's are the moments defined by (7). Prove that

$$T_2(k) \geq 0 \quad \text{for all} \quad k = 2, 3, 4 \ldots. \quad (15)$$

As we shall see below, one way to prove (15) is to establish the following concavity property of the kernel Φ.

Problem 4 *Let $s(t) := \Phi(\sqrt{t})$, where Φ is defined by (1.5). Set $f(t) := s'(t)^2 - s(t)s''(t)$. Prove that*

$$(\log f(t))'' < 0 \quad \text{for } t > 0. \quad (16)$$

First, note that by Theorem 2, $f(t) > 0$ for $t > 0$. Second, it follows from a result of Dimitrov and the author ([12], Theorem 2.4), that if (16) is true, then the double Turán inequalities (15) are satisfied. Finally, although we do not know if the iterated Turán inequalities (cf. Problem 2) are satisfied for an arbitrary function in $\varphi \in \mathcal{L}\text{-}\mathcal{P}^+$, on the basis of our numerical experiments, we propose the following problem.

Problem 5 (The iterated Turán inequalities for the Riemann
ξ function ξ_2.)
*Let $\gamma_k := \frac{k!b_k}{(2k)!}$, where the b_k's are defined by (7). Prove that $T_n(k) \geq 0$
for $k \geq n \geq 2$.*

3. Complex zero decreasing sequences and λ-sequences

A particularly important collection of multiplier sequences are the
complex zero decreasing sequences [4]. These sequences are defined as
follows.

Definition 2 *We say that a sequence $\{\gamma_k\}_{k=0}^{\infty}$ is a complex zero decreasing sequence (CZDS), if for any real polynomial $\sum_{k=0}^{n} a_k x^k$,*

$$Z_c\left(\sum_{k=0}^{n} \gamma_k a_k x^k\right) \leq Z_c\left(\sum_{k=0}^{n} a_k x^k\right),\qquad(17)$$

*where $Z_c(p(x))$ denotes the number of nonreal zeros of the polynomial
$p(x)$, counting multiplicities. (The acronym CZDS will also be used in
the plural.)*

Now it is apparent from (17) (see also Remark 1 (iii)) that a sequence
of nonnegative real numbers which is a CZDS is also a multiplier sequence. The existence of a *nontrivial* CZDS is a consequence of the
classical theorem of Laguerre (see, for example, ([4], Theorem 1.4 (3))
which asserts that if $\varphi \in \mathcal{L}\text{-}\mathcal{P}$ *and if φ has only nonpositive zeros*, then
the sequence $\{\varphi(k)\}_{k=0}^{\infty}$ is a CZDS, that is (17) holds. With the terminology adopted here, the Karlin-Laguerre problem ([4],[1]) alluded to in
the abstract, can be formulated as follows.

Problem 6 (The Karlin-Laguerre problem.) *Characterize all the multiplier sequences which are complex zero decreasing sequences (CZDS).*

This fundamental problem in the theory of multiplier sequences has
eluded the attempts of researchers for over four decades. In order to elucidate some of the subtleties involved, we need to introduce yet another
family of sequences related to CZDS.

Definition 3 *A sequence of nonzero real numbers,* $\Lambda = \{\lambda_k\}_{k=0}^{\infty}$, *is called a* λ-sequence, if

$$\Lambda[p(x)] := \Lambda\left[\sum_{k=0}^{n} a_k x^k\right] := \sum_{k=0}^{n} \lambda_k a_k x^k > 0 \quad \text{for all } x \in \mathbb{R}, \quad (18)$$

whenever $p(x) = \sum_{k=0}^{n} a_k x^k > 0$ *for all* $x \in \mathbb{R}$.

Remark 2 It should be pointed out that λ-sequences are precisely the positive definite sequences and that condition (18) is just one of the many ways of characterizing positive definite sequences (see, for example, [4], Theorem 1.7).

The importance of λ-sequences in the investigation of CZDS stems from the fact that a *necessary condition* for a sequence $\{\gamma_k\}_{k=0}^{\infty}$, $\gamma_k > 0$, to be a CZDS is that the sequence of reciprocals $\Lambda = \{\frac{1}{\gamma_k}\}_{k=0}^{\infty}$ be a λ-sequence. Thus, for example, if $\gamma_k = \varphi(k)$, $k = 0, 1, 2, \ldots$, for $\varphi \in \mathcal{L}\text{-}\mathcal{P}$, where φ possesses only negative zeros, then $\{1/\gamma_k\}_{k=0}^{\infty}$ is a λ-sequence. On the other hand, there are multiplier sequences whose reciprocals are not λ-sequences. For example, it is not difficult to check that the sequence $\{1 + k + k^2\}_{k=0}^{\infty}$ is a multiplier sequence. However, the sequence of reciprocals, $\{\frac{1}{1+k+k^2}\}_{k=0}^{\infty}$ is not a λ-sequence ([4], p. 423). In light of the foregoing discussion, the following natural problem arises.

Problem 7 (Reciprocals of multiplier sequences.) *Characterize those multiplier sequences* $\{\gamma_k\}_{k=0}^{\infty}$, $\gamma_k > 0$, *for which the sequences of reciprocals,* $\{1/\gamma_k\}_{k=0}^{\infty}$, *are* λ-sequences.

In general, if a sequence, $\{\gamma_k\}_{k=0}^{\infty}$, of positive real numbers grows sufficiently rapidly, then it is a λ-sequence. For example, recently it was proved by T. Craven and the author that if $\lambda_k > 0$, $\lambda_0 = 1$, and if $4.07\lambda_k^2 \leq \lambda_{k-1}\lambda_{k+1}$, then $\{\lambda_k\}_{k=0}^{\infty}$ is a positive definite sequence [5]. (The question whether or not the constant 4.07 is best possible remains open.) Thus, applying this criterion to sequences of the form $\{e^{k^p}\}_{k=0}^{\infty}$, where p is a positive integer, $p \geq 3$, it is easy to see that such sequences are positive definite sequences. Furthermore, it is known ([5], p. 438) that the sequence of reciprocals $\{e^{-k^p}\}_{k=0}^{\infty}$, (where p is a positive integer, $p \geq 3$) is a multiplier sequence. However, it is not known whether or not these multiplier sequences are CZDS. For ease of reference, and to tantalize the interested reader, we pose here the following concrete question.

Problem 8 *Is the sequence* $\{e^{-k^3}\}_{k=0}^{\infty}$ *a CZDS?*

The principal source of the difficulty is that, today, the only known, essentially nontrivial CZDS are the multiplier sequences that can be interpolated by functions in $\mathcal{L}\text{-}\mathcal{P}$. We use the terms "essentially non-trivial" advisedly to circumvent trivial examples of the following sort. Let $f(x) := 2 - \sin(\pi x)$. Then, the sequence $\{2, 2, 2, \ldots\}$ is clearly a CZDS, but $f(x) \notin \mathcal{L}\text{-}\mathcal{P}$. More sophisticated examples fostered a renewed scrutiny of the Karlin-Laguerre problem and the investigations have led to the following two theorems.

Theorem 4 ([2], Theorem 2) *Let* $\{\gamma_k\}_{k=0}^{\infty}$, $\gamma_k > 0$, *be a CZDS. If*

$$\overline{\lim}_{k \to \infty} \gamma_k^{1/k} > 0, \tag{19}$$

then there is a function $\varphi(z) \in \mathcal{L}\text{-}\mathcal{P}$ *of the form*

$$\varphi(z) := b e^{az} \psi(z) := b e^{az} \prod_{n=1}^{\infty} \left(1 + \frac{z}{x_n}\right),$$

where $a, b \in \mathbb{R}$, $b \neq 0$, $x_n > 0$ *and* $\sum_{n=1}^{\infty} 1/x_n < \infty$, *such that* $\varphi(z)$ *interpolates the sequence* $\{\gamma_k\}_{k=0}^{\infty}$; *that is,* $\gamma_k = \phi(k)$ *for* $k = 0, 1, 2, \ldots$.

Theorem 5 ([1], Theorem 3.6) *Let* $f(z)$ *be an entire function of exponential type. Suppose that* $\{f(k)\}_{k=0}^{\infty}$ *is a CZDS, where* $f(0) = 1$. *Let* $h_f(\theta)$ *denote the (Phragmén–Lindelöf) indicator function of* $f(z)$, *that is,*

$$h_f(\theta) := h(\theta) := \overline{\lim}_{r \to \infty} \frac{\log |f(re^{i\theta})|}{r}, \tag{20}$$

where $\theta \in [-\pi, \pi]$. *If* $h_f(\pm \pi/2) < \pi$, *then* $f(z)$ *is in* $\mathcal{L}\text{-}\mathcal{P}$ *and* $f(z)$ *can be expressed in the form*

$$f(z) = e^{az} \prod_{n=1}^{\infty} \left(1 + \frac{z}{x_n}\right),$$

where $a \in \mathbb{R}$, $x_n > 0$ *and* $\sum_{n=1}^{\infty} 1/x_n < \infty$.

These theorems are complementary results in the following sense. Theorem 4 asserts that if a CZDS (of positive terms) does *not decay too fast* (cf. (19)), then the sequence can be interpolated by function in $\mathcal{L}\text{-}\mathcal{P}$ having only real negative zeros. In contrast, Theorem 5 says that if for some entire function, f, of exponential type, the sequence $\{f(k)\}_{k=0}^{\infty}$ is a CZDS and if f does *not grow too fast* along the imaginary axis (cf. (20)), then f has only real negative zeros. If a multiplier sequence does decay rapidly (cf. (21)), then the question whether or not such a sequence can be a CZDS remains an open problem.

Problem 9 *If* $\varphi(x) := \sum_{k=0}^{\infty} \frac{\gamma_k}{k!} x^k \in \mathcal{L}\text{-}\mathcal{P}^{+}$ *(so that* $\gamma_k \geq 0$*) and if*

$$\varlimsup_{k \to \infty} \gamma_k^{1/k} = 0, \tag{21}$$

then is $\{\gamma_k\}_{k=0}^{\infty}$ *a CZDS?*

4. The Mellin transform, CZDS and the Riemann Hypothesis

The zeta function, $\zeta(s)$, is defined for $\operatorname{Re} s > 1$, by the simplest Dirichlet series; that is, the series $\sum_{n=1}^{\infty} \frac{1}{n^s}$. It arises naturally in connection with the Mellin transform of the special theta function $g(t) := \sum_{n=1}^{\infty} e^{-n^2 \pi t}$. Indeed, for $\operatorname{Re} s > \frac{1}{2}$, we have

$$\int_0^{\infty} g(t) t^{s-1} \, dt = \int_0^{\infty} \left(\sum_{n=1}^{\infty} e^{-n^2 \pi t} \right) t^{s-1} \, dt$$

$$= \sum_{n=1}^{\infty} \frac{1}{n^{2s} \pi^s} \int_0^{\infty} e^{-t} t^{s-1} \, dt = \frac{\Gamma(s)}{\pi^s} \zeta(2s),$$

where the condition $\operatorname{Re} s > \frac{1}{2}$ arises in two ways; one is to ensure the convergence of the integral and the other is to ensure the convergence of the series. With the aid of the functional equation of the theta function one can obtain the analytic continuation of $\zeta(s)$ ([19], p. 21). Here, our interest is in the distribution zeros of Mellin transform of $\Phi(t)$ and its relation to the distribution of zeros of the Riemann ξ-function. Our starting point is the following beautiful theorem of Pólya.

Theorem 6 ([16], Satz II) *Let* $K : [0, \infty) \to \mathbb{R}$ *be absolutely integrable on* $[0, \infty)$. *Suppose that* $|K(t)| = O(e^{-t^{1/2+\epsilon}})$, *for some* $\epsilon > 0$, *as* $t \to \infty$. *If* $K(t)$ *is analytic in a neighborhood of the origin, then*

$$f(z) := \int_0^{\infty} t^{z-1} K(t) \, dt \tag{22}$$

is a meromorphic function. If $f(z)$ *has only real, negative zeros, then for each positive integer* m, *the entire function*

$$\varphi(x) := \int_0^{\infty} K(t^{2m}) \cos(xt) \, dt \tag{23}$$

has only real zeros and, since the order of $\varphi(x)$ is less than 2, $\varphi(x) \in \mathcal{L}\text{-}\mathcal{P}$.

Problem 10 *Under what additional hypotheses is the converse of Theorem 5.1 valid? That is, if $\varphi(x) \in \mathcal{L}\text{-}\mathcal{P}$ (choose $m = 1$ for simplicity), then under what additional assumptions is it true that the meromorphic function $f(z)$ has only real negative zeros?*

Motivated by Theorem 6, we next exam the Mellin transform of $\Phi(u)$.

Proposition 1 *Let*

$$\psi(x) := \frac{1}{\Gamma(x + 1/2)} \int_0^\infty u^{2x} \Phi(u)\, du, \tag{24}$$

where $\Gamma(x)$ denotes the gamma function and $\Phi(u)$ is the theta function defined by (5). Then

(1) $\psi(x)$ is an entire function,

(2) $\psi(x)$ interpolates $\gamma_k := \frac{4^k k! b_k}{\sqrt{\pi}(2k)!}$, where b_k is the kth moment of $\Phi(u)$ (see (7)); that is, $\psi(k) = \gamma_k$ for $k = 0, 1, 2 \ldots$,

(3)

$$\sum_{k=0}^\infty \frac{\psi(k)}{k!} x^k = \frac{1}{\sqrt{\pi}} \int_0^\infty \Phi(u) \cosh(2u\sqrt{x})\, du, \tag{25}$$

(4) and

$$\begin{aligned}
\xi_2(4x) &= \int_0^\infty \Phi(u) \cosh(2u\sqrt{x})\, du \\
&= \tfrac{1}{16}\left(x - \tfrac{1}{4}\right) \pi^{-\sqrt{x}/2 - 1/4} \Gamma\left(\tfrac{1}{4} + \tfrac{\sqrt{x}}{2}\right) \zeta\left(\tfrac{1}{2} + \sqrt{x}\right).
\end{aligned} \tag{26}$$

Proof (1) Since $\Phi(u)$ is analytic in the strip $|\operatorname{Im} z| < \pi/8$, the Taylor series expansion of $\Phi(u)$ about the origin, $\Phi(u) = \sum_{k=0}^\infty c_{2k} u^{2k}$, converges uniformly and absolutely for $|u| \le \rho < \pi/8$, where we have used the fact that $\Phi(u)$ is an even function (see Theorem 3). Hence, by the uniform convergence of this series, we obtain, for $x > 0$,

$$m(x) := \int_0^\infty u^{2x}\,\Phi(u)\,du = \int_0^\rho \sum_{k=0}^\infty c_{2k} u^{2x+2k}\,du + \int_\rho^\infty u^{2x}\,\Phi(u)\,du$$

$$= \rho^{2x} \sum_{k=0}^\infty c_{2k}\frac{\rho^{2k+1}}{2x+2k+1} + \int_{\log\rho}^\infty \Phi(e^v)e^{(2x+1)v}\,dv,$$

$$:= m_1(x) + m_2(x).$$

Now it is easy to verify that $m_2(x)$ is an entire function and that $m_1(x)$ is meromorphic with poles at $x = -(k+1/2)$, $k = 0, 1, 2, \ldots$. Thus, $m(x)$ is a meromorphic function whose poles coincide with the zeros of $1/\Gamma(x+1/2)$ and whence it follows that $\psi(x)$ is an entire function.

(2) Since $\psi(k) = b_k/\Gamma(k+1/2)$ and $\frac{1}{\Gamma(k+1/2)} = \frac{4^k\,\Gamma(k+1)}{\sqrt{\pi}\,\Gamma(2k+1)} = \frac{4^k\,k!}{\sqrt{\pi}(2k)!}$, we see that $\psi(k) = \gamma_k$.

(3) The verification of part (3) is clear, since

$$\sum_{k=0}^\infty \frac{\psi(k)}{k!}x^k = \frac{1}{\sqrt{\pi}} \sum_{k=0}^\infty \frac{4^k\,k!\,b_k}{(2k)!}\frac{x^k}{k!}$$

$$= \frac{1}{\sqrt{\pi}} \int_0^\infty \Phi(u) \sum_{k=0}^\infty \frac{4^k u^{2k} x^k}{(2k)!}\,du$$

$$= \frac{1}{\sqrt{\pi}} \int_0^\infty \Phi(u) \cosh(2u\sqrt{x})\,du.$$

(4) Fix $x > 0$. Then by (3), (4) and (8), we have

$$8\xi_2(4x) = 8\xi_1(2\sqrt{-x}) = \xi(\sqrt{-x}) = \xi(i\sqrt{x})$$

$$= \tfrac{1}{2}\left(x - \tfrac{1}{4}\right)\pi^{-\frac{\sqrt{x}}{2}-\frac{1}{4}}\Gamma\left(\tfrac{1}{4} + \tfrac{\sqrt{x}}{2}\right)\zeta\left(\tfrac{1}{2} + \sqrt{x}\right).$$

Remark 3 Since we know that the order of the entire function ξ_2 is $\frac{1}{2}$, ξ_2 is of exponential type zero. Also, the indicator function of ξ_2 satisfies $h_F(\pm\frac{\pi}{2}) = 0$. Hence, by Theorem 5 and part (4) of Proposition 1, the Riemann Hypothesis is true if and only if the sequence $\{\xi_2(k)\}_{k=0}^\infty$ (cf. (26)) is a CZDS. Also, from part (2) of Proposition 1 we see that the Riemann Hypothesis is true if and only if the sequence $\{\gamma_k\}_{k=0}^\infty$, where $\gamma_k = k!b_k/(2k)!$ (b_k is defined by (7)), is a multiplier sequence. Thus, *if* $\{\gamma_k\}_{k=0}^\infty$ is a multiplier sequence, then for *any* $f(x) = \sum_{k=0}^\infty \frac{a_k}{k!}x^k \in \mathcal{L}\text{-}\mathcal{P}$, the entire function $\sum_{k=0}^\infty \frac{a_k\gamma_k}{k!}x^k \in \mathcal{L}\text{-}\mathcal{P}$. Now, $\cosh(\sqrt{x}) = \sum_{k=0}^\infty \frac{k!}{(2k)!}\frac{x^k}{k!} \in \mathcal{L}\text{-}\mathcal{P}^+$ and consequently the sequence $\{k!/(2k)!\}_{k=0}^\infty$ is a multiplier sequence. Therefore, it follows that if the

Riemann Hypothesis is true, then

$$\int_0^\infty \Phi(u)\,\theta(u^2 x)\,du \in \mathcal{L}\text{-}\mathcal{P}, \tag{27}$$

where $\theta(x) = \sum_{k=0}^\infty \frac{\alpha_k}{(2k)!} x^k$ is any entire function such that $f(x) = \sum_{k=0}^\infty \frac{\alpha_k}{k!} x^k \in \mathcal{L}\text{-}\mathcal{P}$. Note that in the special case when $f(x) = e^x$, then $\theta(x) = \cosh(\sqrt{x})$.

Example 1 We conclude this paper with some numerical examples involving the sequence $\{F_1(k)\}_{k=0}^\infty$, where

$$F_1(k) := \left(k - \frac{1}{4}\right) \pi^{-\sqrt{k}/2} \Gamma\left(\frac{1}{4} + \frac{\sqrt{k}}{2}\right) \zeta\left(\frac{1}{2} + \sqrt{k}\right), \qquad k = 0, 1, 2, \ldots. \tag{28}$$

Observe that F_1 is just the function ξ_2 (cf. (26)) with the inconsequential factor $\pi^{-\frac{1}{4}}/16$ removed. Using the theoretical framework outlined above, we will now show that if any one of the factors in (28) is omitted, then the resulting sequence cannot be be a multiplier sequence and, *a fortiori*, it cannot be a CZDS. To this end, we consider for $k = 0, 1, 2, \ldots$,

$$F_2(k) := \pi^{-\sqrt{k}/2} \Gamma\left(\frac{1}{4} + \frac{\sqrt{k}}{2}\right) \zeta\left(\frac{1}{2} + \sqrt{k}\right)$$

$$F_3(k) := \left(k - \tfrac{1}{4}\right) \pi^{-\sqrt{k}/2} \Gamma\left(\frac{1}{4} + \frac{\sqrt{k}}{2}\right)$$

$$F_4(k) := \left(k - \frac{1}{4}\right) \Gamma\left(\frac{1}{4} + \frac{\sqrt{k}}{2}\right) \zeta\left(\frac{1}{2} + \sqrt{k}\right)$$

$$F_5(k) := \left(k - \tfrac{1}{4}\right) \pi^{-\sqrt{k}/2} \zeta\left(\tfrac{1}{2} + \sqrt{k}\right).$$

Since $F_2(0) = -5.294\ldots < 0, F_2(1) = 1.806\ldots, F_2(2) = 0.792\ldots$ (and in fact $F_2(k) > 0$ for $k = 1, 2, 3 \ldots$) $\{F_2(k)\}_{k=0}^\infty$ is not a multiplier sequence. Similarly, we find that $F_3(0) = -0.906\ldots < 0, F_3(1) = 0.518\ldots, F_3(2) = 0.799\ldots$ and whence $\{F_3(k)\}_{k=0}^\infty$ is also not a multiplier sequence. The sequences $\{F_4(k)\}_{k=0}^\infty$ and $\{F_5(k)\}_{k=0}^\infty$ are *not* multiplier sequences, since they fail to satisfy the double Turán inequalities. Indeed, with the notation adopted in Section 1 (see Theorem 3), if $\gamma_k = F_4(k)$, then $T_2(2) = -0.405\ldots$ and if $\gamma_k = F_5(k)$, then $T_2(2) = -0.0309\ldots$.

References

[1] Bakan, A., Craven, T., Csordas ,G.: Interpolation and the Laguerre-P'olya class. Southwest J. Pure and Appl. Math. 1 (2001), 38–53.

[2] Bakan, A., Craven, T., Csordas, G., Golub, A.: Weakly increasing zero-diminishing sequences. Serdica 22 (1996), 547–570.

[3] Craven, T., Csordas, G.: Jensen polynomials and the Turán and Laguerre inequalities. Pacific J. Math. 136 (1989), 241–260.

[4] Craven, T., Csordas, G.: Complex zero decreasing sequences. Methods Appl. Anal. 2 (1995), 420–441.

[5] Craven, T., Csordas, G.: A sufficient condition for strict total positivity of a matrix. LAMA 45 (1998), 19–34.

[6] Csordas, G., Norfolk, T.S., Varga, R.S.: The Riemann hypothesis and the Turán inequalities Trans. Amer. Math. Soc. 296 (1986), 521–541.

[7] Csordas, G., Norfolk, T.S., Varga, R.S.: A lower bound for the de Bruijn-Newman constant Λ. Numer. Math. 52 (1988), 483–497.

[8] Csordas, G., Smith, W., Varga, R.S.: Level sets of real entire functions and the Laguerre inequalities. Analysis 12 (1992), 377–402.

[9] Csordas, G., Varga, R.S.: Moment inequalities and the Riemann Hypothesis. Constr. Approx. 4 (1988), 175–198.

[10] Csordas, G., Varga, R.S.: Necessary and sufficient conditions and the Riemann hypothesis. Adv. Appl. Math. 1 (1990), 328–357.

[11] Dimitrov, D.K.: Higher order Turán inequalities. Proc. Amer. Math. Soc. 126 (1998), 2033–2037.

[12] Csordas, G., Dimitrov, D.K.: Conjectures and theorems in the theory of entire functions. Numerical Algorithms 25 (2000), 109–122.

[13] Levin, B.Ja.: Distribution of Zeros of Entire Functions. Transl. Math. Mono. 5, Amer. Math. Soc. Providence, RI 1964; revised ed. 1980.

[14] Obreschkoff, N.: Verteilung und Berechnung der Nullstellen reeller Polynome. VEB Deutscher Verlag der Wissenschaften, Berlin 1963.

[15] Pólya, G.: Über die algebraisch-funktionentheoretischen Untersuchungen von J.L.W.V. Jensen. Klg. Danske Vid. Sel. Math.-Fys. Medd. l7 (1927), 3–33.

[16] Pólya, G.: Über trigonometrische Integrale mit nur reellen Nullstellen. J. Reine Angew. Math. 158 (1927), 6–18.

[17] Pólya, G.: Collected Papers, Vol. II Location of Zeros. (R.P. Boas, ed.) MIT Press. Cambridge, MA 1974.

[18] Pólya, G., Schur, J.: Über zwei Arten von Faktorenfolgen in der Theorie der algebraischen Gleichungen. J. Reine Angew. Math. 144 (1914), 89–113.

[19] Titchmarsh, E.C.: The Theory of the Riemann Zeta-function. 2nd. ed. (revised by D.R. Heath-Brown). Oxford University Press, Oxford 1986.

THEORY OF REPRODUCING KERNELS

Saburou Saitoh

Department of Mathematics
Faculty of Engineering
Gunma University, Kiryu 376-8515, Japan
ssaitoh@math.sci.gunma-u.ac.jp

Abstract   In this survey article, we would like to show that the theory of reproducing kernels is fundamental, is beautiful and is applicable widely in mathematics. At the same time, we shall present some operator versions of our fundamental theory in the general theory of reproducing kernels, as original results.

1.   Reproducing kernel spaces and representation formulas

We consider any positive matrix $K(p,q)$ on E; that is, for an abstract set E and for a complex–valued function $K(p,q)$ on $E \times E$, it satisfies that for any finite points $\{p_j\}$ of E and for any complex numbers $\{C_j\}$,

$$\sum_j \sum_{j'} C_j \overline{C_{j'}} K(p_{j'}, p_j) \geqq 0.$$

Then, by the fundamental theorem by Moore–Aronszajn, we have:

Proposition 1([5]) *For any positive matrix $K(p,q)$ on E, there exists a uniquely determined functional Hilbert space H_K comprising functions $\{f\}$ on E and admitting the reproducing kernel $K(p,q)$ (RKHS H_K) satisfying and characterized by*

$$K(\cdot, q) \in H_K \text{ for any } q \in E \tag{1}$$

and, for any $q \in E$ and for any $f \in H_K$

$$f(q) = (f(\cdot), K(\cdot, q))_{H_K}. \tag{2}$$

H.G.W. Begehr et al. (eds.), Analysis and Applications - ISAAC 2001, 135–150.

In general, a functional Hilbert space on a set E admits a reproducing kernel if and only if point evaluations on the set E are bounded linear functionals, and then the reproducing kernel is constructed and represented by using a complete orthonormal system in the Hilbert space. In this sense, a reproducing kernel is constructive and computable. Two famous reproducing kernels are the Bergman kernel and the Szegö kernel in complex analysis. Their reproducing kernels were introduced by S. Bergman and G. Szegö in 1922 and 1921, respectively in their dissertations at Berlin, in order to construct the Riemann mapping function. For the theory of reproducing kernels in complex analysis, see the books [6], [20] and the ISAAC Congress Proceedings [10] on recent various reproducing kernels.

The classical general theory of reproducing kernels was typically presented by N. Aronszajn [5] and L. Schwartz [39]. The recent general theory of reproducing kernels was presented in [26]. For the history of reproducing kernels, see N. Aronszajn [5], [20] and [30] in order.

We shall assume that H_K is separable. Then, the functions $\{K(\cdot, q); q \in E\}$ generate H_K and there exists a countable set S of E such that $\{K(\cdot, q_j); q_j \in S\}$ is a family of linearly independent functions forming a dense set for H_K. We set $S_n = \{q_1, q_2, \cdots, q_n\} \subset S$ and $\|\Gamma_{jj'n}\|_{1 \leq j, j' \leq n}$ is the inverse of $\|K(q_j, q_{j'})\|_{1 \leq j, j' \leq n}$. Then, we obtain

Proposition 2 ([15] and see Chapter 2, Section 5 in [26])   *For any $f \in H_K$, the sequence of functions f_n defined by*

$$f_n(p) = \sum_{j,j'=1}^{n} f(q_j) \Gamma_{jj'n} K(p, q_{j'}) \tag{3}$$

converges to f as $n \to \infty$ in both the senses in norm of H_K and everywhere on E.

Furthermore, for any function f defined on E satisfying

$$\lim_{n \to \infty} \sum_{j,j'=1}^{n} f(q_j) \Gamma_{jj'n} \overline{f(q_{j'})} < \infty, \quad q_j \in S, \tag{4}$$

the sequence of functions f_n defined by (3) is a Cauchy sequence in H_K whose limit coincides with f on E. Conversely, any member f of H_K is obtained in this way in terms of $\{f(q_j)\}$.

We see in Proposition 2 that extensibility and representation of f in terms of $f(q_j)$, $q_j \in S$ are established by means of the reproducing kernel $K(p, q)$.

In particular, reproducing kernels establish a fundamental relationship between continuum and discrete in reproducing kernel Hilbert spaces. More direct and effective relations will be referred in sampling theory. Note that the theory of reproducing kernels reduces to matrix theory, if the set E is a finite point set.

2.   Connection with linear transforms

Let us connect linear transforms in the framework of Hilbert spaces with reproducing kernels ([18]).

For an abstract set E and for any Hilbert (possibly finite–dimensional) space $\mathcal{H}$, we shall consider an $\mathcal{H}$–valued function h on E

$$h: \quad E \longrightarrow \mathcal{H} \tag{5}$$

and the linear transform for $\mathcal{H}$

$$f(p) = (f, h(p))_{\mathcal{H}} \quad \text{for} \quad f \in \mathcal{H} \tag{6}$$

into a linear space comprising functions on E. For this linear transform (6), we form the positive matrix $K(p, q)$ on E defined by

$$K(p, q) = (h(q), h(p))_{\mathcal{H}} \quad \text{on } E \times E. \tag{7}$$

Then, we have the following fundamental results:

(I) For the RKHS H_K admitting the reproducing kernel $K(p,q)$ defined by (7), the images $\{f(p)\}$ by (6) for $\mathcal{H}$ are characterized as the members of the RKHS H_K.

(II) In general, we have the inequality in (6)

$$\|f\|_{H_K} \leqq \|f\|_{\mathcal{H}}, \tag{8}$$

however, for any $f \in H_K$ there exists a uniquely determined $f^* \in \mathcal{H}$ satisfying

$$f(p) = (f^*, h(p))_{\mathcal{H}} \quad \text{on } E \tag{9}$$

and

$$\|f\|_{H_K} = \|f^*\|_{\mathcal{H}}. \tag{10}$$

In (8), the isometry holds if and only if $\{h(p); p \in E\}$ is complete in $\mathcal{H}$.

(III) We can obtain the inversion formula for (6) in the form

$$f \longrightarrow f^*, \tag{11}$$

by using the RKHS H_K.

However, this inversion formula will depend on, case by case, the realizations of the RKHS H_K.

(IV) Conversely, if we have an isometric mapping $\tilde{L}$ from a RKHS H_K admitting a reproducing kernel $K(p, q)$ on E onto a Hilbert space $\mathcal{H}$, then the mapping $\tilde{L}$ is linear and the isometric inversion $\tilde{L}^{-1}$ is represented in the form (2.2), where the Hilbert space $\mathcal{H}$–valued function h satisfying (5) and (6) is given by

$$h(p) = \tilde{L}K(\cdot, p) \quad \text{on} \quad E \tag{12}$$

and, then $\{h(p); p \in E\}$ is complete in $\mathcal{H}$.

We shall state some general applications of the results (I)∼(IV) to several wide subjects and their basic references:

(1) Linear transforms ([11],[23]). The fact that the image spaces of linear transforms in the framework of Hilbert spaces are characterized as the reproducing kernel Hilbert spaces defined by the beautiful form (2.3) is the most important one in the general theory of reproducing kernels. Therefore, the fact will mean that the theory of reproducing kernels is fundamental and a general concept in mathematics. To look for the characterization of the image space is a starting point when we consider the linear equation (2.2). (II) gives a generalization of the Pythagorean theorem (see also [17]) and means that in the general linear mapping (2.2) there exists essentially an isometric identity between the input and the output. (III) gives a generalized (natural) inverse (solution) of the linear mapping (equation) (2.2). (IV) gives a general method determining and constructing the linear system from an isometric relation between outputs and inputs by using the reproducing kernel in the output space.

(2) Integral transforms among smooth functions ([30]). We considered linear mappings in the framework of Hilbert spaces, however, we can consider linear mappings in the framework of Hilbert spaces comprising smooth functions, similarly. Conversely, reproducing kernel Hilbert spaces are considered as the images of some Hilbert spaces by considering some decomposed representations (2.3) of the reproducing kernels - such decomposition of the positive matrix is, in general, possible. This idea is important in [30] and also in the following items (6) and (7).

(3) Nonharmonic integral transforms ([20]). If the linear system vectors $h(p)$ moved in a small way (perturbation of the linear system) in the Hilbert space $\mathcal{H}$, then we can not caluculate the related positive matrix (2.3), however, we can discuss the inversion formula and an isometric identity of the linear mapping. The prototype result for this idea is the Paley-Wiener theorem for nonharmonic Fourier series.

(4) Various norm inequalities ([20],[24]). Relations among positive matrices correspond to those of the associated reproducing kernel Hilbert spaces, by the minimum principle. So, we can derive various norm inequalities among reproducing kernel Hilbert spaces. We were able to derive many beautiful norm inequalities.

(5) Nonlinear transforms ([24],[28]). In a very general nonlinear transform of a reproducing kernel Hilbert space, we can look for a natural reproducing kernel Hilbert space containing the image space and furthermore, we can derive a natural norm inequality in the nonlinear transform. How to catch nonlinearity in connection with linearity? It seems that the theory of reproducing kernels gives a fundamental and interesting answer for this question.

(6) Linear integral equations ([32]).

(7) Linear differential equations with variable coefficients ([32]). In linear integro-differential equations with general variable coefficients, we can discuss the existence and construction of the solutions, if the solutions exist. This method is called a backward transformation method and by reducing the equations to Fredholm integral equations of the first type - (2.2) - and we can discuss the classical solutions, in very general linear equations.

(8) Approximation theory ([9],[8]). Reproducing kernel Hilbert spaces are very nice function spaces, because the point evaluations are continuous. Then, the reproducing kernels are a fundamental tool in the related approximation theory.

(9) Representations of inverse functions ([25]). For any mapping, we discussed the problem of representing its inverse in term of the direct mapping and we derived a unified method for this problem. As a simple example, we can represent the Taylor coefficients of the inverse of the Riemann mapping function on the unit disc on the complex plane in terms of the Riemann mapping function. This fact was important in the representation of analytic functions in terms of local data in ([37],[38]).

(10) Various operators among Hilbert spaces ([29]). Among various abstract Hilbert spaces, we can introduce various operators of sum, product, integral and derivative by using the linear mapping (2.2) or very general nonlinear transforms. The prototype operator is convolution and we discussed it from a wide and general viewpoint with concrete examples.

(11) Sampling theorems ([26], Chapter 4, Section 2, [12]). The Whittaker-Kotel'nikov-Shannon sampling theorem may be interpretated by (I) and (II), very well and we can discuss the truncation errors in the sampling theory. J. R. Higgins [12] established a fully general theory for [26].

(12) Interpolation problems of Pick-Nevanlinna type ([20],[21]). General and abstract theory of interpolation problems of Pick-Nevanlinna type may be discussed by using the general theory of reproducing kernels.

(13) Analytic extension formulas and their applications ([26],[34]). As shown in Section 3 as typical applications of our theory, we were able to obtain various analytic extension formulas from various isometric identities (II) and their applications. In this survey article, we shall present also new results on

(14) Inversions of a family of bounded linear operators on a Hilbert space into various Hilbert spaces.

By our theory, we were able to derive entirely new inversion formulas for many classical and important integral transforms such that the Cauchy integral formula, the Poisson integral, and integral transforms by many Green's functions, see [23] and [26], but we shall show here only the results on the Weierstrass transform and the Laplace transform, as typical results.

3. Weierstrass transform

We shall consider the Weierstrass transform as a typical integral transform and in order to see typical applications

$$u(x,t) = \frac{1}{\sqrt{4\pi t}} \int_{\boldsymbol{R}} F(\xi) \exp\left[-\frac{(x-\xi)^2}{4t}\right] d\xi \tag{13}$$

for functions $F \in L_2(\boldsymbol{R}, d\xi)$. Then, by using (I) and (II) we obtained in [19] simply and naturally the isometric identity

$$\int_{\boldsymbol{R}} |F(\xi)|^2 d\xi = \frac{1}{\sqrt{2\pi t}} \int\int_{\boldsymbol{R}^2} |u(z,t)|^2 \exp\left[-\frac{y^2}{2t}\right] dx dy \tag{14}$$

for the analytic extension $u(z, t)$ of $u(x, t)$ to the entire complex $z = x + iy$ plane.

Of course, the image $u(x, t)$ of (13) is the solution of the heat equation

$$u_{xx}(x, t) = u_t(x, t) \quad \text{on } \boldsymbol{R} \times \{t > 0\} \tag{15}$$

satisfying the initial condition

$$\lim_{t \to +0} \|u(\cdot, t) - F(\cdot)\|_{L_2(\boldsymbol{R}, dx)} = 0.$$

On the other hand, by using the properties of the solution $u(x, t)$ of (15), we derived the identity

$$\int_{\boldsymbol{R}} |F(\xi)|^2 d\xi = \sum_{j=0}^{\infty} \frac{(2t)^j}{j!} \int_{\boldsymbol{R}} |\partial_x^j u(x, t)|^2 dx \tag{16}$$

in [11].

The two identities (14) and (16) were a starting point for obtaining our various analytic extension formulas and their applications.

As to the equality of (14) and (16), we obtained directly based on (I) in [26]

Theorem 1 ([11]) *For any analytic function $f(z)$ on the strip $S_r = \{|\mathrm{Im}\, z| < r\}$ with a finite integral*

$$\int\int_{S_r} |f(z)|^2 dx dy < \infty,$$

we have the identity

$$\int\int_{S_r} |f(z)|^2 dx dy = \sum_{j=0}^{\infty} \frac{(2r)^{2j+1}}{(2j+1)!} \int_{\boldsymbol{R}} |\partial_x^j f(x)|^2 dx. \tag{17}$$

Conversely, for a smooth function $f(x)$ with a convergence sum (17) on $\boldsymbol{R}$, there exists an analytic extension $f(z)$ onto S_r satisfying (17).

Theorem 2 ([11]) *For any $\alpha > 0$ and for an entire function $f(z)$ with a finite integral*

$$\int\int_{\boldsymbol{R}^2} |f(z)|^2 \exp\left[-\frac{y^2}{\alpha}\right] dx dy < \infty,$$

we have the identity

$$\frac{1}{\sqrt{\alpha\pi}} \int\int_{\boldsymbol{R}^2} |f(z)|^2 \exp\left[-\frac{y^2}{\alpha}\right] dx dy = \sum_{j=0}^{\infty} \frac{\alpha^j}{j!} \int_{\boldsymbol{R}} |\partial_x^j f(x)|^2 dx. \tag{18}$$

142

Conversely, for a smooth function $f(x)$ with a convergence sum (18) on $\mathbf{R}$, there exists an analytic extension $f(z)$ onto $\mathbf{C}$ satisfying the identity (18).

In the Weierstrass transform (13), we obtained the isometric identity, for any fixed $x \in \mathbf{R}$,

$$\int_{-\infty}^{\infty} |F(\xi)|^2 d\xi$$

$$= 2\pi \sum_{j=0}^{\infty} \frac{1}{j!\Gamma\left(j + \frac{3}{2}\right)} \int_{0}^{\infty} \left| \partial_t^j \left[t\partial_t u(x,t) \right] \right|^2 t^{2j - \frac{1}{2}} dt$$

$$+ 2\pi \sum_{j=0}^{\infty} \frac{1}{j!\Gamma\left(j + \frac{5}{2}\right)} \int 0^{\infty} \left| \partial_t^j \left[t\partial_t \partial_x u(x,t) \right] \right|^2 t^{2j + \frac{1}{2}} dt, \quad (19)$$

in [16] based on (I) and (II). ¿From this identity, we can obtain the inversion formula

$$u(x,t) \longrightarrow F(\xi) \quad \text{for any fixed } x. \tag{20}$$

We, in general, in the multi–dimensional Weierstrass transform, established an exact and analytical representation formula of the initial heat distribution F by means of the observations

$$u(x_1, x', t) \quad \text{and} \quad \frac{\partial u(x_1, x', t)}{\partial x_1} \tag{21}$$

for $x' = (x_2, x_3, \cdots, x_n) \in \mathbf{R}^{n-1}$ and $t > 0$, at any fixed point x_1, in [16].

We set

$$\sigma_F = \{\sup |x|, \ x \in \operatorname{supp} F\} \tag{22}$$

and $\operatorname{supp} F$ denotes the smallest closed set outside which F vanishes almost everywhere. By using the isometric identities (14), (16) and (19), we can solve the inverse source problem of determining the size σ_F of the initial heat distribution F from the heat flow $u(x,t)$ observed either at any fixed time t or at any fixed position x. See [40].

Our typical results of another type were obtained from the integral transform

$$v(x,t) = \frac{1}{t} \int_{0}^{t} F(\xi) \frac{x \exp\left\{ \frac{-x^2}{4(t-\xi)} \right\}}{2\sqrt{\pi} \, (t-\xi)^{\frac{3}{2}}} \xi d\xi \tag{23}$$

in connection with the heat equation (15) for $x > 0$ satisfying the conditions, for $u(x,t) = tv(x,t)$

$$u(0,t) = tF(t) \quad \text{for } t \geq 0$$

and

$$u(x,0) = 0 \quad \text{on} \quad x \geq 0.$$

Then, we obtained

Theorem 3 ([1]) *Let $\Delta(\frac{\pi}{4})$ denote the sector $\{|\arg z| < \frac{\pi}{4}\}$. Then, for any analytic function $f(z)$ on $\Delta\left(\frac{\pi}{4}\right)$ with a finite integral*

$$\int\int_{\Delta(\frac{\pi}{4})} |f(z)|^2 dx dy < \infty,$$

we have the identity

$$\int\int_{\Delta(\frac{\pi}{4})} |f(z)|^2 dx dy = \sum_{j=0}^{\infty} \frac{2^j}{(2j+1)!} \int_0^{\infty} x^{2j+1} \left|\partial_x^j f(x)\right|^2 dx. \quad (24)$$

Conversely, for any smooth function $f(x)$ on $\{x > 0\}$ with a convergence sum in (24), there exists an analytic extension $f(z)$ onto $\Delta\left(\frac{\pi}{4}\right)$ satisfying (24).

Let $\Delta(\alpha)$ be the sector $\{|\arg z| < \alpha\}$. Then, by using the conformal mapping e^z, we examined the relation between Theorem 3.1 and Theorem 3.3. Then, we used the Mellin transform and some expansion of Gauss' hypergeometric series $F(\alpha, \beta; \gamma; z)$ and we obtained a general version of Theorem 3.3 and a version for the Szegö space:

Theorem 4 ([2]) *Let $0 < \alpha < \frac{\pi}{2}$. Then, for any analytic function $f(z)$ on $\Delta(\alpha)$ with a finite integral*

$$\int\int_{\Delta(\alpha)} |f(z)|^2 dx dy < \infty,$$

we have the identity

$$\int\int_{\Delta(\alpha)} |f(z)|^2 dx dy = \sin(2\alpha) \sum_{j=0}^{\infty} \frac{(2\sin\alpha)^{2j}}{(2j+1)!}$$
$$\int_0^{\infty} x^{2j+1} |\partial_x^j f(x)|^2 dx. \quad (25)$$

Conversely, for a smooth function $f(x)$ with a convergence sum on $x > 0$ in (25), there exists an analytic extension $f(z)$ onto $\Delta(\alpha)$ satisfying the identity (25).

Theorem 5 ([2]) *Let $0 < \alpha < \frac{\pi}{2}$. Then, for any analytic function $f(z)$ on $\Delta(\alpha)$ satisfying*

$$\int_{|\theta|<\alpha} |f(re^{i\theta})|^2 dr < \infty,$$

we have the identity

$$\int_{\partial\Delta(\alpha)} |f(z)|^2 |dz| = 2\cos\alpha \sum_{j=0}^{\infty} \frac{(2\sin\alpha)^{2j}}{(2j)!} \int_0^{\infty} x^{2j} |\partial_x^j f(x)|^2 dx \qquad (26)$$

where $f(z)$ mean Fatou's nontangential boundary values of f on $\partial\Delta(\alpha)$.

Conversely, for a smooth function $f(x)$ on $x > 0$ with a convergence sum in (26), there exists an analytic extension $f(z)$ onto $\Delta(\alpha)$ satisfying the identity (26).

We obtained various analytic extension formulas on the above line, see the survey article [34]. For their applications to nonlinear partial differential equations, see the survey article [10] by N. Hayashi.

4. Real inversion formulas of the Laplace transform

The inversion formulas of the Laplace transform are, in general, given by complex forms. The observation in many cases however gives us real data only and so, it is important to establish the real inversion formula of the Laplace transform, because we have to extend the real data analytically onto a half complex plane. The analytic extension formula is, in general, very involved and makes the stability unclear. In particular, in the Reznitskaya transform combining the solutions of hyperbolic and parabolic partial differential equations, we need the real inversion formula, because the observation data of the solutions of hyperbolic partial differential equations are real-valued. See [27].

However, if we know a Taylor expansion around any fixed point on the real positive line, then the analytic extension to the complex half plane, which is needed to apply the complex inversion formula, can be done easily. This simple analytic extension formula was very recently obtained in [37]. In a Taylor expansion, however we need, in general, infinite order derivatives, and so we do not like to use the Taylor expansion for the real inversion formula.

Since the image functions of the Laplace transform are, in general, analytic on a half–plane on the complex plane, in order to obtain the real inversion formula, we need a half plane version $\Delta\left(\frac{\pi}{2}\right)$ of Theorem 3.4 and Theorem 3.5, which is a crucial case $\alpha = \frac{\pi}{2}$ in those theorems. By using the famous Gauss summation formula and transformation properties in the Mellin transform we obtained, in a very general version containing the Bergman and the Szegö spaces:

Theorem 6 ([22]) *For any $q > 0$, let $H_{K_q}(R^+)$ denote the Bergman–Selberg space admitting the reproducing kernel*

$$K_q(z, \bar{u}) = \frac{\Gamma(2q)}{(z + \bar{u})^{2q}}$$

on the right half plane $R^+ = \{z; \operatorname{Re} z > 0\}$. Then, we have the identity

$$
\begin{aligned}
\|f\|^2_{H_{K_q}(R^+)} &= \left(\frac{1}{\Gamma(2q-1)\pi} \in t \int_{R^+} |f(z)|^2 (2x)^{2q-2} dx dy \,,\, q > \frac{1}{2}\right) \\
&= \sum_{n=0}^{\infty} \frac{1}{n!\Gamma(n+2q+1)} \\
&\quad \cdot \int_0^{\infty} |\partial_x^n \left(x f'(x)\right)|^2 x^{2n+2q-1} dx.
\end{aligned}
$$ (27)

Conversely, any smooth function $f(x)$ on $\{x > 0\}$ with a convergence summation in (27) can be extended analytically onto R^+ and the analytic extension $f(z)$ satisfying $\lim_{x\to\infty} f(x) = 0$ belongs to $H_{K_q}(R^+)$ and the identity (27) is valid.

For the Laplace transform

$$f(z) = \int_0^{\infty} F(t) e^{-zt} dt,$$ (28)

we have, immediately, the isometric identity, for any $q > 0$

$$
\begin{aligned}
\|f\|^2_{H_{K_q}(R^+)} &= \int_0^{\infty} |F(t)|^2 t^{1-2q} dt \\
(&:= \|F\|^2_{L_q^2})
\end{aligned}
$$ (29)

from (I) and (II). By using (29) and (27), we obtain

Theorem 7 ([7]) *For the Laplace transform (28), we have the inversion formula*

$$F(t) = s - \lim_{N\to\infty} \int_0^{\infty} f(x) e^{-xt} P_{N,q}(xt) dx \quad (t > 0)$$ (30)

where the limit is taken in the space L_q^2 and the polynomials $P_{N,q}$ are given by

$$P_{N,q}(\xi) = \sum_{0 \leq \nu \leq n \leq N} \frac{(-1)^{\nu+1}\Gamma(2n+2q)}{\nu!(n-\nu)!\Gamma(n+2q+1)\Gamma(n+\nu+2q)}\xi^{n+\nu+2q-1}$$

$$\cdot \left\{ \frac{2(n+q)}{n+\nu+2q}\xi^2 - \left(\frac{2(n+q)}{n+\nu+2q} + 3n + 2q \right)\xi \right.$$

$$\left\{ +(n+\nu+2q) \right\}. \tag{31}$$

The truncation error is estimated by the inequality

$$\left\| F(t) - \int_0^\infty f(x)e^{-xt}P_{N,q}(xt)\,dx \right\|_{L_q^2}^2$$

$$\leq \sum_{n=N+1}^\infty \frac{1}{n!\Gamma(n+2q+1)} \int_0^\infty \left| \partial_x^n \left[xf'(x) \right] \right|^2 x^{2n+2q-1}\,dx. \tag{32}$$

Note that in our inversion formula (4.4), we use only the values $f(x)$ on the positive real line.

In order to obtain an inversion formula which converges pointwisely in (30), we considered an inversion formula of the Laplace transform for the Sobolev space satisfying

$$\int_0^\infty \left(|F(t)|^2 + |F'(t)|^2 \right)dt < \infty,$$

in [3]. In some subspaces of $H_{K_q}(R^+)$ and L_q^2, we established an error estimate for the inversion formula (30) in [4]. Conditional stability for the real inversion formula (4.4) of the Laplace transform for some function spaces was also established in [35]. Some characteristics of the strong singularity of the polynomials $P_{N,q}(\xi)$ and some effective algorithms for the real inversion formula (30) are examined by J. Kajiwara and M. Tsuji [[13],[14]]. Furthermore, they gave numerical experiments by using computers.

5.   Operator versions

We shall give operator versions of the fundamental theory (I)~(IV) which may be expected to have many concrete applications. In particular, for the application to the full generalizations of the Pythagorean theorem with geometric meanings, see [17]. Some special versions were given in [36].

For an abstract set Λ, we shall consider an operator-valued function L_λ on Λ,

$$\Lambda \longrightarrow L_\lambda \tag{33}$$

where L_λ are bounded linear operators from a Hilbert space $\mathcal{H}$ into various Hilbert spaces $\mathcal{H}_\lambda$,

$$L_\lambda : \mathcal{H} \longrightarrow \mathcal{H}_\lambda. \tag{34}$$

In particular, we are interested in the inversion formula

$$L_\lambda x \longrightarrow x, \quad x \in \mathcal{H}. \tag{35}$$

Here, we consider $\{L_\lambda x; \lambda \in \Lambda\}$ as informations obtained from x and we wish to determine x from the informations. However, the informations $L_\lambda x$ belong to various Hilbert spaces $\mathcal{H}_\lambda$, and so, in order to unify the informations in a sense, we shall take fixed elements $\mathbf{b}_{\lambda,\omega} \in \mathcal{H}_\lambda$ and consider the linear mapping from $\mathcal{H}$

$$
\begin{aligned}
X_\mathbf{b}(\lambda,\omega) &= (L_\lambda x, \mathbf{b}_{\lambda,\omega})_{\mathcal{H}_\lambda} \\
&= (x, L_\lambda^* \mathbf{b}_{\lambda,\omega})_H, \quad x \in H
\end{aligned} \tag{36}
$$

into a linear space comprising functions on $\Lambda \times \Omega$. For the informations $L_\lambda x$, we shall consider $X_\mathbf{b}(\lambda,\omega)$ as observations (measurements, in fact) for x depending on λ and ω. For this linear transform (5.4), we form the positive matrix $K_\mathbf{b}(\lambda,\omega;\lambda',\omega')$ on $\Lambda \times \Omega$ defined by

$$
\begin{aligned}
K_\mathbf{b}(\lambda,\omega;\lambda',\omega') &= (L_{\lambda'}^* \mathbf{b}_{\lambda',\omega'}, L_\lambda^* \mathbf{b}_{\lambda,\omega})_{\mathcal{H}} \\
&= (L_\lambda L_{\lambda'}^* \mathbf{b}_{\lambda',\omega'}, \mathbf{b}_{\lambda,\omega})_{\mathcal{H}_\lambda} \quad \text{on} \quad \Lambda \times \Omega.
\end{aligned} \tag{37}
$$

Then, as in (I)~(IV), we have the following fundamental results:

(I') For the RKHS $H_{K_\mathbf{b}}$ admitting the reproducing kernel $K_\mathbf{b}(\lambda,\omega;\lambda',\omega')$ defined by (37), the images $\{X_\mathbf{b}(\lambda,\omega)\}$ by (36) for $\mathcal{H}$ are characterized as the members of the RKHS $H_{K_\mathbf{b}}$.

(II') In general, we have the inequality in (36)

$$\|X_\mathbf{b}\|_{H_{K_\mathbf{b}}} \leqq \|x\|_{\mathcal{H}}, \tag{38}$$

however, for any $X_\mathbf{b} \in H_{K_\mathbf{b}}$ there exists a uniquely determined $x' \in \mathcal{H}$ satisfying

$$X_\mathbf{b}(\lambda,\omega) = (x', L_\lambda^* \mathbf{b}_{\lambda,\omega})_{\mathcal{H}} \quad \text{on} \quad \Lambda \times \Omega \tag{39}$$

and

$$\|X_{\mathbf{b}}\|_{H_{K_{\mathbf{b}}}} = \|x'\|_{\mathcal{H}}. \tag{40}$$

In (38), the isometry holds if and only if $\{L_\lambda^* \mathbf{b}_{\lambda,\omega}; (\lambda, \omega) \in boldsymbol\Lambda \times \Omega\}$ is complete in $\mathcal{H}$.

(III') We can obtain the inversion formula for (36) and so, for the mapping (5.3) as in (III), in the form

$$L_\lambda x \longrightarrow (L_\lambda x, \mathbf{b}_{\lambda,\omega})_{\mathbf{H}_\lambda} = X_{\mathbf{b}}(\lambda, \omega) \longrightarrow x', \tag{41}$$

by using the RKHS $H_{K_{\mathbf{b}}}$.

(IV') Conversely, if we have an isometric mapping $\tilde{L}$ from a RKHS $H_{K_{\mathbf{b}}}$ admitting a reproducing kernel $K_{\mathbf{b}}(\lambda, \omega; \lambda', \omega')$ on $\Lambda \times \Omega$ in the form (5.5) using bounded linear operators L_λ and fixed vectors $\mathbf{b}_{\lambda,\omega}$ onto a Hilbert space $\mathcal{H}$, then the mapping $\tilde{L}$ is linear and the isometric inversion $\tilde{L}^{-1}$ is represented in the form (5.4) by using

$$L_\lambda^* \mathbf{b}_{\lambda,\omega} = \tilde{L} K_{\mathbf{b}}(\cdot, \cdot; \lambda, \omega) \quad \text{on} \quad \Lambda \times \Omega. \tag{42}$$

Further, then $\{L_\lambda^* \mathbf{b}_{\lambda,\omega}; (\lambda, \omega) \in \Lambda \times \Omega\}$ is complete in $\mathcal{H}$.

Acknowledgments

The author wishes to express his deep thanks the organizing committee and all the related stuff, in particular, Professor H. G. W. Begehr and Ms. B. Wengel for their great contributions to the big 3rd ISAAC Congress at Berlin.

References

[1] Aikawa, H., Hayashi, N., Saitoh, S.: The Bergman space on a sector and the heat equation. Complex Variables, Theory Appl. 15 (1990), 27–36.

[2] Aikawa, H., Hayashi, N., Saitoh, S.: Isometrical identities for the Bergman and the Szegö spaces on a sector. J. Math. Soc. Japan 43 (1991), 196–201.

[3] Amano, K., Saitoh, S., Syarif, A.: A real inversion formula for the Laplace transform in a Sobolev space. Zeitschrift für Analysis und ihre Anwendungen 18 (1999), 1031–1038.

[4] Amano, K., Saitoh, S., Yamamoto, M.: Error estimates of the real inversion formulas of the Laplace transform. Graduate School of Math. Sci. The University of Tokyo, Preprint Series 1998, 98–29. Integral Transforms and Special Functions 10 (2000), 1–14 .

[5] Aronszajn, N.: Theory of reproducing kernels. Trans. Amer. Math. Soc. 68(1950), 337–404.

[6] Bergman, S.: The kernel function and conformal mapping. Amer. Math. Soc., Providence, R. I. 1970.

[7] Byun, D.-W., Saitoh, S.: A real inversion formula for the Laplace transform. Zeitschrift für Analysis und ihre Anwendungen 12 (1993), 597–603.

[8] Byun, D.-W., Saitoh, S.: Approximation by the solutions of the heat equation. J. Approximation Theory 78 (1994), 226–238.

[9] Byun, D.-W., Saitoh, S.: Best approximation in reproducing kernel Hilbert s-paces. Proc. of the 2th International Colloquium on Numerical Analysis, VSP-Holland, 1994, 55–61.

[10] Hayashi, N.: Analytic function spaces and their applications to nonlinear evolution equations. Analytic Extension Formulas and their Applications, Kluwer Academic Publishers, 2001, 59–86.

[11] Hayashi, N., Saitoh, S.: Analyticity and smoothing effect for the Schrödinger equation. Ann. Inst. Henri Poincaré 52 (1990), 163–173.

[12] Higgins, J.R.: A sampling principle associated with Saitoh's fundamental theory of linear transformations. Analytic Extension Formulas and their Applications, Kluwer Academic Publishers, Dordrecht 2001, 73–86.

[13] Kajiwara, J., Tsuji, M.: Program for the numerical analysis of inverse formula for the Laplace transform. Proceedings of the Second Korean-Japanese Colloquium on Finite or Infinite Dimensional Complex Analysis, 1994, 93–107.

[14] Kajiwara, J., Tsuji, M.: Inverse formula for Laplace transform. Proceedings of the 5th International Colloquium on Differential Equations, VHP-Holland 1995, 163–172.

[15] Körezlioğlu, H.: Reproducing kernels in separable Hilbert spaces. Pacific J. Math. 25 (1968), 305–314.

[16] Nakamura, G., Saitoh, S., Syarif, A.: Representations of initial heat distributions by means of their heat distributions as functions of time. Inverse Problems 15 (1999), 1255–1261.

[17] Rassias, Th.M., Saitoh, S.: The Pythagorean theorem and linear mappings. PanAmerican Math. J. 12 (2002), 1–10.

[18] Saitoh, S.: Hilbert spaces induced by Hilbert space valued functions. Proc. Amer. Math. Soc. 89 (1983), 74–78.

[19] Saitoh, S.: The Weierstrass transform and an isometry in the heat equation. Applicable Analysis 16 (1983), 1–6.

[20] Saitoh, S.: Theory of Reproducing Kernels and its Applications. Pitman Research Notes in Mathematics Series 189. Longman Scientific & Technical, Harlow 1998.

[21] Saitoh, S.: Interpolation problems of Pick-Nevanlinna type. Pitman Research Notes in Mathematics Series 212 (1989), 253–262.

[22] Saitoh, S.: Representations of the norms in Bergman-Selberg spaces on strips and half planes. Complex Variables, Theory Appl. 19 (1992); 231–241.

[23] Saitoh, S.: One approach to some general integral transforms and its applications. Integral Transforms and Special Functions 3 (1995), 49–84.

[24] Saitoh, S.: Natural norm inequalities in nonlinear transforms. General Inequalities 7 (1997), 39–52.

[25] Saitoh, S.: Representations of inverse functions. Proc. Amer. Math. Soc. 125 (1997), 3633–3639.

[26] Saitoh, S.: Integral Transforms, Reproducing Kernels and their Applications. Pitman Research Notes in Mathematics Series 369 . Addison Wesley Longman, Harlow 1997.

[27] Saitoh, S., Yamamoto, M.: Stability of Lipschitz type in determination of initial heat distribution. J. of Inequa. & Appl. 1 (1997), 73–83.

[28] Saitoh, S.: Nonlinear transforms and analyticity of functions. Nonlinear Mathematical Analysis and Applications. Hadronic Press, Palm Harbor, 1998, 223–234.

[29] Saitoh, S.: Various operators in Hilbert space induced by transforms. International J. of Applied Math. 1 (1999), 111–126.

[30] Saitoh, S.: Applications of the general theory of reproducing kernels. Reproducing Kernels and their Applications. Kluwer Academic Publishers, Dordrecht 1999, 165–188.

[31] Saitoh, S., Yamamoto, M.: Integral transforms involving smooth functions. Reproducing Kernels and their Applications. Kluwer Academic Publishers, Dordrecht 1999, 149–164.

[32] Saitoh, S.: Linear integro-differential equations and the theory of reproducing kernels. Volterra Equations and Applications. C. Corduneanu and I.W. Sandberg (eds), Gordon and Breach Science Publishers, Amsterdam 2000.

[33] Saitoh, S.: Representations of the solutions of partial differential equations of parabolic and hyperbolic types by means of time observations. Applicable Analysis 76 (2000), 283–289.

[34] Saitoh, S.: Analytic extension formulas, integral transforms and reproducing kernels. Analytic Extension Formulas and their Applications. Kluwer Academic Publishers, Dordrecht, 2001, 207–232.

[35] Saitoh, S., Tuan, V.K., Yamamoto, M.: Conditional Stability of a Real Inverse Formula for the Laplace Transform. Zeitschrift für Analysis und ihre Anwendungen 20(2001), 193–202.

[36] Saitoh, S.: Applications of the reproducing kernel theory to inverse problems. Comm. Korean Math. Soc. 16 (2001), 371–383.

[37] Saitoh, S., Mori, M.: Representations of analytic functions in terms of local values by means of the Riemann mapping function. Complex Variables, Theory Appl. 45 (2002), 387–393.

[38] Saitoh, S.: Principle of telethoscope. Proceedings of the Graz Workshop on "Functional-analytic and complex methods, their interaction and applications to Partial Differential Equations", World Scientific, Singapore, 2001, 101–117.

[39] Schwartz, L.: Sous-espaces hilbertiens d'espaces vectoriels topologiques et noyaux associés (noyaux reproduisants). J. Analyse Math. 13(1963), 115–256.

[40] Tuan, V.K., Saitoh, S., Saigo, M.: Size of support of initial heat distribution in the 1D heat equation. Applicable Analysis 74 (2000), 439–446.

ANALYTIC FUNCTIONS AND ANALYTIC FUNCTIONALS ON SOME BALLS IN THE COMPLEX EUCLIDEAN SPACE

Keiko Fujita

Faculty of Culture and Education

Saga University

Saga 840-8502, Japan

keiko@cc.saga-u.ac.jp

Mitsuo Morimoto

International Christian University

Osawa 3-10-2, Mitaka

Tokyo 181-8585, Japan

mitsuo@icu.ac.jp

Abstract   By generalizing the Lie norm, the Euclidean norm and the dual Lie norm, we could define a series of norms $\{N_p\}_{1 \leq p \leq \infty}$ on $\mathbf{C}^{n+1}$ (see [5] and [8]). Our results in [1], [2] and [3] for the Lie ball, the complex Euclidean ball and the dual Lie ball can be generalized for the N_p-balls. In this note, following our paper [4], we consider analytic functions and analytic functionals on the N_p-balls $\tilde{B}_p(r)$, and characterize them by their growth behavior of their harmonic components in their double series expansion.

1.   Lie ball and N_p-ball

Let us denote $\mathbf{E} = \mathbf{R}^{n+1}$ and $\tilde{\mathbf{E}} = \mathbf{C}^{n+1}$. Let $\|x\| = \sqrt{x_1^2 + \cdots + x_{n+1}^2}$ be the Euclidean norm of $x = (x_1, x_2, \cdots, x_{n+1}) \in \mathbf{E}$. For $z = (z_1, \cdots, z_{n+1}) \in \tilde{\mathbf{E}}$, the cross norm $\|z\|_C$ of $\|x\|$ is defined by

$$\|z\|_C = \inf \left\{ \sum_{j=1}^{m} |a_j| \|x_j\|; z = \sum_{j=1}^{m} a_j x_j \in \mathbf{E}, a_j \in \mathbf{C}, x_j \in \mathbf{E}, m \in \mathbf{Z}_+ \right\}.$$

H.G.W. Begehr et al. (eds.), Analysis and Applications - ISAAC 2001, 151–159.

It is known that $\|z\|_C$ equals to the Lie norm $L(z)$ (see, e.g., [7]):

$$\|z\|_C = L(z) = \sqrt{\|z\|^2 + \sqrt{\|z\|^4 - |z^2|^2}},$$

where $\|z\| = \sqrt{|z_1|^2 + \cdots + |z_{n+1}|^2}$ and $z^2 = z_1^2 + \cdots + z_{n+1}^2$.
For a norm $N(z)$, the dual norm $N^*(z)$ of $N(z)$ is defined by

$$N^*(z) = \sup\{|z \cdot \zeta|; N(\zeta) \leq 1\}.$$

Then the dual Lie norm $L^*(z)$ is given by $L^*(z) = \sqrt{(\|z\|^2 + |z^2|)/2}$.
Further, $L^*(z)$ can be represented by using the Lie norm:

$$L^*(z) = \frac{1}{2}\left(L(z) + \left(\frac{|z^2|}{L(z)}\right)\right). \tag{1}$$

Noting $|z^2|/L(z) = \sqrt{\|z\|^2 - \sqrt{\|z\|^4 - |z^2|^2}}$, the complex Euclidean norm $\|z\|$ is also represented by using the Lie norm:

$$\|z\| = \left(\frac{1}{2}\left(L(z)^2 + \left(\frac{|z^2|}{L(z)}\right)^2\right)\right)^{\frac{1}{2}}. \tag{2}$$

Now, for $p \geq 1$, consider the function

$$N_p(z) = \left(\frac{1}{2}\left(L(z)^p + \left(\frac{|z^2|}{L(z)}\right)^p\right)\right)^{\frac{1}{p}}.$$

Then by (1) and (2), we have $N_1(z) = L^*(z)$ and $N_2(z) = \|z\|$. Because

$$\lim_{p \to \infty} N_p(z) = L(z), \tag{3}$$

we interpret $N_\infty(z) = L(z)$.

Moreover, $N_p(z)$ has the following properties (see [5] and [8]):

1. $N_p(z)$ is monotone increasing in p; that is, $N_s(z) \leq N_t(z)$ for $s < t$.

2. $N_p(z)$ is a norm on $\tilde{\mathbf{E}}$.

3. $N_p^*(z) = N_q(z)$ if $1/p + 1/q = 1, 1 \leq p, q \leq \infty$.

For $p \geq 1$ we define the open and the closed N_p-balls on $\tilde{\mathbf{E}}$ by

$$\tilde{B}_p(r) = \left\{z \in \tilde{\mathbf{E}}; N_p(z) < r\right\}, \quad \tilde{B}_p[r] = \left\{z \in \tilde{\mathbf{E}}; N_p(z) \leq r\right\}.$$

We note that $\tilde{B}_1[r]$ is the dual Lie ball, $\tilde{B}_2[r]$ is the complex Euclidean ball, and $\tilde{B}_\infty[r] = \bigcap_{p>0} \tilde{B}_p[r]$ is the Lie ball.

2. Double series expansion

We denote by $\mathcal{P}_\Delta^k(\tilde{\mathbf{E}})$ the space of homogeneous harmonic polynomials of degree k on $\tilde{\mathbf{E}}$ and by $N(k, n)$ the dimension of $\mathcal{P}_\Delta^k(\tilde{\mathbf{E}})$. It is known that

$$N(k, n) = \frac{(2k + n - 1)(k + n - 2)!}{k!(n - 1)!}.$$

The harmonic extension of the Legendre polynomial $P_{k,n}(t)$ of degree k and of dimension $n + 1$ is defined by

$$\tilde{P}_{k,n}(z, w) = (\sqrt{z^2})^k (\sqrt{w^2})^k P_{k,n}\left(\frac{z}{\sqrt{z^2}} \cdot \frac{w}{\sqrt{w^2}}\right).$$

Then $\tilde{P}_{k,n}(z, w) = \tilde{P}_{k,n}(w, z)$ and

$$\tilde{P}_{k,n}(\cdot, w) \in \mathcal{P}_\Delta^k(\tilde{\mathbf{E}}). \tag{4}$$

Let $\mathcal{O}(\{0\})$ be the space of holomorphic functions in a neighborhood of 0. For $f \in \mathcal{O}(\{0\})$, we define

$$f_k(z) = \frac{1}{2\pi i} \int_{|t|=\rho} \frac{f(tz)}{t^{k+1}} dt,$$

where ρ is sufficiently small. Then $f \in \mathcal{O}(\{0\})$ can be expanded locally into the homogeneous polynomials f_k: $f = \sum_{k=0}^{\infty} f_k$. Moreover, the homogeneous polynomial f_k can be expanded into the homogeneous harmonic polynomials:

$$f_k(z) = \sum_{l=0}^{[k/2]} (z^2)^l f_{k,k-2l}(z), \quad f_{k,k-2l} \in \mathcal{P}_\Delta^{k-2l}(\tilde{\mathbf{E}}). \tag{5}$$

$f_{k,k-2l}$ is uniquely determined by f_k as follows:

$$f_{k,k-2l}(z) = N(k - 2l, n) \int_S f_k(\omega) \tilde{P}_{k-2l,n}(z, \omega) d\omega, \tag{6}$$

where $d\omega$ is the normalized invariant measure on the unit real sphere S. Then by (4), $f_{k,k-2l} \in \mathcal{P}_\Delta^{k-2l}(\tilde{\mathbf{E}})$. Thus $f \in \mathcal{O}(\{0\})$ can be expanded locally by means of homogeneous harmonic polynomials:

$$f(z) = \sum_{k=0}^{\infty} \sum_{l=0}^{[k/2]} (z^2)^l f_{k,k-2l}(z).$$

We call the right-hand side *the double series expansion* of f and $f_{k,k-2l}$ the $(k, k - 2l)$-*harmonic component* of f.

Since $\mathcal{P}_\Delta^k(\tilde{\mathbf{E}})$ and $\mathcal{P}_\Delta^j(\tilde{\mathbf{E}})$ are orthogonal with respect to the integral over S, we have

$$f_{k,k-2l}(z) = N(k - 2l, n) \int_S f_{k,k-2l}(\omega) \tilde{P}_{k-2l,n}(z, \omega) d\dot{\omega} \qquad (7)$$

by (5) and (6).

3. Holomorphic functions on the N_p-ball

Let $\mathcal{O}(\tilde{B}_p(r))$ be the space of holomorphic functions on $\tilde{B}_p(r)$ with the topology of uniform convergence on compact sets. Put

$$\mathcal{O}(\tilde{B}_p[r]) = \lim_{r'>r} \text{ind}\, \mathcal{O}(\tilde{B}_p(r'))$$

and we equip it with the inductive limit locally convex topology.

We can characterize $f \in \mathcal{O}(\tilde{B}_p(r))$ by the growth behavior of their harmonic components $f_{k,k-2l}$.

Theorem 1 ([4]) *Let* $f = \sum_{k=0}^{\infty} \sum_{l=0}^{[k/2]} (z^2)^l f_{k,k-2l}(z)$. *If* $f \in \mathcal{O}(\tilde{B}_p(r))$, *then the sequence* $\{f_{k,k-2l}\}$ *of* $(k, k-2l)$-*harmonic components of* f *satisfies*

$$\limsup_{k\to\infty} \left(\left(\frac{2^k l! (k - l)!}{k!} \right)^{\frac{1}{p}} \|f_{k,k-2l}\|_{C(S)} \right)^{\frac{1}{k}} \leq \frac{1}{r}, \qquad (8)$$

where $\|f\|_{C(S)} = \sup_{z \in S} |f(z)|$ *is the supremum norm on* S.

Conversely, if a sequence $\{f_{k,k-2l}\}$ *of harmonic polynomials satisfies* (8), *then the double series*

$$\sum_{k=0}^{\infty} \sum_{l=0}^{[k/2]} (z^2)^l f_{k,k-2l}(z)$$

converges to a holomorphic function f *in* $\mathcal{O}(\tilde{B}_p(r))$.

This theorem for $p = 1$ and ∞ was proved in [1], [2] and [3], and for $p = 2$ in [3].

For the sake of brevity, we state such a kind of theorem as follow:

THEOREM   Let $r > 0$ and $f(z) = \sum\limits_{k=0}^{\infty} \sum\limits_{l=0}^{[k/2]} (z^2)^l f_{k,k-2l}(z)$.

(i) $f \in \mathcal{O}(\tilde{B}_p(r)) \iff \limsup\limits_{k\to\infty} \left(\left(\dfrac{2^k l! (k-l)!}{k!} \right)^{\frac{1}{p}} \|f_{k,k-2l}\|_{C(S)} \right)^{\frac{1}{k}} \leq \dfrac{1}{r}$,

(ii) $f \in \mathcal{O}(\tilde{B}_p[r]) \iff \limsup\limits_{k\to\infty} \left(\left(\dfrac{2^k l! (k-l)!}{k!} \right)^{\frac{1}{p}} \|f_{k,k-2l}\|_{C(S)} \right)^{\frac{1}{k}} < \dfrac{1}{r}$.

4.   Entire functions of exponential type

For an entire function f on $\tilde{\mathbf{E}}$ and $N_p(z)$, put

$$\|f\|_{X(r,N_p)} = \sup\{|f(z)| \exp(-rN_p(z)); z \in \tilde{\mathbf{E}}\},$$

$$X(r, N_p) = \{f \in \mathcal{O}(\tilde{\mathbf{E}}); \|f\|_{X(r,N_p)} < \infty\}.$$

Then $X(r, N_p)$ is a Banach space with respect to the norm $\|\cdot\|_{X(r,N_p)}$. We set

$$\begin{aligned} \mathrm{Exp}\,(\tilde{\mathbf{E}}; (r, N_p)) &= \bigcap_{r'>r} X(r', N_p) \\ &= \{f \in \mathcal{O}(\tilde{\mathbf{E}}); \forall r' > r, \|f\|_{X(r',N_p)} < \infty\}, \\ \mathrm{Exp}\,(\tilde{\mathbf{E}}; [r, N_p]) &= \bigcup_{r'<r} X(r', N_p) \\ &= \{f \in \mathcal{O}(\tilde{\mathbf{E}}); \exists r' < r, \|f\|_{X(r',N_p)} < \infty\}. \end{aligned}$$

We equip $\mathrm{Exp}\,(\tilde{\mathbf{E}}; (r, N_p))$ with the projective limit locally convex topology and $\mathrm{Exp}\,(\tilde{\mathbf{E}}; [r, N_p])$ with the inductive limit locally convex topology.

Theorem 2 ([4]) *Let* $F(z) = \sum\limits_{k=0}^{\infty} \sum\limits_{l=0}^{[k/2]} (z^2)^l F_{k,k-2l}(z)$. *Suppose p and q satisfy $1/p + 1/q = 1$. Then we have*

(i) $F \in \mathrm{Exp}\,(\tilde{\mathbf{E}}; (r, N_p))$

$$\iff \limsup\limits_{k\to\infty} \left(\left(2^k l!(k-l)!\right)^{\frac{1}{p}} (k!)^{\frac{1}{q}} \|F_{k,k-2l}\|_{C(S)} \right)^{\frac{1}{k}} \leq r,$$

(ii) $F \in \mathrm{Exp}\,(\tilde{\mathbf{E}}; [r, N_p])$

$$\iff \limsup\limits_{k\to\infty} \left(\left(2^k l!(k-l)!\right)^{\frac{1}{p}} (k!)^{\frac{1}{q}} \|F_{k,k-2l}\|_{C(S)} \right)^{\frac{1}{k}} < r.$$

This theorem for $p = 1$ and ∞ was proved in [1] and [2].

5.   Analytic functionals on the N_p-ball

We denote by $\mathcal{O}'(\tilde{B}_p(r))$ the dual space of $\mathcal{O}(\tilde{B}_p(r))$ and by $\mathrm{Exp}\,'(\tilde{\mathbf{E}};(r,N_p$
the dual space of $\mathrm{Exp}\,(\tilde{\mathbf{E}};(r,N_p))$. An element of $\mathcal{O}'(\tilde{B}_p(r))$ is called an
analytic functional on $\mathcal{O}(\tilde{B}_p(r))$ and an element of $\mathrm{Exp}\,'(\tilde{\mathbf{E}};(r,N_p))$ is
called an entire functional on $\tilde{\mathbf{E}}$.

Let $T \in \mathrm{Exp}\,'(\tilde{\mathbf{E}};(0)) = \mathrm{Exp}\,'(\tilde{\mathbf{E}};(0,N_p))$ and put

$$T_{k,k-2l}(w) = \left\langle T_z, N(k-2l,n)(z^2)^l \tilde{P}_{k-2l,n}(z,w) \right\rangle.$$

Then $T_{k,k-2l} \in \mathcal{P}_\Delta^{k-2l}(\tilde{\mathbf{E}})$. We call $T_{k,k-2l}$ the $(k,k-2l)$-harmonic component of T.

For $f(z) = \displaystyle\sum_{k=0}^{\infty}\sum_{l=0}^{[k/2]} (z^2)^l f_{k,k-2l}(z) \in \mathrm{Exp}\,(\tilde{\mathbf{E}};(0))$, by (7), we have

$$
\begin{aligned}
\langle T_z, f(z) \rangle &= \left\langle T_z, \sum_{k=0}^{\infty}\sum_{l=0}^{[k/2]} (z^2)^l f_{k,k-2l}(z) \right\rangle \\
&= \left\langle T_z, \sum_{k=0}^{\infty}\sum_{l=0}^{[k/2]} (z^2)^l N(k-2l,n) \int_S f_{k,k-2l}(\omega)\tilde{P}_{k-2l,n}(z,\omega)d\dot\omega \right\rangle \\
&= \sum_{k=0}^{\infty}\sum_{l=0}^{[k/2]} \int_S T_{k,k-2l}(\omega) f_{k,k-2l}(\omega)d\dot\omega.
\end{aligned}
$$

Theorem 3 ([4]) *Let $T \in \mathcal{O}'(\tilde{B}_p(r))$ and $T_{k,k-2l}$ be the $(k,k-2l)$-harmonic component of T. Then the sequence $\{T_{k,k-2l}\}$ of $(k,k-2l)$-harmonic components of T satisfies*

$$\limsup_{k\to\infty} \left(\left(\frac{k!}{2^k l!(k-l)!}\right)^{\frac{1}{p}} \|T_{k,k-2l}\|_{C(S)} \right)^{\frac{1}{k}} < r, \qquad (9)$$

and we have

$$\langle T_z, f(z) \rangle = \sum_{k=0}^{\infty}\sum_{l=0}^{[k/2]} \int_S T_{k,k-2l}(\omega) f_{k,k-2l}(\omega)d\dot\omega, \qquad f \in \mathcal{O}(\tilde{B}_p(r)).$$

Conversely, if a sequence $\{T_{k,k-2l}\}$ of harmonic polynomials satisfies
(9), then the mapping

$$T : f \mapsto \sum_{k=0}^{\infty}\sum_{l=0}^{[k/2]} \int_S T_{k,k-2l}(\omega) f_{k,k-2l}(\omega)d\dot\omega$$

defines an analytic functional T on $\tilde{B}_p(r)$ and the given $T_{k,k-2l}$ is the $(k, k-2l)$-harmonic component of T.

This theorem for $p = 1$ and ∞ was proved in [1] and [2].

Similarly, we have

$$T \in \mathcal{O}'(\tilde{B}_p[r]) \iff \limsup_{k \to \infty} \left(\left(\frac{k!}{2^k l! (k-l)!} \right)^{\frac{1}{p}} \|T_{k,k-2l}\|_{C(S)} \right)^{\frac{1}{k}} \leqq r.$$

Theorem 4 ([4]) *Let $T \in \mathrm{Exp}'(\tilde{\mathbf{E}}; (0))$ and $T_{k,k-2l}$ be the $(k, k-2l)$-harmonic component of T. Suppose p and q satisfy $1/p + 1/q = 1$. Then we have*

(i) $T \in \mathrm{Exp}'(\tilde{\mathbf{E}}; (r, N_p))$

$$\iff \limsup_{k \to \infty} \left(\left(\frac{1}{2^k l! (k-l)!} \right)^{\frac{1}{p}} \left(\frac{1}{k!} \right)^{\frac{1}{q}} \|T_{k,k-2l}\|_{C(S)} \right)^{\frac{1}{k}} < \frac{1}{r},$$

(ii) $T \in \mathrm{Exp}'(\tilde{\mathbf{E}}; [r, N_p])$

$$\iff \limsup_{k \to \infty} \left(\left(\frac{1}{2^k l! (k-l)!} \right)^{\frac{1}{p}} \left(\frac{1}{k!} \right)^{\frac{1}{q}} \|T_{k,k-2l}\|_{C(S)} \right)^{\frac{1}{k}} \leqq \frac{1}{r}.$$

This theorem for $p = 1$ and ∞ was proved in [1] and [2].

6.   The Fourier-Borel transform

The Fourier-Borel transform $\mathcal{F}T$ of $T \in \mathrm{Exp}'(\tilde{\mathbf{E}}; (0))$ is defined by

$$\mathcal{F}T(\zeta) = \langle T_z, \exp(z \cdot \zeta) \rangle,$$

where ζ is sufficiently small. We call the mapping $\mathcal{F} : T \mapsto \mathcal{F}T$ the Fourier-Borel transformation.

For general norms, A. Martineau proved the following theorem:

Theorem 5 ([6]) *Let $N(z)$ be a norm on $\tilde{\mathbf{E}}$. The Fourier-Borel transformation $\mathcal{F}$ establishes the following topological linear isomorphisms:*

$$\mathcal{F} : \ \mathcal{O}'(\tilde{B}_N[r]) \xrightarrow{\sim} \mathrm{Exp}\,(\tilde{\mathbf{E}}; (r, N^*)), \quad 0 \leq r < \infty,$$
$$\mathcal{F} : \ \mathcal{O}'(\tilde{B}_N(r)) \xrightarrow{\sim} \mathrm{Exp}\,(\tilde{\mathbf{E}}; [r, N^*]), \quad 0 < r \leq \infty,$$
$$\mathcal{F} : \ \mathrm{Exp}'(\tilde{\mathbf{E}}; (r, N^*)) \xrightarrow{\sim} \mathcal{O}(\tilde{B}_N[r]), \quad 0 \leq r < \infty,$$
$$\mathcal{F} : \ \mathrm{Exp}'(\tilde{\mathbf{E}}; [r, N^*]) \xrightarrow{\sim} \mathcal{O}(\tilde{B}_N(r)), \quad 0 < r \leq \infty,$$

where $\mathcal{O}'(\tilde{B}_N[r])$, $\mathrm{Exp}\,(\tilde{\mathbf{E}}; (r, N^*))$, etc., are defined as $\mathcal{O}'(\tilde{B}_p[r])$, $\mathrm{Exp}\,(\tilde{\mathbf{E}}; (r, N_p))$ etc..

In case of $N(z) = N_p(z)$, we can prove Martineau's theorem by our results on the double series expansion. Note that, for $p = 1$ and ∞, it was done in [1] and [2].

Theorem 6 ([4]) *Suppose p and q satisfy $1/p + 1/q = 1$. Then the Fourier-Borel transformation $\mathcal{F}$ establishes the following topological linear isomorphisms:*

$$\mathcal{F}: \; \mathcal{O}'(\tilde{B}_p[r]) \xrightarrow{\sim} \operatorname{Exp}(\tilde{\mathbf{E}}; (r, N_q)), \quad 0 \le r < \infty,$$

$$\mathcal{F}: \; \mathcal{O}'(\tilde{B}_p(r)) \xrightarrow{\sim} \operatorname{Exp}(\tilde{\mathbf{E}}; [r, N_q]), \quad 0 < r \le \infty,$$

$$\mathcal{F}: \; \operatorname{Exp}'(\tilde{\mathbf{E}}; (r, N_p)) \xrightarrow{\sim} \mathcal{O}(\tilde{B}_q[r]), \quad 0 \le r < \infty,$$

$$\mathcal{F}: \; \operatorname{Exp}'(\tilde{\mathbf{E}}; [r, N_p]) \xrightarrow{\sim} \mathcal{O}(\tilde{B}_q(r)), \quad 0 < r \le \infty.$$

PROOF.　Since

$$\exp(z \cdot \zeta) = \sum_{k=0}^{\infty} \sum_{l=0}^{[k/2]} \frac{\Gamma(\frac{n+1}{2}) N(k - 2l, n)}{2^k l! \Gamma(k - l + \frac{n+1}{2})} (z^2)^l (\zeta^2)^l \tilde{P}_{k-2l,n}(z, \zeta),$$

for $T \in \operatorname{Exp}'(\tilde{\mathbf{E}}; (0))$ and sufficiently small ζ, we have

$$\mathcal{F}T(\zeta) = \sum_{k=0}^{\infty} \sum_{l=0}^{[k/2]} \frac{\Gamma(\frac{n+1}{2})}{2^k l! \Gamma(k - l + \frac{n+1}{2})} (\zeta^2)^l T_{k,k-2l}(\zeta). \tag{10}$$

Because the dual norm of $N_p(z)$ is $N_q(z)$, considering their growth behavior of their harmonic components, the first and the second isomorphisms result from Theorems 2 and 3, and the third and the fourth isomorphisms from Theorems 1 and 4.　　　　　q.e.d.

References

[1] Fujita, K.: On the double series expansion with harmonic components. Finite or Infinite Dimensional Complex Analysis, Lecture Notes in Pure and Applied Mathematics, Marcel Dekker, New York, 2000, 77–86.

[2] Fujita, K.: Entire eigenfunctions of the Laplacian of exponential type with respect to the Lie norm and the dual Lie norm, Acta Mathematica Vietnamica 26 (2001), 81–94.

[3] Fujita, K., Morimoto, M.: Holomorphic functions on the dual Lie ball. Proceedings of the Second ISAAC Congress, Kluwer Academic Publishers, 2000, 771–780.

[4] Fujita, K., Morimoto, M.: On the double series expansion of holomorphic functions, to appear in J. Math. Anal. Appl.

[5] Kimura, T.: Norm related to the Lie norm. Tokyo J. Math. 24 (2001), 319–322.

[6] Martineau, A.: Sur les fonctionnelles analytiques et la transformation de Fourier-Borel. J. d'analyse Math. de Jérusalem 11 (1963), 1–164.

[7] Morimoto, M.: Analytic Functionals on the Sphere. Translation of Math. Monographs 178, AMS, 1998.

[8] Morimoto, M., Fujita, K.: Between Lie norm and dual Lie norm. Tokyo J. Math. 24 (2001), 499–507.

BATTERIES AND ENERGY MINIMIZATION PROBLEMS

Promarz M. Tamrazov *

Institute of Mathematics
of National Ukrainian Academy of Sciences
Tereshchenkivska str., 3
Kiev, 01601 MSP, Ukraine

ptamraz@capt.kiev.ua

Abstract   We consider energy minimization problems in classes of real signed measures consistent with a finite or infinite collection of signed sets that are allowed to intersect each other, and with nonzero real numbers attached to these sets.

1.   Introduction

In 70-th the author had introduced a notion called a *battery*. It is a collection of signed sets (called *plates*) *which may intersect each other* We had posed energy minimization problems in classes of charges (real signed measures) consistent with a battery and a given collection of nonzero real numbers of the corresponding signs attached to plates of the battery. By 1978 the author had studied such problems for logarithmic energy at batteries in the Euclidean plane with compact plates which were allowed to intersect each other.

The work contains results on sufficient conditions for solvability of energy minimization problems at general batteries in Euclidean spaces, on uniqueness of extremals and on advanced properties of minimizing charges and their potentials.

*This research was supported in part by INTAS-99-00089 project.

H.G.W. Begehr et al. (eds.), Analysis and Applications - ISAAC 2001, 161–169.
© 2003 *Kluwer Academic Publishers.*

2.   Main assumptions, notations and problems

Let Ω be a topological space, and $\vec{E} := \{E^i\}_{i \in I}$ be a family of non-empty sets $E^i \subset \Omega$ for which the subscript i runs over an abstract non-empty set I decomposed on two non-intersecting subsets I^+, I^- (each of them is allowed to be either empty or non-empty, finite or infinite, even uncountable).

A family $\vec{E}$ is called a *battery*. The sets E^i are called *plates*. A plate E^i is called *positive* or *negative* if i belongs to I^+ or I^-, respectively.

Plates of the same sign may intersect each other (in arbitrary way).

We assume that $E^i \neq E^j$ if $i \neq j$ for $i, j \in I$.

Denote

$$E^+ := \cup_{i \in I^+} E^i,$$

$$E^- := \cup_{i \in I^-} E^i.$$

Suppose that Ω is a locally compact Hausdorff space.

Signed Radon measures on Ω will be called *charges*, and we consider only those charges ν for which $\nu(K)$ is finite for every compact set K.

Suppose that for some charges μ in Ω the energy $\mathcal{E}(\mu)$ is defined in some a way. Denote $\|\mu\| := \mathcal{E}(\mu)^{1/2}$, and let $\|\mu\| \geq 0$ when $\mathcal{E}(\mu) \geq 0$.

Let $\mathfrak{E}$ be the class of all charges in Ω with a finite energy.

Suppose that the sets E^i ($i \in I$) are measurable with respect to every charge with a finite energy.

Let $\vec{r} := \{r^i\}_{i \in I}$ be a family of real numbers r^i, corresponding to all $i \in I$ and satisfying the conditions

$$r^i > 0 \quad \forall\, i \in I^+,$$

$$r^i < 0 \quad \forall\, i \in I^-.$$

For a charge ν in Ω, let $S(\nu)$ denote its support.

Let $\mathfrak{H}_*(\vec{E}, \vec{r})$ denote the class of all charges with the Jordan decomposition $\nu = \nu^+ - \nu^-$ for which

$$S(\nu^+) \subset E^+, \quad S(\nu^-) \subset E^-,$$

$$\nu(E^i) \geq r^i \quad \forall\, i \in I^+,$$

$$\nu(E^i) \leq r^i \quad \forall\, i \in I^-.$$

Let $\mathfrak{N}(\vec{E},\vec{r})$ denote the subclass, consisting of all those $\nu \in \mathfrak{H}_*(\vec{E},\vec{r})$, for which

$$\nu(E^i) = r^i \quad \forall\, i \in I.$$

If the class $\mathfrak{H}_*(\vec{E},\vec{r}) \cap \mathfrak{E}$ is non-empty, let us introduce the quantity

$$\mathcal{H}_*(\vec{E},\vec{r}) := \inf\{\mathcal{E}(\nu) : \nu \in \mathfrak{H}_*(\vec{E},\vec{r}) \cap \mathfrak{E}\},$$

and if the class $\mathfrak{N}(\vec{E},\vec{r}) \cap \mathfrak{E}$ is non-empty, then we introduce the quantity

$$V(\vec{E},\vec{r}) := \inf\{\mathcal{E}(\nu) : \nu \in \mathfrak{N}(\vec{E},\vec{r}) \cap \mathfrak{E}\}.$$

Let us consider the following variational problems.

Problem I To solve the question on the existence and uniqueness of a charge $\lambda_{\vec{E},\vec{r}} := \lambda \in \mathfrak{N}(\vec{E},\vec{r}) \cap \mathfrak{E}$ for which

$$\mathcal{E}(\lambda) = V(\vec{E},\vec{r}).$$

Problem II To solve the question on the existence and uniqueness of a charge $\eta_{\vec{E},\vec{r},*} := \eta_* \in \mathfrak{H}_*(\vec{E},\vec{r}) \cap \mathfrak{E}$ for which

$$\mathcal{E}(\eta_*) = \mathcal{H}_*(\vec{E},\vec{r}).$$

If there is $i \in I$ for which E^i coincides with E^+ (E^-), then we denote this i via i^+ (i^-, respectively) and denote the corresponding r^i via r^+ (r^-, respectively). In this case let $\mathfrak{H}(\vec{E},\vec{r})$ denote the subclass of all those $\nu \in \mathfrak{H}_*(\vec{E},\vec{r})$ for which

$$\nu(E^+) = r^+, \quad \nu(E^-) = r^-,$$

and if the class $\mathfrak{H}(\vec{E},\vec{r}) \cap \mathfrak{E}$ is non-empty, then we define

$$\mathcal{H}(\vec{E},\vec{r}) := \inf\{\mathcal{E}(\nu) : \nu \in \mathfrak{H}(\vec{E},\vec{r}) \cap \mathfrak{E}\}$$

and pose the following variational problem.

Problem III To solve the question on the existence and uniqueness of a charge $\eta_{\vec{E},\vec{r}} := \eta \in \mathfrak{H}(\vec{E},\vec{r}) \cap \mathfrak{E}$ for which

$$\mathcal{E}(\eta) = \mathcal{H}(\vec{E},\vec{r}).$$

Each of the variational problems is complemented with the following problem:

Problem A. To descript extremals and their potentials.

From now on we consider only batteries with relatively compact plates.

3.   Results for batteries with compact plates

In this section we consider batteries with compact plates.

Before 70-th the following cases of the formulated problems were studied.

Case 1 The index set I consists of a single element.

In this case classical results solve the formulated problems and Problem A in Euclidean spaces [5, p. 167 – 221] with classical kernels; B. Fuglede [3] solved such problems for locally compact spaces and a wide class of kernels.

Case 2 Each of the sets I^+, I^- consists of one element, and $r^+ = -r^- = 1$.

This is the case of a usual condenser. For this case M. Ohtsuka [6] studied Problem I for locally compact spaces and a wide class of kernels, with a particular investigation of Problem A for it, and T. Bagby [1] established the following results for a condenser $\{E^+, E^-\}$ in the complex plane $\mathbb{C}$ with the connected complement $\mathbb{C} \setminus (E^+ \cup E^-)$ and the logarithmic kernel: properties of classes of associated charges, sufficient conditions for solvability of the energy minimization problem for normalized charges, uniqueness of the minimizing charge, some immediate properties of its support, description of the potential of the minimizing charge in terms of its support, a discrete description of the capacity of the condenser, characterization of the minimal energy problem via other potential-theoretic problems.

Case 3 The set I is finite and the plates E^i are mutually disjoint.

For this case M. Ohtsuka [6] studied Problem I for locally compact spaces and a wide class of kernels under some additional assumptions, with particular investigation of Problem A for it.

In 70-th the author had obtained a series of results given in my manuscript "Theory of logarithmic potential for plane condensers and batteries" (in Russian). Below as for "Manuscript", I refer to a handwritten copy (82 pages) of that my work made in 1979 from my original text by my former research student N.V. Zorii.

In "Manuscript", batteries with intersecting plates and batteries with an infinite number of plates were considered, some problems for them in

$\mathbb{C}$ were solved and the following results in logarithmic potential theory were established:

(a) Under conditions of Case 2 (see above), the connectedness require-ment for the complement of the condenser assumed in [1] was avoided (without any compensation), and for such general condensers analogues of all above mentioned results of [1] were established.

(b) For Case 2 the complete descriptions of the minimizing charge and of its potential in terms of the geometry of the condenser itself were obtained.

(c) For Case 3 the description of the potential of the minimizing charge was essentially deepened.

(d) For batteries with intersecting plates and with any (finite or infinite) number of plates results on sufficient conditions of solvability of Problems II and III were obtained, and this was done both for positive and signed measures.

(e) In the search of proper formulations of problems and results, and in proofs the essential role was played by set theoretic and potential-theoretic constructions.

Some of results mentioned in the items (a) and (b) were published in the preprint [7], and results mentioned in the item (d) are exposed in the preprint [8] (see also [9, 10, 11]).

In the Euclidean spaces $\mathbb{R}^d$ $(d \geq 2)$ we consider the classical kernels:

$$k(x, y) := -\log |x - y|, \ d = 2,$$

$$k(x, y) := |x - y|^{2-d}, \ d \geq 3,$$

and define the mutual energy of chargesv $\mu, \nu \in \mathfrak{E}$ by the formula

$$\mathcal{E}(\mu, \nu) := \int k(x, y) d\mu(x) d\nu(y)$$

when it makes sense. The energy of a charge $\mu \in \mathfrak{E}$ is defined as $\mathcal{E}(\mu) := \mathcal{E}(\mu, \mu)$.

Let $\vec{E}$ be a battery in $\mathbb{R}^d$ $(d \geq 2)$ with a finite or infinite (even un-countable) index set I. In the following statements plates with coinciding signs are allowed to intersect each other (in arbitrary way).

Assume that the sets E^+, E^- are compact and the class $\mathfrak{H}(\vec{E}, \vec{r}) \cap \mathfrak{E}$ is non-empty. If $d = 2$, let us additionaly assume that either $r^+ + r^- = 0$ or $E^+ \cup E^- \subset \{x : |x| < 1\}$.

We refer to [3, 6] concerning the *vague* and *weak* convergence of charges, and the *strong* convergence is the convergence with respect to $\| \cdot \|$.

Under these assumptions Problem III makes sense and the following results are true.

Theorem 0 *The set $\mathfrak{H}(\vec{E}, \vec{r})$ is convex and vaguely compact.*

Theorem 1 *The variational problem on minimization the energy $\mathcal{E}(\nu)$ in the class of charges $\nu \in \mathfrak{H}(\vec{E}, \vec{r}) \cap \mathfrak{E}$ has a solution $\eta \in \mathfrak{H}(\vec{E}, \vec{r}) \cap \mathfrak{E}$, and the quantity $\mathcal{H}(\vec{E}, \vec{r}) = \mathcal{E}(\eta)$ is finite and (strictly) positive. Moreover, a minimizing charge η is unique.*

Concerning Problem II we confine ourself to the following remarks. This problem makes sense in the case when a kernel is non-negative in the region of integration (for the logarithmic kernel it is the case when $E^+ \cup E^- \subset \{x : |x| < 1/2\}$). In this case analogues of the corresponding results of Problem III are true.

For $d = 2$ these results for Problems III and II were given in "Manuscript" (see also [8]).

The mentioned results for Problem II are true without requirement that the sets E^+, E^- are compact.

In 70-th the author posed to N.V. Zorii the problem to extend his results of "Manuscript" to higher dimensions and appropriate kernels. For batteries with a finite number of plates, she extended a part of results of "Manuscript" and [7] to other spaces and other kernels, sometimes with some changes (see [12, 13]).

4.   Minimizing sequences

Let $T = (T^+, T^-)$ be a pair of sets T^+, T^- in the Euclidean space $\mathbb{R}^d$, $d \geq 2$, and let $T^+ \neq \emptyset$, $T^+ \cap T^- = \emptyset$. We say that a charge μ *is associated with* T, if its Jordan decomposition $\mu = \mu^+ - \mu^-$ satisfies the conditions: $\mu^+(\mathbb{R}^d \setminus T^+) = 0$, $\mu^-(\mathbb{R}^d \setminus T^-) = 0$.

Let $\mathfrak{N}$ be some non-empty class of charges with a finite energy associated with T. Denote $V(\mathfrak{N}) := \inf\{\mathcal{E}(\mu) : \mu \in \mathfrak{N}\}$. If $d = 2$ then additionally assume that $V(\mathfrak{N}) \geq 0$ and for every charges $\mu, \nu \in \mathfrak{N}$ there holds $\mathcal{E}(\mu - \nu) \geq 0$, while $\mathcal{E}(\mu - \nu) = 0$ holds only under $\mu = \nu$.

A sequence S of charges $\mu_n \in \mathfrak{N}$ ($n = 1, 2, \ldots$) is called *minimizing (in $\mathfrak{N}$)*, if $\mathcal{E}(\mu_n) \to V(\mathfrak{N})$ ($n \to \infty$). A charge $\lambda \in \mathfrak{N}$ is called *minimizing (for $\mathfrak{N}$)*, if $\mathcal{E}(\lambda) = V(\mathfrak{N})$.

We have established the following result.

Theorem 2 *Suppose that $\frac{1}{2}(\mu + \nu) \in \mathfrak{N} \; \forall \mu, \nu \in \mathfrak{N}$. Let S be a minimizing sequence of charges in $\mathfrak{N}$. Then there exists a subsequence $S' \subset S$ that vaguely and strongly converges to some charge μ_* associated with T. This charge must satisfy the condition $\mathcal{E}(\mu_*) \leq V(\mathfrak{N})$, and S' converges to μ_* also weakly.*

5. Results for batteries with finely closed plates

Concerning notions of the fine topology see [2, 5, 4, 6].

For a battery consisting of a single finely closed plate, the energy minimization problem was investigated by Fuglede in general setting (see [4]).

Below we give some results on energy minimization problem for general batteries with finely closed plates in the Euclidean spaces and advanced results on minimizing charges.

Let $T = (T^+, T^-)$ be a pair of sets T^+, T^- in $\mathbb{R}^d$, $d \geq 2$, with the same properties as in Section 4.

Let $\vec{I}$, $\vec{r}$, r^+, r^- be the same as in Section 2.

Let $\vec{E}$ be the same battery as in Section 2, with $\Omega := \mathbb{R}^d$ ($d \geq 2$) and finely closed plates E^i for all $i \in I$.

Assume that the set E^i is contained in T^+ for every $i \in I^+$, and is contained in T^- for every $i \in I^-$.

If $d = 2$, let us additionaly assume that either $r^+ + r^- = 0$ or $T^+ \cup T^- \subset \{x : |x| < 1\}$.

Let $\mathfrak{H}_*(\vec{E}, \vec{r})$ be the class of all charges μ with a finite energy associated with T, for which

$$\mu(E_i) \geq r_i \; \forall \, i \in I^+, \quad \mu(E_i) \leq r_i \; \forall \, i \in I^-.$$

The class $\mathfrak{H}_*(\vec{E}, \vec{r})$ is convex.

Suppose that the class $\mathfrak{H}_*(\vec{E}, \vec{r}) \cap \mathfrak{E}$ is non-empty.

In the following statements, plates with coinciding signs are allowed to intersect each other (in arbitrary way).

Under these assumptions the following results are true.

Theorem 3 *Let S be a minimizing sequence of charges in $\mathfrak{H}_*(\vec{E},\vec{r})$. Then there exists a subsequence $S' \subset S$ that strongly and vaguely converges to some charge $\mu_* \in \mathfrak{H}_*(\vec{E},\vec{r})$. This charge is minimizing for $\mathfrak{H}_*(\vec{E},\vec{r})$, and S' converges to μ_* also weakly. Moreover, in $\mathfrak{H}_*(\vec{E},\vec{r})$ there is no other minimizing (for $\mathfrak{H}_*(\vec{E},\vec{r})$) charge.*

If $Q \subset \mathbb{R}^d$, then we set $CQ := \mathbb{R}^d \setminus Q$, and as $\tilde{Q} = Q^\sim, \partial_f Q$ and $b(Q)$ we denote *the fine closure, fine boundary* and the *base*

$$b(Q) := C((C\tilde{Q})^\sim \setminus \partial_f((C\tilde{Q})^\sim))$$

of the set Q, respectively (another definition of the base see in [4]). Let us denote $G := Cb(\bigcup_{i \in I} b(\partial_f E_i))$.

Theorem 4 *The class $\mathfrak{H}_*(\vec{E},\vec{r})$ contains (the unique) minimizing charge λ, and the restriction of this charge to G equals zero.*

Let $i \in I^+$. The set $Cb(E_i)$ is finely open. Let Y_i^+ denote the union of all relatively compact, finely connected components of the set $Cb(E_i)$ which don't intersect T^-. Let us denote $E_i' := Y_i^+ \cup b(E_i)$. In symmetric way for $i \in I^-$, let us define the sets Y_i^- and E_i'. Now let us denote $G' := Cb(\bigcup_{i \in I} b(\partial_f E_i'))$ and sharpen Theorem 4 (since $G \subset G'$) :

Theorem 5 *Under the conditions of Theorem 4, the restriction of the charge λ to G' equals zero.*

References

[1] Bagby, T.: The modulus of a plane condenser. J. Math. and Mech. 17, no 4 (1967), 315–329.

[2] Brelot, M.: On Topologies and Boundaries in Potential Theory. Lecture Notes in Mathematics, 175, Springer, Berlin etc. 1971.

[3] Fuglede, B.:On the theory of potentials in locally compact spaces. Acta Math. 103 (1960), 139–215.

[4] Fuglede, B.: Finely Harmonic Functions. Lecture Notes in Mathematics 289, Springer, Berlin etc. 1972.

[5] Landkof, N.S.: Foundations of Modern Potential Theory. Nauka, Moscow 1966, (Russian).

[6] Ohtsuka, M.: On potentials in locally compact spaces. J. Sci. Hiroshima Univ. Ser. A-1. 25, no 2 (1961), 135–352.

[7] Tamrazov, P.M.: On variational problems of the logarithmic potential theory. Investigations on Potential Theory. Preprint 80.25 of Inst. of Mathem. of Acad. of Sci. of Ukrainian SSR. Kiev, 1980, 3–13 (Russian).

[8] Tamrazov, P.M.: Batteries and variational problems of the logarithmic potential theory. Preprint 01.1 of Inst. of Mathem. of National Acad. of Sci. of Ukraine. Kiev 2001 (Russian).

[9] Tamrazov, P.M.: Batteries and energy minimization problems. International Conference on Complex Analysis and Potential Theory, Kiev, 7–12 August 2001. Abstracts, 54–56 (Russian).

[10] Tamrazov, P.M.: Strengthened results on batteries with intersecting plates. International Conference on Complex Analysis and Potential Theory, Kiev, 7–12 August 2001. Abstracts, 82–84.

[11] Tamrazov, P.M.: Batteries and energy minimization problems of potential theory. 3-th International ISAAC Congress, Freie Universität Berlin, Germany, August 20–25, 2001.

[12] Zorii, N.V.: On some properties of spatial condensers. Preprint 80.26 of Inst. of Mathem. of Acad. of Sci. of Ukrainian SSR. Kiev 1980 (Russian).

[13] Zorii, N.V.: Extremal problems in the theory of capacities of condensers in locally compact spaces. I, II, III. Ukrain. Math. Zh. 53 (2001), no 2, 168–189; no 4, 466–488; no 6, 758–782 (Russian).

VECTOR-VALUED EXTENTIONS OF SOME CLASSICAL THEOREMS IN HARMONIC ANALYSIS

Maria Girardi *
Department of Mathematics
University of South Carolina
Columbia, SC 29208
U.S.A.
girardi@math.sc.edu

Lutz Weis †
Mathematisches Institut I
Universität Karlsruhe
Englerstraße 2
76128 Karlsruhe
Germany
Lutz.Weis@math.uni-karlsruhe.de

Abstract   This paper surveys some recent results on vector-valued Fourier multiplier theorems and pseudo differential operators, which have found important application in the theory of evolution equations. The approach used combines methods from Fourier analysis and the geometry of Banach spaces, such as R-boundedness.

Keywords: R-boundedness, Mihlin multiplier theorem, pseudo differential operators, Fourier type, Littlewood-Paley decomposition, vector-valued Besov spaces
Subject Classification: Primary 42B15, 46E40, 46B09. Secondary 46B20, 46E35.

*Girardi is supported in part by the Alexander von Humboldt Foundation.
†Weis is supported in part by *Landesforschungsschwerpunkt Evolutionsgleichungen des Landes Baden-Württenberg.*

H.G.W. Begehr et al. (eds.), Analysis and Applications - ISAAC 2001, 171–185.

1.   Introduction

Boundedness theorems for Fourier multiplier operators, singular integral operators, and pseudo differential operators play an important role in analysis. In recent years it has become apparent that one needs not only the classical theorems but also vector-valued extensions with operator-valued multiplier functions or symbols. These extensions allow one to treat certain problems for evolution equations with partial differential operators in an elegant and efficient manner in analogy to ordinary differential equations. For example, such theorems are used: in studying maximal regularity of parabolic equations (see, e.g., [1, 4, 5, 9, 17, 18, 19, 27, 28]), in stability theory (see, e.g., [21, 26]), in the theory of pseudo differential operators on manifolds with singularities (see, e.g., [23]),and for elliptic operators on infinite dimensional state spaces (see, e.g., [3, 10]).

This paper surveys some recent results in harmonic analysis of Banach space valued functions and tries to elucidate the interesting interplay with the geometry of the underlying Banach space, which in the end leads to significant applications to evolution equations.

The first Fourier multiplier theorem for *operator-valued* multiplier functions was J. Schwartz's version of Mihlin's theorem.

Theorem 1 (J. Schwartz) *Let H be a Hilbert space and*

$$m \colon \mathbb{R}^N \setminus \{0\} \to \mathcal{B}(H)$$

be such that the sets

$$\left\{ |t|^{|\alpha|} D^\alpha m(t) \colon t \in \mathbb{R}^N \setminus \{0\} \right\} \tag{1}$$

are norm bounded for each multi-index $\alpha \in \mathbb{N}_0^N$ with $|\alpha| \leq [N/2] + 1$. Then

$$T_m(f) := \mathcal{F}^{-1}[m(\mathcal{F}f)] \quad for \quad f \in \mathcal{S}(\mathbb{R}^N, H)$$

defines a bounded operator on $L_q(\mathbb{R}^N, H)$ for each $q \in (1, \infty)$.

Here $\mathcal{F}$ is the Fourier transform and $\mathcal{S}(\mathbb{R}^N, H)$ is the Schwartz class of rapidly decreasing functions from $\mathbb{R}^N$ to H. Does Theorem 1 formally generalize by replacing the Hilbert space H by an arbitrary Banach space X? No; G. Pisier showed that (isomorphic images of) Hilbert spaces are the **only** Banach spaces for which Theorem 1 holds in the above form. In recent years, two approaches were found to circumvent this difficulty.

(i) Replace Bochner spaces $L_q\left(\mathbb{R}^N, X\right)$ by Besov spaces $B_{q,r}^s\left(\mathbb{R}^N, X\right)$. Using the characterization of Besov spaces in terms of the Paley-Littlewood decomposition, one can prove a Mihlin-type theorem as well as boundedness results for pseudo differential operators for *any* Banach space X. Section 4 elaborates on this approach. Note that in this setting one needs only norm boundedness, and not R-boundedness, of the sets in (1).

(ii) In the case of the Bochner spaces $L_q\left(\mathbb{R}^N, X\right)$, it makes sense to consider only those Banach spaces X for which the simplest multiplier function, i.e. $M\left(\cdot\right) = \operatorname{sign}\left(\cdot\right)\chi_X$, is a Fourier multiplier (or equivalently, for which the Hilbert transform is bounded) on $L_q\left(\mathbb{R}^N, X\right)$. Such Banach spaces are called **UMD spaces**. Subspaces of $L_q\left(\Omega, \mathbb{C}\right)$-spaces, for $1 < q < \infty$, are examples of UMD spaces. There are many results showing the UMD spaces form the proper class of Banach spaces for vector-valued harmonic analysis (see, e.g., [6, 7, 29]). For starters, if X is a UMD space, then there is a Paley-Littlewood decompositition for $L_q\left(\mathbb{R}^N, X\right)$. But this decomposition is more delicate than the corresponding decomposition for Besov spaces. Therefore one has to replace the norm bounded condition in (1) by an R-bounded condition. This leads to boundedness results for Fourier multipliers and pseudo differential operators. Section 5 elaborates on this approach. Since large classes of classical operators are R-bounded (cf. [12, and references therein]), the assumptions in this approach are not too restrictive for applications.

2.   Definitions and Notation

Notation is standard; consult [14, 15] for the needed definitions and notations. Here some basics are recalled.

Schwartz used Plancherel's identity for $L_2\left(\mathbb{R}^N, H\right)$ in his proof of Theorem 1. Since Plancherel's identity holds only for Hilbert space valued Bochner spaces $L_2\left(\mathbb{R}^N, X\right)$, the following concept from Banach space theory is needed.

Definition 2 ([22]) *Let $1 \le p \le 2$. A Banach space X has **Fourier type** p provided the Fourier transform $\mathcal{F}$ defines a bounded linear operator from $L_p\left(\mathbb{R}^N, X\right)$ to $L_{p'}\left(\mathbb{R}^N, X\right)$ for some (and thus then for each) $N \in \mathbb{N}$.*

The simple estimate $\|\mathcal{F}f(t)\|_X \le \|f\|_{L_1(X)}$ shows that each Banach space X has Fourier type 1. The notion becomes more restrictive as p increases to 2. A Banach space has Fourier type 2 if and only if X

is isomorphic to a Hilbert space [20]. A space $L_q(\Omega, \mathbb{R})$ has Fourier type $p = \min(q, q')$ [22]. Each closed subspace, the dual, and quotient space of a Banach space X has the same Fourier type as X; the first fact holds by definition, the last by duality.

The Fourier type is connected with the minimal smoothness assumptions on the multiplier function. For example, the condition (1) holding for $|\alpha| \leq [N/2] + 1$, instead of $|\alpha| \leq N$ for the Bochner case and $|\alpha| \leq N + 1$ for the Besov case, expresses the fact that a Hilbert space has Fourier type 2.

To define Besov spaces, first consider a partition of unity $\{\varphi_k\}_{k \in \mathbb{N}_0}$ of functions from $\mathcal{S}(\mathbb{R}^N, \mathbb{R})$ as follows. Take a nonnegative function ψ in $\mathcal{S}(\mathbb{R}, \mathbb{R})$ with support in $[2^{-1}, 2]$ that satisfies

$$\sum_{k=-\infty}^{\infty} \psi(2^{-k} s) = 1 \text{ for } s \in \mathbb{R} \setminus \{0\}$$

and let, for $t \in \mathbb{R}^N$,

$$\varphi_k(t) = \psi\left(2^{-k}|t|\right) \quad \text{for } k \in \mathbb{N} \quad \text{and} \quad \varphi_0(t) = 1 - \sum_{k=1}^{\infty} \varphi_k(t) .$$

To simplify notation, let $\varphi_k \equiv 0$ if $k < 0$. Note that φ_k and φ_j have overlapping support if and only if $|k - j| \leq 1$.

Among the many equivalent descriptions of Besov spaces, the most useful one in this context is given in terms of the so-called *Littlewood-Paley decomposition*. Roughly speaking this means that one considers $f \in \mathcal{S}'(X)$ as a distributional sum

$$f = \sum_{k=0}^{\infty} \left[\varphi_k \hat{f}\right]^{\vee} = \sum_{k=0}^{\infty} \check{\varphi}_k * f = \sum_{k=0}^{\infty} f_k , \quad \text{where} \quad f_k := \check{\varphi}_k * f ,$$

of analytic functions f_k whose Fourier transforms have support in the (slightly overlapping dyadic-like) intervals $\{\operatorname{supp} \varphi_k\}_{k \in \mathbb{N}_0}$ and then one defines the Besov norm in terms of the *blocks* $\{f_k\}_{k \in \mathbb{N}_0}$ of the Littlewood-Paley decomposition of f.

Definition 3 *The **Besov space*** $B_{q,r}^s \left(\mathbb{R}^N, X \right)$, *where* $1 \leq q, r \leq \infty$ *and* $s \in \mathbb{R}$, *is the space of all* $f \in \mathcal{S}' \left(\mathbb{R}^N, X \right)$ *for which*

$$\|f\|_{B_{q,r}^s(\mathbb{R}^N,X)} := \left\| \left\{ 2^{ks} \|\check{\varphi}_k * f\|_{L_q(X)} \right\}_{k=0}^\infty \right\|_{\ell_r} \tag{2}$$

$$\equiv \begin{cases} \left[\displaystyle\sum_{k=0}^\infty 2^{ksr} \|\check{\varphi}_k * f\|_{L_q(X)}^r \right]^{1/r} & \text{if } r \neq \infty \\[2em] \displaystyle\sup_{k \in \mathbb{N}_0} \left[2^{ks} \|\check{\varphi}_k * f\|_{L_q(X)} \right] & \text{if } r = \infty \end{cases}$$

is finite; q *is the **main index** while* s *is the **smoothness index**. The space* $B_{q,r}^s \left(\mathbb{R}^N, X \right)$, *together with the norm in* (2), *is a Banach space.*

3. A weak Fourier multiplier theorem for Bochner spaces

The following theorem is a *weak* Fourier multiplier theorem in the sense that its assumption (3) is quite *strong*; indeed, (3) implies that the Fourier multiplier function m is in $L_p \left(\mathbb{R}^N, \mathcal{B}(X,Y) \right)$.

Theorem 4 ([14]) *Let* X *and* Y *have Fourier type* $p \in [1,2]$ *and*

$$m \in B_{p,1}^{N/p} \left(\mathbb{R}^N, \mathcal{B}(X,Y) \right) . \tag{3}$$

Then m *is a Fourier multiplier from* $L_q(\mathbb{R}^N, X)$ *to* $L_q(\mathbb{R}^N, Y)$ *for each* $q \in [1, \infty]$; *furthermore,*

$$\|T_m\|_{L_q(X) \to L_q(Y)} \leq C \, \mathcal{M}_p(m) , \tag{4}$$

where C *is a constant independent of* m *and*

$$\mathcal{M}_p(m) := \inf \left\{ \|m(a\cdot)\|_{B_{p,1}^{N/p}(\mathbb{R}^N, \mathcal{B}(X,Y))} : a > 0 \right\} .$$

Theorem 4 leads to vector-valued extensions of classical multiplier theorems (such as Mihlin-, Hörmanders-, and Lipschitz-type theorems) for Besov spaces (see Section 4) and Bochner spaces (see Section 5) by considering a Littlewood-Paley decomposition of these spaces and then applying Theorem 4 to the blocks of the decomposition with the multiplier function m *restricted* to the support of the blocks. This is the charm behind using Littlewood-Paley decompositions: the classical (weak) assumptions on m imply the (strong) assumption of Theorem 4 *when* restricted to the support of the blocks.

The first step in proving Theorem 4 is to extend two other classical results to the vector-valued setting. The first extension shows that, for spaces with Fourier type p, the Sobolev Embedding (SE) factors through L_1 via the Fourier transform $\mathcal{F}$.

$$B_{p,1}^{N/p}\left(\mathbb{R}^N, X\right) \xrightarrow[\text{SE}]{i} L_\infty\left(\mathbb{R}^N, X\right)$$

$$\mathcal{F} \searrow \qquad \nearrow \mathcal{F}^{-1}$$

$$L_1\left(\mathbb{R}^N, X\right)$$

Lemma 5 ([14]) *Let X have Fourier type $p \in [1,2]$. Then the Fourier transform defines bounded operator from $B_{p,1}^{N/p}\left(\mathbb{R}^N, X\right)$ to $L_1\left(\mathbb{R}^N, X\right)$.*

The next lemma extends a well-known boundedness result for classical integral operators.

Lemma 6 ([13]) *Let $F \subseteq Y^*$ be a subspace that norms Y. Let*

$$k\colon \mathbb{R}^N \to \mathcal{B}\left(X, Y\right)$$

be such that k and k^* are srongly measurable and satisfy

$$\int_{\mathbb{R}^N} \|k(s)x\|_Y \, ds \leq C_0 \|x\|_X \qquad \text{for each} \quad x \in X$$

$$\int_{\mathbb{R}^N} \|k(s)^* y^*\|_{X^*} \, ds \leq C_1 \|y^*\|_{Y^*} \qquad \text{for each} \quad y^* \in F$$
(5)

for some constants C_i. Then the convolution operator K, defined for finitely-valued functions $f\colon \mathbb{R}^N \to X$ with finite support by

$$(Kf)(t) = \int_{\mathbb{R}^N} k(t-s)f(s)\, ds \qquad \text{for } t \in \mathbb{R}^N \ ,$$

extends to a bounded operator $K\colon L_q\left(\mathbb{R}^N, X\right) \to L_q\left(\mathbb{R}^N, Y\right)$ for each $q \in [1,\infty)$; furthermore, $\|K\|_{L_q \to L_q} \leq C_0^{\frac{1}{q}} C_1^{1-\frac{1}{q}}$. If, in addition, Y does not contain c_0, then the same holds true for $q = \infty$.

To prove Theorem 4, first consider a function m in the Schwartz class. Applying Lemma 5 to the functions

$$t \to m(t)x \quad \text{for } x \in X \qquad \text{and} \qquad t \to m^*(t)y^* \quad \text{for } y^* \in Y^*$$

gives that $k := \check{m}$ satisfies the assumptions in (5) and so one has that the corresponding operator T_m satisfies (4) (even if Y contains c_0). Now, thanks to the bound in (4) on the norm of the T_m's, a density argument finishes the job.

4. Fourier multiplier theorems for Besov spaces

One can think of a Besov space as a direct sum

$$B^s_{q,r}\left(\mathbb{R}^N, X\right) = \sum_{k \in \mathbb{N}_0} Z_k \quad \text{where} \quad Z_k := \left\{\check{\varphi}_k * f : f \in B^s_{q,r}\left(\mathbb{R}^N, X\right)\right\}.$$

To see how a Fourier multiplier operator T_m, for a multiplier function $m \colon \mathbb{R}^N \to \mathcal{B}(X, Y)$, formally behaves on a *blocks* Z_k of the above Littlewood-Paley decomposition, fix an $f \in \mathcal{S}\left(\mathbb{R}^N, X\right)$. Since the function $\psi_k := \varphi_{k-1} + \varphi_k + \varphi_{k+1}$ is 1 on $\operatorname{supp}\varphi_k$,

$$T_m\left(\check{\varphi}_k * f\right) = \left[\psi_k\, m\, \varphi_k\, \widehat{f}\,\right]^\vee = \check{\varphi}_k * \left[m\, \psi_k\, \widehat{f}\,\right]^\vee = \check{\varphi}_k * T_m\left(\check{\psi}_k * f\right) \in Z_k.$$

Thus T_m leaves the blocks Z_k invariant. Furthermore,

$$\left[\check{\varphi}_k * T_m f\right]^\widehat{} = \psi_k\, \varphi_k\, m\, \widehat{f} = m\, \psi_k\, \left(\check{\varphi}_k * f\right)^\widehat{} = \left[T_{m\psi_k}\left(\check{\varphi}_k * f\right)\right]^\widehat{}, \qquad (6)$$

and so

$$f = \sum_{k \in \mathbb{N}_0} \check{\varphi}_k * f \quad \text{and} \quad T_m f = \sum_{k \in \mathbb{N}_0} \check{\varphi}_k * T_m f = \sum_{k \in \mathbb{N}_0} T_{m\psi_k}\left(\check{\varphi}_k * f\right).$$

Thus T_m behaves as a Fourier multiplier operator $T_{\psi_k m}$ on *each* block Z_k of the Littlewood-Paley decomposition; furthermore, the operator $T_{\psi_k m}$ depends only on the values of m on the supports of ψ_k. This suggests the following approach to boundedness results for T_m.

(1st) Estimate $\|T_{m\psi_k}\|_{L_q \to L_q}$ on each block of the Littlewood-Paley decomposition *separably*. For this, apply Theorem 4 to the multiplier function $m\,\psi_k$.

(2nd) Sum over the blocks; with the help of (6):

$$\|T_m f\|_{B^s_{p,r}(X)} = \left[\sum_{k=0}^{\infty} 2^{ksr}\, \|T_{m\psi_k}\left(\check{\varphi}_k * f\right)\|^r_{L_q(X)}\right]^{1/r}$$

$$\leq \sup_k \|T_{m\psi_k}\|_{L_q \to L_q}\, \|f\|_{B^s_{p,r}(X)}.$$

This gives the heuristic idea behind the proof of the main result of this section (additional considerations are necessary if q or r is ∞):

Theorem 7 [14] *Let X and Y be Banach spaces with Fourier type p. Let $m : \mathbb{R}^N \to \mathcal{B}(X, Y)$ satisfy, for each $k \in \mathbb{N}_0$,*

$$\varphi_k \cdot m \in B_{p,1}^{N/p}\left(\mathbb{R}^N, \mathcal{B}(X, Y)\right) \qquad and \qquad \mathcal{M}_p(\varphi_k \cdot m) \leq A \,.$$

Then m is a Fourier multiplier from $B_{q,r}^s(\mathbb{R}^N, X)$ to $B_{q,r}^s(\mathbb{R}^N, Y)$ for each $q, r \in [1, \infty]$ and $s \in \mathbb{R}$. Furthermore, $\|T_m\|_{B_{q,r}^s \to B_{q,r}^s} \leq CA$ for some constant C that is independent of m.

Note that each Banach space has Fourier type 1 and each uniformly convex Banach space has Fourier type p for some $p > 1$. Our result shows that the required smoothness of the multiplier function m depends not only on the dimension of $\mathbb{R}^N$ but also on the geometry of the Banach spaces X and Y. It follows from results in [26] that the smoothness N/p is sharp for the Besov scale.

An advantage of the rather general formulation of the assumptions in Theorem 7 is that one can deduce from them, by simple estimates, several multiplier theorems with *classical* assumptions. For example, the Mihlin-type multiplier theorem below follows easily; it was the first multiplier theorem of this kind and its parts i) and iii) are due independently to H. Amann [2] and L. Weis [26], respectively.

Corollaries of Theorem 7. [14]   Let $q, r \in [1, \infty]$ and $s \in \mathbb{R}$.

<u>Mihlin condition</u> If $m \colon \mathbb{R}^N \to \mathcal{B}(X, Y)$ satisfies, for some constant A, the estimate

$$\sup_{t \in \mathbb{R}^N} \left\| (1 + |t|)^{|\alpha|} D^\alpha m(t) \right\|_{\mathcal{B}(X,Y)} \leq A$$

for each multi-index $\alpha \in \mathbb{N}_0^N$ with $|\alpha| \leq l$, then m is a Fourier multiplier from $B_{q,r}^s(\mathbb{R}^N, X)$ to $B_{q,r}^s(\mathbb{R}^N, Y)$ provided one of the following conditions hold:

i) X and Y are arbitrary Banach spaces and $l = N + 1$

ii) X and Y are uniformly convex Banach spaces and $l = N$

iii) X and Y have Fourier type p and $l = [\frac{N}{p}] + 1$.

<u>Hörmander condition</u> Let X and Y have Fourier type p and $l = \left[\frac{N}{p}\right] + 1$. Let $m \colon \mathbb{R}^N \to \mathcal{B}(X, Y)$ satisfy, for some constant A and each $R \in [1, \infty)$,

the estimates:

$$\left[\int_{|t| \leq 2} \| D^\alpha m(t) \|^p \, dt \right]^{1/p} \leq A$$

$$\left[R^{-N} \int_{R < |t| < 4R} \| D^\alpha m(t) \|^p \, dt \right]^{1/p} \leq A R^{-|\alpha|}$$

for each α with $|\alpha| \leq l$. Then m is a Fourier multiplier from $B^s_{q,r}(\mathbb{R}^N, X)$ to $B^s_{q,r}(\mathbb{R}^N, Y)$.

<u>Lipschitz condition</u> Let X and Y have Fourier type p and $l \in (1/p, 1)$. Assume that $m \colon \mathbb{R} \to \mathcal{B}(X, Y)$ satisfies, for some constant A, the estimates:

$$\| m(t) \| \leq A \quad \text{for} \quad t \in \mathbb{R}$$

$$(1 + |t|)^l \left\| \frac{m(t + u) - m(t)}{|u|^l} \right\| \leq A \quad \text{for} \quad u, t \in \mathbb{R}, \ u \neq 0.$$

Then m is a Fourier multiplier from $B^s_{q,r}(X)$ to $B^s_{q,r}(Y)$.

It is also useful to consider pseudo differential operators with operator-valued symbols. A pseudo differential operators Ψ_a with symbol a is formally defined by

$$\Psi_a f(t) := (2\pi)^{-N} \int_{\mathbb{R}^N} e^{it \cdot s} a(t, s) \hat{f}(s) \, ds \quad , \quad f \in \mathcal{S}(\mathbb{R}^N, X).$$

In analogy to classical symbol classes, for $\delta \in [0, 1)$ and an $r > 0$, let $S^0_{1,\delta}(r, X)$ be the the class of symbols $a \colon \mathbb{R}^N \times \mathbb{R}^N \to \mathcal{B}(X)$ so that for all multi-indices α there is a constant C_α with

$$\left\| (1 + |s|)^{|\alpha|} \partial^\alpha_s a(t, s) \right\| \leq C_\alpha \quad \text{for each} \quad t, s \in \mathbb{R}^N \tag{7}$$

$$\| \partial^\alpha_s a(\cdot, s) \|_{B^r_{\infty,\infty}} \leq C_\alpha (1 + |s|)^{\delta r - |\alpha|} .$$

By extending the Coifman-Meyer decomposition of symbols to the operator-valued case, Z. Štrkalj showed the following theorem.

Theorem 8 [24] *Let X be a separable Banach space. Let $q, r \in [1, \infty]$ and $-(1 - \delta) r < s < r$. If $a \in S^0_{1,\delta}(r, X)$ then Ψ_a is bounded on $B^s_{q,r}(\mathbb{R}^N, X)$.*

5. Fourier multiplier theorems for Bochner spaces

This section presents Fourier multiplier theorems on Bochner spaces. The methods are similar to those in the Besov case: one uses Theorem 4 and a Littlewood-Paley decomposition for Bochner spaces. For this decomposition, one needs to decompose $\mathbb{R}^N$ not only for $|t| \to \infty$ but also for $|t| \to 0$.

So consider a *partition of unity* $\{\phi_k\}_{k \in \mathbb{Z}}$ of functions from $S\left(\mathbb{R}^N, \mathbb{R}\right)$ defined as follows. Take a nonnegative function $\phi_0 \in C^\infty(\mathbb{R}^N, \mathbb{R})$ that has support in $\{t : 2^{-1} \leq |t| \leq 2\}$ and satisfies, for $\phi_k(t) := \phi_0(2^{-k}t)$ for each $k \in \mathbb{Z}$, that $\sum_{k \in \mathbb{Z}} \phi_k(t) = 1$ for each $t \neq 0$. Note that

$$\|\check{\phi}_k\|_{L_1} = \|\check{\phi}_0\|_{L_1} \quad \text{and} \quad \operatorname{supp} \phi_k \subset \left\{t : 2^{k-1} \leq |t| \leq 2^{k+1}\right\}$$

for each $k \in \mathbb{Z}$.

Bourgain [7, $N = 1$] and Zimmermann [29, $N > 1$] proved that if a scalar-valued function $m \colon \mathbb{R}^N \setminus \{0\} \to \mathbb{C}$ satisfied a certain Mihlin-type smoothness condition and X is a UMD space, then $m(\cdot) I_X$ is a Fourier multiplier from $L_q(\mathbb{R}^N, X)$ to $L_q(\mathbb{R}^N, X)$ for each $q \in (1, \infty)$. Their result leads to a Littlewood-Paley decomposition for Bochner spaces. Henceforth, $\{r_k\}_{k \in \mathbb{Z}}$ is just any enumeration of the Rademacher functions.

Corollary 9 ([15]) *Let X be a UMD space and $1 < q < \infty$. There is a constant C so that*

$$\frac{1}{C}\|f\|_{L_q(\mathbb{R}^N, X)} \leq \int_{[0,1]} \left\|\sum_{k \in \mathbb{Z}} r_k(t)\left(\check{\phi}_k * f\right)\right\|_{L_q(\mathbb{R}^N, X)} dt \leq C\|f\|_{L_q(\mathbb{R}^N, X)} \quad (8)$$

for each $f \in L_q\left(\mathbb{R}^N, X\right)$.

To see the fundamental difference between the Besov- and Bochner-space case (for $1 < q < \infty$), let's compare the norms. If $f \in B_{q,2}^0(X)$ (which is *closest* to $L_q(X)$), then

$$\|f\|_{B_{q,2}^0(X)} = \left[\sum_{k=0}^\infty \|\check{\varphi}_k * f\|_{L_q(X)}^2\right]^{1/2} ; \quad (9)$$

thus, $\{\check{\varphi}_k * f\}_{k \in \mathbb{N}_0}$ is <u>absolutely 2-summable</u> in $L_q(X)$. If $f \in L_q(X)$, then Corollary 9 gives not only that $\{\check{\phi}_k * f\}_{k \in \mathbb{Z}}$ is <u>almost uncondi-</u>

tionally summable in $L_q(X)$ but also (in the scalar case) that

$$\|f\|_{L_q(\mathbb{C})} \sim \left\| \left[\sum_{k \in \mathbb{Z}} |\check{\phi}_k * f|^2 \right]^{1/2} \right\|_{L_q(\mathbb{C})}, \tag{10}$$

with the help of Kahane's and Khintchine's inequalities. Compare (9) and (10)! In the Besov case, one can estimate the Bochner norm of each block $\check{\varphi}_k * f$ of f *separately* (via Theorem 4) and then sum over the blocks. But this approach is not possible in the Bochner case since one sums over the blocks inside the Bochner norm. Therefore one needs tools to estimate the blocks simultaneously as an unconditionally summable sequence. Definition 10 is the first tool; Definition 12 is the second tool.

Definition 10 *Let X be a Banach space. Then the space $\boldsymbol{Rad}(X)$, or simply $\widetilde{X}$, is*

$$Rad(X) \stackrel{or}{=} \widetilde{X} := \{ \{x_k\}_{k \in \mathbb{Z}} \in X^{\mathbb{Z}} :$$
$$\sum_{k=-n}^{n} r_k(\cdot)\, x_k : [0,1] \to X \text{ is convergent in } L_1([0,1], X) \}.$$

For $1 \le p < \infty$, when equipped with one of the following norms, which are equivalent by Kahane's inequality,

$$\|\{x_k\}_{k \in \mathbb{Z}}\|_{\mathrm{Rad}_p(X)} := \left\| \sum_{k \in \mathbb{Z}} r_k(\cdot)\, x_k \right\|_{L_p([0,1],X)},$$

$\mathrm{Rad}_p(X)$ becomes a Banach space. Much can be found about $\mathrm{Rad}(X)$ in the literature (see, e.g. [11]).

Condition (8) can thus be reformulated as follows:

$$\|f\|_{L_q(\mathbb{R}^N, X)} \sim \left\| \{\check{\phi}_k * f\}_{k \in \mathbb{Z}} \right\|_{L_q(\mathbb{R}^N, \widetilde{X})}.$$

It is now possible to estimate the blocks simultaneously in the following way: for a given function $m : \mathbb{R}^N \setminus \{0\} \to B(X, Y)$, simultaneously *roll in* all the blocks to the 0th block by defining the corresponding mapping $M : \mathbb{R}^N \setminus \{0\} \to B\left(\widetilde{X}, \widetilde{Y}\right)$ by

$$M(s) := \left\{ \phi_0(s) m(2^k s) \right\}_{k \in \mathbb{Z}}. \tag{11}$$

Next (if possible), apply Theorem 4 to the function M to get a Fourier multiplier operator $T_M : L_q\left(\widetilde{X}\right) \to L_q\left(\widetilde{Y}\right)$, which then can be *rolled back out* to a Fourier multiplier operator $T_m\colon L_q(X) \to L_q(Y)$. This approach leads to the following theorem (note that if X has Fourier type p (resp. UMD), then so does $\widetilde{X}$).

Theorem 11 [15] *Let X and Y be UMD Banach spaces with Fourier type $p \in (1,2]$ and $1 < q < \infty$. Let $m : \mathbb{R}^N \setminus \{0\} \to \mathcal{B}(X,Y)$ be a measurable function so that the corresponding mapping M, as defined in (11), satisfies that $M \in B_{p,1}^s\left(\mathbb{R}^N, \mathcal{B}\left(\widetilde{X},\widetilde{Y}\right)\right)$ for some $s > N/p$. Then m is a Fourier multiplier from $L_q(\mathbb{R}^N, X)$ to $L_q(\mathbb{R}^N, Y)$.*

The assumption in Theorem 11 may look awkward; however, it is general enough to yield *classical* multiplier theorems, with the help of our second tool.

Definition 12 *A subset τ of $\mathcal{B}(X,Y)$ is **R-bounded** provided there is a constant C_p so that for each $n \in \mathbb{N}$ and subset $\{T_j\}_{j=1}^n$ of τ and subset $\{x_j\}_{j=1}^n$ of X*

$$\left\|\sum_{j=1}^n r_j(\cdot)\, T_j(x_j)\right\|_{L_p(\Omega,Y)} \leq C_p \left\|\sum_{j=1}^n r_j(\cdot)\, x_j\right\|_{L_p(\Omega,X)} \tag{12}$$

*for some (and thus then, by Kahane's inequality, for each) $p \in [1,\infty)$. The **R-bound** of τ, $R(\tau)$, is smallest constant C_1 for which (12) holds.*

Note the following connection between our two tools: Definition 10 and Definition 12.

Remark 13 *A sequence $\{T_j\}_{j\in\mathbb{Z}}$ from $\mathcal{B}(X,Y)$ is R-bounded if and only if the mapping*

$$\widetilde{X} \ni \{x_j\}_{j\in\mathbb{Z}} \xrightarrow{\;\widetilde{T}\;} \{T_j x_j\}_{j\in\mathbb{Z}} \in \widetilde{Y}$$

defines an element in $\mathcal{B}\left(\widetilde{X},\widetilde{Y}\right)$ for some (or equiv., for each) $p \in [1,\infty)$.

The statements, and proofs, of the following corollaries to Theorem 11 are similar to the corresponding corollaries to Theorem 7.

Corollaries of Theorem 11. [15]
Let X and Y be UMD spaces with Fourier type p and $1 < q < \infty$.
<u>Mihlin condition</u> If for $m\colon \mathbb{R}^N \setminus \{0\} \to \mathcal{B}(X,Y)$ the set

$$\left\{ |t|^{|\alpha|}\, D^\alpha m(t) : t \in \mathbb{R}^N \setminus \{0\} \right\}$$

is R-bounded for each multi-index $\alpha \in \mathbb{N}_0^N$ with $|\alpha| \leq [\frac{N}{p}] + 1$, then m is a Fourier multiplier from $L_q(\mathbb{R}^N, X)$ to $L_q(\mathbb{R}^N, Y)$.

<u>Hörmander condition</u> If for $m \colon \mathbb{R}^N \setminus \{0\} \to \mathcal{B}(X, Y)$ the term

$$\left[\int_{\frac{1}{2} < |t| < 2} R\left(\left\{ \left|2^k t\right|^{|\alpha|} D^\alpha m \left(2^k t\right) \right\}_{k \in \mathbb{Z}} \right)^p dt \right]^{1/p}$$

is finite for each multi-index $\alpha \in \mathbb{N}_0^N$ with $|\alpha| \leq [\frac{N}{p}] + 1$, then m is a Fourier multiplier from $L_q(\mathbb{R}^N, X)$ to $L_q(\mathbb{R}^N, Y)$.

<u>Lipschitz condition</u> If for $m \colon \mathbb{R} \to \mathcal{B}(X, Y)$ the set

$$\left\{ m(t), |t|^l \frac{m(t+s) - m(t)}{|s|^l} : t, s \in \mathbb{R} \setminus \{0\} \right\}$$

is R-bounded for some $l \in (1/p, 1)$, then m is a Fourier multiplier from $L_q(\mathbb{R}^N, X)$ to $L_q(\mathbb{R}^N, Y)$.

For the connection between R-boundedness and multiplier theorems with scalar multiplier functions, see [8]. The operator-valued Mihlin multiplier theorem was first proved for $N = 1$ in [28] and for higher dimensions in [25]. For variants of the proof of the Mihlin-type result, see [4, 5, 9, 10, 16].

Now returning to pseudo differential operators, define the symbol class $\mathfrak{R}S_{1,\delta}^0(r, X)$ similarly to the symbol class $S_{1,\delta}^0(r, X)$: just replace (7) by the condition that

$$R\left(\left\{ (1 + |s|)^{|\alpha|} \partial_s^\alpha a(t, s) : s \in \mathbb{R}^N \right\} \right) \leq C_\alpha \quad \text{for each} \quad t \in \mathbb{R}^N .$$

In this context, using the same tools and the Coifman-Meyer decomposition of symbols, Z. Štrkalj has shown, among other things, the following boundedness result.

Theorem 14 [24] *Let X be a separable UMD Banach space, $\delta \in [0, 1)$, and $r > 0$. If $a \in \mathfrak{R}S_{1,\delta}^0(r, X)$ then Ψ_a is bounded on $L_q(\mathbb{R}^N, X)$ for each $q \in (1, \infty)$.*

References

[1] Amann, H.: Linear and quasilinear parabolic problems. I: Abstract linear theory. Birkhäuser Boston Inc., Boston, MA, 1995.

[2] Amann, H.: Operator-valued Fourier multipliers, vector-valued Besov spaces, and applications. Math. Nachr. 186 (1997), 5–56.

[3] Amann, H.: Elliptic operators with infinite-dimensional state spaces. J. Evol. Equ. 1 (2001), no 2, 143–188.

184

[4] Arendt, W., Bu, S.: The operator-valued Marcinkiewicz multiplier theorem and maximal regularity. Math. Z. (to appear).

[5] Blunck, S.: Maximal regularity of discrete and continuous time evolution equations. Studia Math. 146 (2001), no 2, 157–176.

[6] Bourgain, J.: Some remarks on Banach spaces in which martingale difference sequences are unconditional. Ark. Mat. 21 (1983), no 2, 163–168.

[7] Bourgain, J.: Vector-valued singular integrals and the H^1-B M O duality. Probability theory and harmonic analysis (Cleveland, Ohio, 1983), Dekker, New York, 1986, 1–19.

[8] Clément, P., de Pagter, B., Sukochev, F.A., Witvliet, H.: Schauder decomposition and multiplier theorems. Studia Math. 138 (2000), no 2, 135–163.

[9] Clément, P., Prüss, J.: An operator-valued transference principle and maximal regularity on vector-valued L_p-spaces. Evolution equations and their applications in physical and life sciences (Bad Herrenalb, 1998), Dekker, New York, 2001, 67–87.

[10] Denk,R., Hieber, M., Prüß, J.: R-boundedness, Fourier multipliers and problems of elliptic and parabolic type. Mem. Amer. Math. Soc. (to appear).

[11] Diestel, J., Jarchow, H., Tonge, A.: Absolutely summing operators. Cambridge University Press, Cambridge 1995.

[12] Girardi, M., Weis, L.: Criteria for R-boundedness of operator families. Recent Contributions to Evolution Equations. Marcel Dekker Lecture Notes in Math. (to appear).

[13] Girardi, M., Weis, L.: Integral operators with operator-valued kernels. In preparation.

[14] Girardi, M., Weis, L.: Operator-valued Fourier multiplier theorems on Besov spaces. Mathematische Nachrichten (to appear).

[15] Girardi, M., Weis, L.: Operator-valued Fourier multiplier theorems on $L_p(X)$ and geometry of Banach spaces (submitted).

[16] Haller, R., Heck, H., Noll, A.: Mikhlin's theorem for operator-valued Fourier multipliers on n-dimensional domains. Math. Nachr. (to appear).

[17] Kalton, N.J., Lancien, G.: A solution to the problem of L^p-maximal regularity. Math. Z. 235 (2000), 559–568.

[18] Kunstmann, P.C.: Maximal L_p-regularity for second order elliptic operators with uniformly continuous coefficients on domains (submitted).

[19] Kunstmann, P.C., Weis, L.: Perturbation theorems for maximal L_p-regularity. Ann. Sc. Norm. Sup. Pisa 30 (2001), 415–435.

[20] Kwapień, S.: Isomorphic characterizations of inner product spaces by orthogonal series with vector valued coefficients. Collection of articles honoring the completion by Antoni Zygmund of 50 years of scientific activity, VI. Studia Math. 44 (1972), 583–595.

[21] Latushkin, Y., Räbiger, F.: Fourier multipliers in stability and control theory. Preprint.

[22] Peetre, J.: Sur la transformation de Fourier des fonctions à valeurs vectorielles. Rend. Sem. Mat. Univ. Padova 42 (1969), 15–26.

[23] Schulze, B.-W.: Boundary value problems and singular pseudo-differential operators. John Wiley & Sons Ltd., Chichester, 1998.

[24] Štrkalj, Ž.: $\mathcal{R}$-Beschränktheit, Summensätze abgeschlossener Operatoren und operatorwertige Pseudodifferentialoperatoren. Ph.D. thesis, Universität Karlsruhe, Karlsruhe, Germany, 2000.

[25] Štrkalj, Ž., Weis, L.: On operator-valued Fourier multiplier theorems (submitted).

[26] Weis, L.: Stability theorems for semi-groups via multiplier theorems. Differential equations, asymptotic analysis, and mathematical physics (Potsdam 1996), Akademie Verlag, Berlin 1997, 407–411.

[27] Weis, L.: A new approach to maximal L_p-regularity. Evolution equations and their applications in physical and life sciences (Bad Herrenalb, 1998), Dekker, New York 2001, 195–214.

[28] Weis, L.: Operator-valued Fourier multiplier theorems and maximal L_p-regularity. Math. Ann. 319 (2001), 735–758.

[29] Zimmermann, F.: On vector-valued Fourier multiplier theorems. Studia Math. 93 (1989), 201–222.

CHAOTIC ZONE IN THE BOGDANOV-TAKENS BIFURCATION FOR DIFFEOMORPHISMS

Vassili Gelfreich

Institut für Mathematik I

Freie Universität Berlin

Arnimallee 2-6

14195 Berlin, Germany

gelf@math.fu-berlin.de

Abstract   We consider a two-parametric analytic family of diffeomorphisms near the Bogdanov-Takens bifurcation. It is known that if the parameters belong to a homoclinic zone, the map has homoclinic points. The width of the homoclinic zone is exponentially small. We derive an asymptotic formula for the width of the homoclinic zone. An analytic invariant associated with a parabolic fixed point is an important ingredient of this formula. The proof of the asymptotic formula is not complete. Additionally we provide results of computations of the invariant for model families.

Keywords: Bogdanov-Takens bifurcation, homoclinic orbits, exponentially small phenomena

1.   Introduction

There is a remarkable similarity between bifurcations of equilibria for planar vector fields and bifurcations of periodic orbits for 3 dimensional vector fields. This similarity is due to the fact that in many cases the Poincaré map near a bifurcating periodic orbit can be formally written as a time-1 map of a planar vector field. Although all Taylor coefficients of the interpolating vector field can be determined, there is no reason to expect convergence of the corresponding series since the dynamics in the 3 dimensional space can be much richer than the dynamics on the plane. On the other hand the interpolating planar vector field can be used to construct normal forms for bifurcations of the periodic orbits. Thus the difference between these two different types of bifurcations is

H.G.W. Begehr et al. (eds.), Analysis and Applications - ISAAC 2001, 187–197.

moved beyond all algebraic orders. In real-analytic theory this difference is typically exponentially small [2, 6, 9].

The Bogdanov-Takens bifurcation [1, 16, 5] has codimension 2. In the case of a planar vector field it corresponds to an equilibrium with two vanishing eigenvalues. In the case of a periodic orbit in the 3 dimensional space both multipliers of the periodic orbit equal 1. It is more convenient to study two dimensional Poincaré maps instead of 3 dimensional vector fields.

The Bogdanov-Takens bifurcation for planar diffeomorphisms has been intensively studied in the recent time [3, 2]. It was shown that, similar to planar vector fields, there are both Hopf and homoclinic bifurcations in a neighborhood of the Bogdanov-Takens bifurcation. In the vector field case the values of parameters, which correspond to the homoclinic bifurcations, form a line on the plane of parameters. In the case of diffeomorphisms transversal homoclinic orbits are possible. The latter are stable with respect to small changes of parameters, so one can expect that there is a sector (instead of the line) on the parameter plane, where the map has homoclinic points. We call this sector a *homoclinic zone*. An exponentially small upper bound for its width was established in [2]. The aim of this paper is to present a lower bound. Of course this lower bound is exponentially small, and we conjecture that it is positive for a generic map. In order to prove this conjecture it is necessary to show that a certain constant Θ (see Section 3) does not vanish. We checked this conjecture numerically for several examples.

Let us formulate our main result more precisely. We consider a two-parametric analytic family $F_{\mu\nu} : (x, y) \mapsto (x_1, y_1)$:

$$
\begin{aligned}
x_1 &= x + y, \\
y_1 &= y + f(x, y, \mu, \nu)
\end{aligned}
\tag{1}
$$

with analytic f. We assume f to be of the form

$$
f(x, y, \mu, \nu) = -\mu + x^2 + \nu y + \gamma xy + \ldots
\tag{2}
$$

It is convenient to order terms of Taylor series according to values of the calibration function, which assigns orders to monomials:

$$
\delta(\mu^j \nu^k x^l y^m) = 4j + 2k + 2l + 3m.
$$

The coefficients μ, ν, and γ are invariant under close-to-identity transformations, which preserve the form of the map $F_{\mu\nu}$.

This form of the map is not exceptional: a generic two-parametric family of maps near its Bogdanov-Takens bifurcation can be transformed to the form (1), (2) by a change of coordinates and parameters [3].

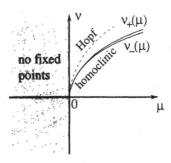

Figure 1.   Bifurcation diagram

If $\mu = \nu = 0$ the map $F_{\mu\nu}$ has a parabolic fixed point at the origin. Following [3, 2] we study study the bifurcation of this point on the plane of (μ, ν). The corresponding bifurcation diagram is shown on Fig. 1.

If a point (μ, ν) is situated between the lines $\nu = \nu_-(\mu)$ and $\nu = \nu_+(\mu)$, $0 < \mu < \mu_0$, then the map $F_{\mu\nu}$ has a homoclinic trajectory. On the lower and upper boundary the homoclinic trajectory disappears. The first and the last homoclinic points are shown on Fig. 2. The pictures correspond to $\nu = \nu_-(\mu)$ and $\nu = \nu_+(\mu)$ respectively.

It is not very difficult to show that $\nu_\pm(\mu) = \frac{5}{7}(\gamma - 2)\mu^{1/2} + O(\mu^{3/4})$. A much more difficult problem is to prove the following estimate:

$$\nu_+(\mu) - \nu_-(\mu) = \frac{A_\gamma}{\mu^{5/4}} \cdot e^{-\sqrt{2}\pi^2 / \sqrt[4]{\mu}}, \tag{3}$$

$$A_\gamma = \frac{5|\Theta|}{6\sqrt{2}} e^{-6\pi^2(\gamma-2)/7} + O(\mu^{1/4} \log \mu). \tag{4}$$

Here Θ is a constant defined later in Section 3. It depends on γ and on all other Taylor coefficients of the function $f(x, y, 0, 0)$. If the map F_{00} is exactly the time-1 map of a planar flow, the corresponding constant Θ vanishes. Each F_{00} can be interpolated by a flow up to an error, which is less than any power of x and y. Thus all Taylor coefficients of $f(x, y, 0, 0)$ are relevant for Θ.

In [14] a similar asymptotic formula was proposed after an accurate analysis of direct numerical measurements for the width of the homoclinic zone.

The formula (3) implies a lower bound for the width of the zone provided Θ does not vanish. It was proved analytically that Θ does not vanish in the cases of the Hénon family [10]; numerically $|\Theta(\text{Hénon})| = 2.47 \ldots \cdot 10^6$. One should expect that Θ does not vanish generically, since

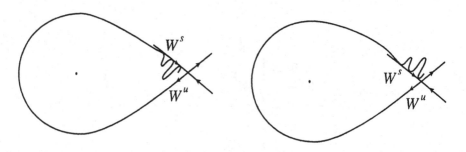

Figure 2. The first and last homoclinic points of the map $F_{\mu\nu}$

it depends analytically on F_{00}. We recall that the Hénon map

$$u_1 = v,$$
$$v_1 = av^2 - bu + 1$$

can be transformed to the form (1) with

$$f(x, y, \mu, \nu) = (x + y)^2 - \mu + \nu y.$$

The coefficients μ and ν are in one-to-one correspondence with the original parameters:

$$a = (1 + \nu/2)^2 - \mu, \qquad b = 1 + \nu.$$

The Hénon family has the special property: it is area-preserving for $\nu = 0$. The separatrices of the area-preserving Hénon map split transversally [9, 10, 4].

The rest of the paper contains an informal derivation of the asymptotic formula (3). It can not be considered as a complete proof. The approach is based on Lazutkin's method originally proposed for a study of the standard map [13, 8]. Section 5 provides a summary of numerical experiments used to evaluate the splitting constant Θ for some model families and to check the correctness of the asymptotic formula (3).

2. Formal interpolation by a flow

In this section we discuss the formal interpolation of the family $F_{\mu\nu}$ by an autonomous vector-field

$$X = X^1 \partial_x + X^2 \partial_y,$$

where ∂_x and ∂_y stand for derivatives in x and y respectively; X^1 and X^2 are power series in x, y, μ and ν. The vector field is a differential operator and can be expanded into a sum of homogeneous operators:

$$X = \sum_{k \geq 1} X_k,$$

where X_k increases the order of a δ-homogeneous polynomial by k.

It is not too difficult to check that there is a unique formal vector-field, such that

$$F_{\mu\nu}(\underline{x}) = e^X \underline{x}, \qquad \underline{x} = (x, y)^T.$$

In general one expects that the series for X diverges. Comparing the first two orders of the series on both sides we obtain

$$
\begin{aligned}
X_1 &= y\partial_x + (x^2 - \mu)\partial_y, \\
X_2 &= -\tfrac{1}{2}(x^2 - \mu)\partial_x + (\nu + (\gamma - 1)x)y\partial_y.
\end{aligned}
$$

The first oder vector-field X_1 is Hamiltonian with the Hamiltonian function

$$h_0 = \frac{y^2}{2} + \mu x - \frac{x^3}{3}.$$

If $\mu > 0$ it has a homoclinic loop defined by $h_0(x, y) = \tfrac{2}{3}\mu^{3/2}$. The perturbation by X_2 can destroy this loop. Let us study the homoclinic bifurcation of the vector field $X_1 + X_2$ using the Melnikov method [11]. The unperturbed homoclinic orbit can be find explicitly:

$$
\begin{aligned}
x_0(t, \mu) &= \sqrt{\mu}\left(1 - \frac{3}{\cosh^2(t\sqrt[4]{\mu}/\sqrt{2})}\right), \\
y_0(t, \mu) &= \dot{x}_0 = 3\sqrt{2}\mu^{3/4}\frac{\sinh(t\sqrt[4]{\mu}/\sqrt{2})}{\cosh^3(t\sqrt[4]{\mu}/\sqrt{2})}.
\end{aligned}
$$

The Melnikov function describes the distance between the stable and unstable separatrices of the flow generated by $X_1 + X_2$ measured in units of h_0:

$$
\begin{aligned}
M(\mu, \nu) &= \int_{-\infty}^{+\infty} (X_2 h_0)|_{x_0(t), y_0(t)} \, dt \\
&= (\gamma - 2)\int_{-\infty}^{+\infty} x_0(t)y_0^2(t) \, dt + \nu \int_{-\infty}^{+\infty} y_0^2(t) \, dt.
\end{aligned}
$$

It is not too difficult to calculate this integral explicitly:

$$M(\mu, \nu) = \frac{24\sqrt{2}}{35}\mu^{5/4}\left(-5(\gamma - 2)\sqrt{\mu} + 7\nu\right).$$

The Melnikov function vanishes along the line $\nu = 5(\gamma - 2)\sqrt{\mu}/7$, which gives an asymptotic of the actual line of homoclinic bifurcations. In the following we will need the following derivative:

$$\frac{\partial M}{\partial \nu}(\mu, \nu) = \frac{24\sqrt{2}}{5}\mu^{5/4}, \tag{5}$$

which describes how fast the distance between separatrices changes as a function of ν for a fixed μ.

Of course, Melnikov-type computations do not give the width of the homoclinic zone for the map $F_{\mu\nu}$, because for a flow it reduces to a line.

Let $\nu = 5(\gamma - 2)\sqrt{\mu}/7$ and find the first correction to the unperturbed homoclinic: $x = x_0(t) + x_1(t) + \ldots$, $y = y_0(t) + y_1(t) + \ldots$ We have

$$\begin{aligned} \dot{x}_1 &= y_1 - \tfrac{1}{2}(x_0^2 - \mu), \\ \dot{y}_1 &= 2x_0x_1 + (\gamma - 1)x_0y_0 + \nu y_0. \end{aligned}$$

For the purposes of the present paper it is sufficient to compute x_1 only:

$$x_1(t) = \tfrac{3}{7}(\gamma - 2)y_0(t)\log\left(\sqrt{\mu} - x_0(t)\right).$$

This calculation can be continued and approximations for the separatrix loop for all δ-orders can be constructed. All these approximations give a closed homoclinic loop. In order to detect the exponentially small splitting of separatrices we analytically continue in t and study the approximations of the separatrices near a complex singularity.

It is clear that x_0 and x_1 have a common singularity at $t = \pm i\rho$ with

$$\rho = \frac{\pi}{\sqrt{2}\mu^{1/4}}. \tag{6}$$

Both x_0 and x_1 are analytic in the complex strip $|\Im t| < \rho$. The main parts of x_0 and x_1 at the singularity are:

$$\begin{aligned} x_0(t) &= \frac{6}{(t - i\rho)^2} + O(1), \\ x_1(t) &= \tfrac{36}{7}(\gamma - 2)\frac{i\pi + \log 6 - \log(t - i\rho)^2}{(t - i\rho)^3} + O\left((t - i\rho)^{-1}\right). \end{aligned} \tag{7}$$

We will use this expansions for the complex matching method described later in Section 4.

3.   Splitting constant Θ

Let us consider the map $F_0 \equiv F_{00}$, which has a parabolic fixed point at the origin. The fixed point has stable and unstable invariant manifold, which we parameterize by solutions of the finite-difference equation:

$$x_0^\pm(\tau + 1) = F_0(x_0^\pm(\tau)).$$

We supply this equation with asymptotic boundary conditions for the first components of x_0^+ and x_0^-:

$$x_0^\pm(\tau) = \frac{6}{\tau^2} - \frac{36}{7}(\gamma - 2)\frac{\log \tau^2}{\tau^3} + O\left(\tau^{-4}\log^2 \tau\right). \tag{8}$$

The unstable solution x_0^- is analytic in a sectorial domain

$$D_- = \{\tau \in \mathbb{C} : \delta < \arg(\tau + c) < 2\pi - \delta\}$$

with $\delta \in (0, \frac{\pi}{2})$ and sufficiently large $c > 0$. The stable solution x_0^+ is analytic in the domain D_+, which is a reflection of D_- with respect to the imaginary axis. These conditions define x_0^+ and x_0^- uniquely.

Consider the auxiliary scalar equation

$$w_0(\tau + 1) = J_0(\tau)w_0(\tau), \qquad \text{where } J_0(\tau) = \det F_0'(x_0^-(\tau)).$$

In D_- this linear finite-difference equation has a unique analytic solution, which satisfies the asymptotic boundary condition $w_0(\tau) = 1 + o(\tau^{-1})$. Let us define the splitting function

$$\tilde{\theta}(\tau) = \frac{1}{w_0(\tau)} \det\left(\frac{dx_0^-}{d\tau}(\tau); \; x_0^+(\tau) - x_0^-(\tau)\right) \tag{9}$$

for $\tau \in \Omega_-$, the lower component of $D_- \cap D_+$. Then the splitting constant Θ is defined by

$$\Theta = \lim_{\Im \tau \to -\infty} e^{2\pi i \tau}\tilde{\theta}(\tau).$$

This definition generalizes the definition from [9]. The existence of the limit can be proved by methods of [8, 10]. The expression under the limit sign converges exponentially as $|\Im \tau| \to -\infty$. Given a map F_0, the constant Θ can be numerically computed directly from the definition. Finally we rewrite the definition of Θ in the form

$$\tilde{\theta}(\tau) \approx w_0(\tau)\Theta\, e^{-2\pi i \tau} \tag{10}$$

in oder to use it later in Section 4.

4. Flow box theorem

Now let us come back to the study of separatrices of the original map $F_{\mu\nu}$. We assume that parameters μ and ν are both small and close to the unperturbed homoclinic line $\nu = \frac{5}{7}(\gamma - 2)\mu^{1/2}$. So the stable and unstable separatrices of the map are close to each other.

We parameterize the stable and unstable separatrices of the hyperbolic fixed point by solutions of finite-difference equation

$$\underline{x}^{u,s}(t+1) = F_{\mu\nu}(\underline{x}^{u,s}(t)),$$

which satisfy appropriate boundary conditions [8, 9]. These functions are close to the separatrix solution of the differential equations of Section 2.

In a neighborhood of the first intersection of the curve $\underline{x}^u(t)$ with the x-axes we introduce the time-energy coordinates (E, T) [7]. In these coordinates the map $F_{\mu\nu}$ takes the form of the translation

$$(E, T) \mapsto (E, T + 1).$$

The coordinates can be normalized in such a way that

$$E(\underline{x}^u(t)) = 0, \qquad T(\underline{x}^u(t)) = t.$$

Moreover the energy E is close to the Hamiltonian function h_0. In the (E, T) coordinates a segment of the stable manifold is a graph of a periodic function $E = A(T, \mu, \nu)$, $A(T + 1, \mu, \nu) = A(T, \mu, \nu)$. We can expand this function in Fourier series

$$A(T, \mu, \nu) = A_0(\mu, \nu) + A_1(\mu, \nu) \sin 2\pi(T - T_0(\mu, \nu)) + h.o.t.$$

In [7] it was shown the h.o.t. is an exponentially small quantity of higher order. The zeroes of $A(T, \mu, \nu)$ correspond to homoclinic points. Moreover, if this function have nor zeroes, it can be shown that the separatrices do not intersect at all.

The first term A_0 is approximated by the Melnikov function described in Section 2:

$$A_0(\mu, \nu) = M(\mu, \nu) + O(\mu^2).$$

The amplitude A_1 of the second term is exponentially small. The product $A_1 \sin 2\pi T$ can be large for non-real T. In order to approximate it we use the following complex matching method. We expand $E(\underline{x}^s(t)) - E(\underline{x}^u(t))$ in powers of $\underline{x}^s(t) - \underline{x}^u(t)$ and use

$$\partial_x E(\underline{x}^u(t)) = \frac{\dot{y}^u(t)}{w(t)}, \qquad \partial_y E(\underline{x}^u(t)) = -\frac{\dot{x}^u(t)}{w(t)},$$

where $1/w(t)$ is the Jacobian of the map $(x, y) \mapsto (E, T)$ computed at the point $\underline{x}^u(t)$. We obtain that

$$A(t, \mu, \nu) = \frac{1}{w(t)} \det(\underline{\dot{x}}^u(t); \underline{x}^s(t) - \underline{x}^u(t)) + O(|\underline{x}^s(t) - \underline{x}^u(t)|^2). \quad (11)$$

In oder to approximate this function we consider its analytic continuation in the variable t. The functions $\underline{x}^u(t)$ and $\underline{x}^s(t)$ are approximated by $\underline{x}_0^-(\tau)$ and $\underline{x}_0^+(\tau)$ respectively, where

$$\tau = t - i\rho - \rho_1,$$

ρ is defined in (6); and $\rho_1 = \frac{3}{7}(\gamma - 2)(i\pi + \log 6)$ takes into account the matching of the expansions (8) and (7).

Comparing (10), (9) and (11) we obtain

$$A_1 \approx 2|\Theta|\,|e^{-2\pi i\rho_1}|\,e^{-2\pi\rho} = 2|\Theta|\,e^{-6\pi^2(\gamma-2)/7}\,e^{-\sqrt{2}\pi^2/\sqrt[4]{\mu}}.$$

Now we calculate the width of the homoclinic zone:

$$\nu_+(\mu) - \nu_-(\mu) = \left.\frac{2A_1(\mu,\nu)}{\frac{\partial A_0}{\partial\nu}(\mu,\nu)}\right|_{\nu:A_0(\mu,\nu)=0} = \frac{5|\Theta|e^{-6\pi^2(\gamma-2)/7}}{6\mu^{5/4}}\,e^{-\sqrt{2}\pi^2/\sqrt[4]{\mu}},$$

where we used the approximations for A_0 and A_1. Thus we obtained the desired asymptotic (3) for the width of the homoclinic zone on the parameter plane of the Bogdanov-Takens bifurcation for diffeomorphisms. The error term comes from the consideration of next order corrections to this asymptotic.

5. Numerical experiments

In the most cases there is not known ways to find the splitting constant Θ analyticly. Given a map F_0, this constant is not too difficult to compute with any reasonable precision. For the area-preserving Hénon map this constant was computed with several thousands digits [15]. We did high-precision computations of Θ for the family (1) with $f = x^2 + \gamma xy$ for several γ. The method is essentially the same as in [9]. Numerically the splitting constant Θ does not vanish on the interval $[-0.5, 2]$. Several values are given in the following table:

γ	Θ
0	$5.202169191369226 \cdot 10^{-5}$
1	$6.989873i51682470943...$
2	$1.203229677075980622 \cdot 10^8$

The estimates for the width of the homoclinic zone were derived non-rigorously. So it is interesting to compare them with results of direct numerical computations. We choose the family (1) with

$$f(x, y, \mu, \nu) = -\mu + x^2 + \nu y + \gamma xy,$$

and the Hénon map. There is a reasonably good agreement between our theory and experiments. For $\mu^{1/4} = 0.08$ the relative error of the asymptotic formula is typically around 20%.

It is difficult to perform computations for essentially smaller values of μ. So we took the data for several μ and then extrapolated towards $\mu = 0$. The values of Θ obtained in this way coincide in several digits with the values from the table. This provides numerical evidence for the correctness of the asymptotic formula (3).

Acknowledgments

The author thanks the International Society for Analysis, Applications and Computation for presenting him the "ISAAC Award for Young Scientists".

References

[1] Bogdanov, R.I.: Versal deformations of a singular point on the plane in the case of zero eigenvalues. Functional Analysis and Applications 9(2), 144–145 (1975).

[2] Broer, H.W., Roussarie, R.: Exponential confinement of chaos in the bifurcation set of real analytic diffeomorphisms. Broer H. (ed.) et al., Global Analysis of Dynamical Systems (dedicated to F. Takens for his 60th birthday). IOP, Bristol 2001, 167–210.

[3] Broer, H.W., Roussarie, R., Simó, C.: Invariant circles in the Bogdanov-Takens bifurcation for diffeomorphisms. Erg. Th. Dyn. Syst. 16 (1996), no 6, 1147–1172.

[4] Chernov, V.: Splitting of separatrices for the Hénon map. Preprint, 2001.

[5] Chow, S.-N., Li, C., Wang, D.: Normal forms and bifurcations of planar vector fields. Cambridge University Press: Cambridge, 1994.

[6] Fontich, E., Simó, C.: The splitting of separatrices for analytic diffeomorphisms. Ergodic Theory Dyn. Syst. 10 (1990), no 2, 295–318.

[7] Gelfreich, V.G.: Conjugation to a shift and splitting of separatrices, Applicationes Mathematicae 24 (1996), 2, 127–140.

[8] Gelfreich, V.G.: A proof of the exponentially small transversality of the separatrices for the Standard Map Comm. Math. Phys. 201 (1999), 155–216.

[9] Gelfreich, V.G.: Splitting of a small separatrix loop near the saddle-center bifurcation in area-preserving maps. Physica D 136 (2000), 266–279.

[10] Gelfreich, V.G., Sauzin, D.: Borel summation and the splitting of separatrices for the Hénon map. Annales l'Institut Fourier 51 (2001), 2, 513–567.

[11] Guckenheimer, J., Holmes, Ph.: Nonlinear oscillations, dynamical systems, and bifurcations of vector fields. Appl. Math. Sciences 42. Springer, Berlin etc. 1983.

[12] Kuznetsov, A.Yu.: Elements of applied bifurcation theory. Springer, Berlin etc. 1998.

[13] Lazutkin, V.F.: Splitting of separatrices for the Chirikov's standard map. VINITI no. 6372/84 (1984) (Russian). Revised English version: Mathematical Physics Preprint Archive 98–421, http://www.ma.utexas.edu/mp_arc.

[14] Simó, C., Broer, H., Roussarie, R.: A numerical exploration of the Takens-Bogdanov bifurcation for diffeomorphisms. Mira, C. (ed.) et al., European Conference on Iteration Theory (ECIT 89), Batschuns (Austria). World Scientific, Singapore, 1991, 320–334.

[15] Simó, C.: Analytical and numerical detection of exponentially small phenomena. Fiedler, B. (ed.) et al., International conference on differential equations. Proceedings of the conference, Equadiff '99, Berlin, Germany, August 1–7, 1999. Vol. 2. World Scientific, Singapore, 2000, 967–976.

[16] Takens, F.: Forced oscillations and bifurcations. Applications of global analysis. I (Sympos., Utrecht State Univ., Utrecht, 1973). Comm. Math. Inst. Rijksuniv. Utrecht, no 3, (1974), 1–59.

CARLEMAN ESTIMATES FOR A PLATE EQUATION ON A RIEMANN MANIFOLD WITH ENERGY LEVEL TERMS

I. Lasiecka and R. Triggiani

Department of Mathematics
University of Virginia
Charlottesville, VA 22904
il2v@virginia.edu and rt7u@virginia.edu

P.F. Yao

Institute of Systems Science
Academy of Mathematics and Systems Science
Chinese Academy of Sciences
Beijing 100080, PRC
pfyao@yahoo.com

Abstract   We provide Carleman estimates for an Euler-Bernoulli type plate equation with energy level terms, defined on an open bounded set Ω of a finite-dimensional Riemann manifold (M, g). The energy level for this problem is $H^3(\Omega) \times H^1(\Omega)$. The basic assumption made is the existence of a strictly convex function on Ω. Carleman estimates are also a critical springboard from which one may derive the *a-priori* inequalities of continuous observability/uniform stabilization of interest in control theory of PDEs.

1.   Introduction

Model Throughout this paper M is a finite-dimensional Riemann manifold with metric $g(\,\cdot\,,\,\cdot\,) = \langle\,\cdot\,,\,\cdot\,\rangle$ and norm $|X|^2 = g(X, X)$, and Ω is an *open bounded* set of M with smooth boundary $\Gamma = \Gamma_0 \cup \Gamma_1$. We let n denote the outward unit normal field along the boundary Γ. Further, we denote by Δ the Laplace (Laplace-Beltrami) operator on the manifold M and by D the Levi-Civita connection of M. A summary of the main Riemann geometric notions needed in this paper is provided in Appendix A below.

H.G.W. Begehr et al. (eds.), Analysis and Applications - ISAAC 2001, 199–236.

In this paper, we study the following Euler-Bernoulli plate equation with third-order space derivatives on the displacement w and first-order space derivatives on the velocity w_t on Ω:

$$w_{tt} + \Delta^2 w + F(w) + f = 0 \quad \text{in } (0,T] \times \Omega \equiv Q, \tag{1}$$

under the following assumptions:

(H.1) the energy-level differential term $F(w)$ satisfies the following estimate: there exists a constant $C_T > 0$ such that

$$|F(w)|^2 \leq C_T\{|D^3 w|^2 + |D^2 w|^2 + |Dw|^2 + w^2 + |Dw_t|^2 + w_t^2\}, \quad \forall\, t, x \in Q, \tag{2}$$

where $Dw = \nabla_g w$ for the scalar function w. So Dw is a vector field, $Dw \in \mathcal{X}(M) = $ the set of all vector fields on M. Generally, $D^k w$ is the kth covariant differential of the (scalar) function w in the metric g, which is a k-order tensor field $T^k(\Omega)$ on Ω. For simplicity of notation, two vertical bars $|\cdot|$ denote the norm in the space $T^k(\Omega)$, see definitions (A.8), (A.9) in Appendix A. In particular, the case $k = 2$ is a special case of the covariant differential DH of a vector field $H \in \mathcal{X}(M)$: see (A.3) below and in general see Appendix A, in particular Remark A.1.

Furthermore, we assume throughout that the forcing term f in (1) satisfies

$$f \in L_2(0,T; L_2(\Omega)) \equiv L_2(Q) : \int_Q f^2 dQ < \infty, \tag{3}$$

where $dQ = d\Omega\, dt$, and $d\Omega$ is the volume element of the manifold M in its Riemann metric g.

Main assumption (H.2) We assume that there exists a function $v : \overline{\Omega} \to \mathbb{R}$ of class C^2 which is strictly convex on $\overline{\Omega}$ with respect to the Riemann metric g of M. This means that the Hessian $D^2 v$, as defined in (A.6) below as a 2-order covariant tensor, is strictly positive on Ω: there exists a constant $a > 0$ such that

$$D^2 v(X, X) \geq a|X|^2, \quad \forall\, X \in M_x,\ x \in \Omega, \tag{4}$$

M_x denoting the tangent space of M at x. $\qquad \square$

Without loss of generality we may always translate $v(\cdot)$ as to make it non-negative on $\overline{\Omega}$: $v \geq 0$ on Ω. Under assumption (H.2), we shall work with the vector field

$$h(x) = \nabla_g v(x) = Dv(x) = \sum_{i=1}^{n}\left(\sum_{j=1}^{n} g^{ij} \frac{\partial}{\partial x_j}\right) \frac{\partial}{\partial x_i}. \tag{5}$$

In Appendix A we provide a short summary of the Riemann geometric apparatus needed in this paper. In Appendix B, we show how the present setting covers Euler-Bernoulli plate equations defined on a smooth bounded domain of $\mathbb{R}^n$, $n \geq 2$, with C^1-variable coefficients in space of the principal part, and L_∞-variable coefficients in both time and space of the energy level terms. For the present problem, the energy level is $H^3(\Omega) \times H^1(\Omega)$ for $\{w, w_t\}$.

Accordingly, we introduce the following 'energy' $E(t)$:

$$E(t) \equiv \int_\Omega [|D(\Delta w)|^2 + |Dw_t|^2]d\Omega. \tag{6}$$

Remark 1 Assumption (H.2) = (4) is the same that arises in the case of second-order hyperbolic equations, [18, 19] and Schrödinger equations [28], on a manifold. It is possible to give several general criteria which guarantee that the main assumption (H.2) = (4) is satisfied [5], [29], [30], [4]. These criteria, in turn, allow one to construct many classes of examples [30, 31], [18, 19] covered by (H.2). We refer to Remark B.1 for illustrations involving a differential operator with variable coefficients on an Euclidean domain. Several examples are given in the above references which involve suitable classes of coefficients $a_{ij}(x)$ defining the differential operator $\mathcal{A}$ in (B.1), such that the corresponding Riemann metric g in (B.7) admits a smooth strictly convex function on *all of* $(\mathbb{R}^n, g)$. Accordingly, *any* (sufficiently smooth) bounded domain in $\mathbb{R}^n$ can be chosen. Finally, we note that a most recent result, valid in dim $\Omega = 2$, which yields the required strictly convex function v in (H.2) = (4) on Ω, was given in [4]: it requires that the boundary $\partial\Omega$ be strictly convex in the metric g (second fundamental form positive).   □

Goal The goal of the present paper is to provide Carleman-type estimates for (sufficiently smooth) solutions of Eqn. (1) [and no Boundary Conditions (B.C.)], that involve the energy $E(t)$ in (6) and, in an *explicit way*, the boundary traces as well. This goal is accomplished in Theorems 2.1 and 2.2, over an arbitrary time $T > 0$, under the main geometric assumption (H.2) = (4).

These estimates are at the $H^3(\Omega) \times H^1(\Omega)$-level for $\{w, w_t\}$. Besides being of interest in themselves, these estimates are also a critical springboard, from which one may derive those *a-priori* inequalities that are required in control theory of PDEs, when Eqn. (1) is supplemented by appropriate B.C. These inequalities are usually referred to as 'continuous observability/uniform stabilization' inequalities. They arise when Eqn. (1) is supplemented, respectively, either by suitable *homogeneous*

B.C. or by suitable *dissipative* B.C. As is well-known, by duality, continuous observability inequalities are equivalent (e.g., [12, 13, 14, 15, 16, 17] to exact controllability properties for (dual) corresponding boundary control problems. The form of the continuous observability/uniform stabilization inequalities depends on the specific B.C. and involves 'dual' (complementary) traces. The topological level of these sought-after inequalities also depends on the specific B.C. The $H^3(\Omega) \times H^1(\Omega)$-level estimates in Section 2 are most directly of use when Eqn. (1) is supplemented by *hinged* B.C. However, a full, sharp account even in the case of hinged B.C.—and surely more so for other B.C. such as *clamped* B.C.—requires additional tools [15, 16], [23, 24, 25] to deal with a few remaining issues. These include: (i) the issue of reducing the number of traces needed in the continuous observability inequalities (that is, the issue of reducing the number of controls needed to obtain the corresponding (dual) exact controllability results); (ii) the issue of shifting (in particular, shifting down) the $H^3(\Omega) \times H^1(\Omega)$-level of topologies of the original estimate (32). (The case of *clamped* controls of interest in control theory requires $H^2(\Omega) \times L_2(\Omega)$-energy level in the estimates [12, 13, 14], [20].) For lack of space, these remaining issues will be dealt with in detail in a subsequent paper, along the lines of [15, 16], [23, 24, 25]. Here, we concentrate on the Carleman estimates as well as on *preliminary* continuous observability result, that follows readily from estimate (32). It is given by Theorem 4 below. It involves homogeneous *hinged* B.C. (by duality, hinged boundary controls). A more general treatment of the second part of the present project—beyond Carleman estimates—will be carried out in a successor paper, where a sharpening form of Theorem 4 will also be given.

Literature The emphasis of this paper is twofold and consists of: (i) treating plate equations of Euler-Bernoulli-type with variable coefficients in a bounded Euclidean domain (see Appendix B); and (ii) including, as far as possible in the analysis, energy level terms. In all cases, we require only (essentially) minimum smoothness on the coefficients, see the statement below (5).

The 'classical' literature, beginning with [12, 13, 14], [20] studied only the *canonical* case of an Euler-Bernoulli plate equation, where then $\mathcal{A} = \Delta$ (in the Euclidean domain) and $F(w) \equiv 0$. The energy method (multipliers) used in these classical efforts for $H^3(\Omega) \times H^1(\Omega)$-energy level estimate were: $h \cdot \nabla(\Delta w)$ and $(\Delta w)\text{div } h$, where $h(x)$ is a coercive vector field [12, 13, 14] (in particular, radial [20]). These multipliers are wholly inadequate, however, in treating the present class which has (i) variable coefficient principal part and (ii) energy level terms.

This far broader class was included, instead, in the general treatment for a general evolution equation in [23, 24, 25], which is based on general pseudo-differential multipliers (symbols) derived from pseudo-convex functions [9] for general evolution equations [23, 24, 25]. These techniques with pseudo-differential multipliers proposed in [23, 24, 25], which in prior literature [9] were applied to solutions with compact support (thereby not accounting for boundary traces which are instad critical for continuous observability/uniform stabilization inequalities) are unifying across several evolution equations. However, they require the assumption on the existence of a pseudo-convex function (suitably defined in terms of Poisson brackets [9]), a property that is not readily verified, except for the constant coefficient a_{ij} case.

In [16], the authors specialized the Carleman estimates, and subsequent continuous observability/uniform stabilization inequalities, for the Euler-Bernoulli plate model of Appendix B on a bounded Euclidean domain with general energy level $F(w)$ as in (B.5), but, however, with $\mathcal{A} = \Delta$ (the Euclidean Laplacian). This study used explicit computations in the energy method based on the two main multipliers

$$e^{\tau\phi(x,t)}\langle\nabla v, \nabla(\Delta w)\rangle_{\mathbb{R}^n} \text{ and } \operatorname{div}(e^{\tau\phi}\nabla v), \tag{7}$$

where $v(x) = \|x - x_0\|^2$ in the Euclidean norm, and ϕ is correspondingly defined by (9a). The first use of Riemann geometric methods for Euler-Bernoulli plates was given in [31] (following [30] for waves) under a slightly weaker assumption (the existence of a 'coercive' vector field H in the Riemann metric g given by (B.7)): however, that method—which is the Riemann counterpart of the classical literature on canonical cases [12, 13, 14], [20]—took $\mathcal{A}(w)$ as in (B.1) but required the assumption $F(w) \equiv 0$. In fact, [31] uses the multipliers $\langle H, \nabla_g(\Delta_g w)\rangle_g$ and $(\Delta_g w) \operatorname{div}_g H$, the Riemann version of [12, 13, 14], [20] in the Euclidean case.

The present paper extends the methods and results of both [16] and [31]. Technically, this paper is the Riemann counterpart of [16]. It uses, accordingly, the same multipliers (7) except in the Riemann metric: that is

$$e^{\tau\phi(x,t)}\langle\nabla_g v, \nabla_g(\Delta_g w)\rangle_g \text{ and } \operatorname{div}_g(e^{\tau\phi}\Delta_g w), \tag{8}$$

where now, more generally, v is the function of assumption (H.2) = (4), and ϕ is correspondingly defined by (9a). In particular, our Theorem 4 on continuous observability in the case of hinged B.C. extends [Y.2, Eqn. (2.2.38)] from the case $F(w) \equiv 0$ to the case where F satisfies (H.1) = (2), (H.5) = (29).

Conclusion In conclusion, we mention that advantages of the present approach include the following features: (i) it applies to E-B equations on manifolds; (ii) when specialized to the case of Appendix B with

variable coefficients on a Euclidean domain, it requires essentially minimal regularity of the coefficients: C^1 in space for the principal part;
L_∞ in time and space for energy level terms; (iii) the main assumption
required is in common to other evolution equations (second-order hyperbolic equations [18, 19]; Schrödinger equations [28], etc.) and permits to
construct many classes of non-trivial illustrative examples with variable
coefficients and, often, on an arbitrary (sufficiently smooth) Euclidean
domain; (iv) it allows arbitrarily small time $T > 0$ in the estimates; (v)
it is a fully self-contained energy method, which may be viewed as a far-
reaching generalization of the classical energy methods (multipliers) for
canonical plate equations with $\mathcal{A} = \Delta$ and $F(w) \equiv 0$ [12, 13, 14], [20];
(vi) it identifies explicitly the relevant observability/controllability portion of the boundary (though this need not be minimal); (vii) it combines
well with other techniques to resolve remaining problems (reduction of
the number of traces, shift of topology level).

2. Statement of main results

Consequence of main assumption (H.2) = (4) Let $v : \overline{\Omega} \to \mathbb{R}$ be
the (non-negative) strictly convex function with respect to the Riemann
metric g, provided by the main assumption (H.2) in (4). Define the
function $\phi : \Omega \times \mathbb{R} \to \mathbb{R}$, with Ω an open bounded set of M, by

$$\phi(x,t) = v(x) - c\left(t - \frac{T}{2}\right)^2, \tag{9a}$$

where $c = c_T > 0$ is a constant depending on T chosen as follows. Let
$T > 0$ be arbitrary but fixed, and let $\delta > 0$. Accordingly, let $c = c_T > 0$
be a constant sufficiently large, such that

$$4 \sup_{x \in \Omega} v(x) + 4\delta \leq cT^2, \tag{9b}$$

where the left-hand side is finite. Henceforth, with $T > 0$ and $\delta > 0$
chosen, let ϕ be defined by (9a) with $c = c_T > 0$ given as in (9b). Such
function $\phi(x,t)$ possesses then the following properties:

(i)
$$\phi(x,0) < -\delta \text{ and } \phi(x,T) < -\delta \text{ uniformly in } x \in \Omega, \tag{9c}$$

as it follows from (9b);

(ii) there exist t_0 and t_1 with $0 < t_0 < \frac{T}{2} < t_1 < T$, such that

$$\min_{x \in \Omega, t \in [t_0, t_1]} \phi(x,t) \geq -\frac{\delta}{2} \tag{9d}$$

since $\phi\left(x, \frac{T}{2}\right) = v(x) \geq 0$ for all $x \in \Omega$, as assumed.

The important property (9c) will be invoked in the proof of estimate (89) of Lemma 6 (see Eqn. (90)). The important property (9d) (in fact, only the weaker property: min $\phi(x,t) \geq \sigma > -\delta$ is actually needed) will be invoked in going from (11a) to (11b) in the statement of Theorem 1 (Carleman estimate, first version), but not before obtaining (11a).

Carleman estimates, first version, without B.C.

Theorem 1 Assume (H.1) = (2) on F, (3) on f, and (H.2) = (4). Let w be a solution of Eqn. (1) on $(0,T] \times \Omega$ [with no boundary conditions imposed] within the following class:

$$\begin{cases} \{w, w_t\} \in L_2(0,T; H^3(\Omega) \times H^1(\Omega)); \\ w_t, \ \Delta w, \ \dfrac{\partial \Delta w}{\partial n}, \ Dw_t, \ D(\Delta w) \in L_2(\Sigma). \end{cases} \tag{10}$$

Let $\phi(x,t)$ be the functon defined by (9a-b). Then, for all $\tau > 0$ sufficiently large, there exist constants $C > 0$, $C_T > 0$, $C_{\tau,T} > 0$ such that the following one-parameter family of estimates holds true:

$$B_\Sigma(w)$$

$$+ \frac{2}{\tau} \int_Q e^{\tau\phi} f^2 dQ + C_{\tau,T} \int_Q [|D^2 w|^2 + |Dw|^2 + w^2 + w_t^2] dQ$$

$$\geq \left(a - \frac{C}{\tau}\right) \int_Q e^{\tau\phi} [|Dw_t|^2 + |D(\Delta w)|^2] dQ \tag{11a}$$

$$- C_T e^{-\delta\tau} [P_w(T) + P_w(0)]$$

$$\tag{11b}$$

$$\geq \left(a - \frac{C}{\tau}\right) e^{-\frac{\tau\delta}{2}} \int_{t_0}^{t_1} E(t) dt - C_T e^{-\delta\tau} [E(T) + E(0)]$$

$$- C_T e^{-\delta\tau} \max_{[0,T]} \left\{ \int_\Omega [|D^2 w|^2 + |Dw|^2 + w^2 + w_t^2] d\Omega \right\},$$

where $a > 0$ is the constant in assumption (H.2) = (4), and where:
(i) the boundary terms $B_\Sigma(w)$ over $(0,T] \times \Gamma$ are given by

$$B_\Sigma(w) \equiv B_\Sigma^1(w) + B_\Sigma^2(w), \tag{12}$$

$$B_\Sigma^1(w) \equiv \int_\Sigma \left\{ [H(w_t) + w_t \ \mathrm{div} H] \frac{\partial w_t}{\partial n} + H(\Delta w) \frac{\partial \Delta w}{\partial n} \right\} d\Sigma$$

$$- \int_{\Sigma} \left\{ \left[w_t \Delta w_t + \frac{1}{2}(|Dw_t|^2 + |D(\Delta w)|^2) \right] \right.$$

$$\left. \langle H, n \rangle + \frac{1}{2} w_t^2 \langle D(\operatorname{div} H), n \rangle \right\} d\Sigma \tag{13}$$

$$B_\Sigma^2(w) \equiv \frac{1}{2} \int_\Sigma \left\{ \left[\Delta w \frac{\partial \Delta w}{\partial n} - w_t \frac{\partial w_t}{\partial n} \right] \operatorname{div} H \right.$$

$$\left. + \frac{1}{2} [w_t^2 - (\Delta w)^2] \langle D(\operatorname{div} H), n \rangle \right\} d\Sigma; \tag{14}$$

where H is a vector field constructed from the assumed strictly convex function v in (4) and the function ϕ in (9a), and defined by (recall (5))

$$H = e^{\tau \phi} Dv = e^{\tau \phi} D\phi = e^{\tau \phi} h, \quad h \equiv Dv = D\phi. \tag{15}$$

(ii) Moreover, $E(t)$ and $P_w(t)$ are defined by

$$E(t) \equiv \int_\Omega [|D(\Delta w)|^2 + |Dw_t|^2] d\Omega; \tag{16}$$

$$P_w(t) = \int_\Omega [|D(\Delta w)|^2 + |Dw_t|^2 \tag{17a}$$

$$+ |D^2 w|^2 + |Dw|^2 + w^2 + w_t^2] d\Omega$$

$$= E(t) + \int_\Omega [|D^2 w|^2 + |Dw|^2 + w^2 + w_t^2] d\Omega. \tag{17b}$$

(iii) Finally, $lot(w)$ is defined by

$$lot(w) = \mathcal{O} \left\{ \int_Q [|D^2 w|^2 + |Dw|^2 + w^2 + w_t^2] dQ \right\}, \tag{18}$$

so that $P_w(t) = E(t) + lot(w)$. $\qquad \Box$

The proof of Theorem 1 will be given in Sections 3.1, 3.2.

Carleman estimate, first version, subject to geometrical conditions on Γ_0 and B.C. on w Theorem 1 referred to regular solutions of Eqn. (1), with no Boundary Conditions (B.C.) imposed on them. Next, we impose B.C. To this end, we divide the boundary $\partial \Omega = \Gamma$ of $\Omega \subset M$ into two parts: an uncontrolled part Γ_0 and a controlled part Γ_1:

$$\partial \Omega = \Gamma = \Gamma_0 \cup \Gamma_1, \tag{19}$$

where, on Γ_0, we make the following geometrical assumption:

(H.3) the strictly convex function $v : \overline{\Omega} \to \mathbb{R}$ of class C^2 in assumption (H.2) = (4) satisfies further the following condition on $h = Dv \in \mathcal{X}(M)$:

$$\frac{\partial v}{\partial n} = \langle Dv, n \rangle = \langle h, n \rangle \leq 0, \quad \forall\, x \in \Gamma_0, \tag{20}$$

where n is the unit outward normal field along Γ.

Case 1 We then impose the following 'hinged' B.C. on Γ_0 on regular solutions of (1):

$$w|_{\Sigma_0} = \Delta w|_{\Sigma_0} \equiv 0 \quad \text{on } (0, T] \times \Gamma_0 \equiv \Sigma_0. \tag{21}$$

For this case (2.13), we then have:

Corollary 1 Assume (H.1) = (2) on F, (3) on f, (H.2) = (4), and (H.3) = (20) on Γ_0. Let w be a solution of Eqn. (1) on $(0, T] \times \Omega$, within the class (10), satisfying, moreover, the hinged B.C. (21) on Γ_0. Then, estimates (11a–11b) hold true, where now, recalling (12)–(14), we have

$$\begin{cases} B_{\Sigma_0}(w) & \equiv \; B_{\Sigma_0}^1(w) + B_{\Sigma_0}^2(w) = B_{\Sigma_0}^1(w) \\[2mm] & = \; \dfrac{1}{2} \displaystyle\int_{\Sigma_0} \left[\left(\dfrac{\partial w_t}{\partial n}\right)^2 + \left(\dfrac{\partial(\Delta w)}{\partial n}\right)^2 \right] \langle H, n \rangle d\Sigma_0 \leq 0 \quad (22) \\[4mm] B_{\Sigma_0}^2(w) & = \; 0, \hspace{6.5cm} (23) \end{cases}$$

and hence

$$\begin{aligned} B_\Sigma(w) \; &= \; B_{\Sigma_0}(w) + B_{\Sigma_1}(w) \leq B_{\Sigma_1}(w) \\[2mm] &\leq \; C_T \int_{\Sigma_1} \left[|Dw_t|^2 + w_t^2 + |D(\Delta w)|^2 \right. \hspace{2.5cm} (24) \\[2mm] & \hspace{2cm} \left. + |w_t \Delta w_t| + \left| \Delta w \, \frac{\partial \Delta w}{\partial n} \right| + (\Delta w)^2 \right] d\Sigma_1. \quad \square \end{aligned}$$

The proof of Corollary 1 will be given in Section 3.3.

Case 2 Finally, we impose the following 'clamped' B.C. on all of Γ on regular solutions of (1):

$$w|_\Sigma \equiv 0, \quad \frac{\partial w}{\partial n}\bigg|_\Sigma \equiv 0, \quad \text{on } (0, T] \times \Gamma \equiv \Sigma. \tag{25}$$

For this case we have

Corollary 2 Assume (H.1) = (2) on F, (3) on f, (H.2) = (4). Let w be a solution of Eqn. (1) on $(0, T] \times \Omega$, within the class (10), satisfying, moreover, the clamped B.C. (25) on Γ. Let $p = \operatorname{div} H$.

Then, estimates (11a-b) hold true, where now we have

$$B^1_\Sigma(w) = \int_\Sigma \left[H(\Delta w)\frac{\partial \Delta w}{\partial n} - \frac{1}{2}|D(\Delta w)|^2\langle H, n\rangle \right] d\Sigma, \qquad (26)$$

$$B^2_\Sigma(w) = \frac{1}{2}\int_\Sigma \left[\Delta w\frac{\partial \Delta w}{\partial n}p - \frac{1}{2}(\Delta w)^2\langle Dp, n\rangle \right] d\Sigma, \qquad (27)$$

hence, since $H(\Delta w) = \langle D(\Delta w), H\rangle$, via (A.2):

$$B_\Sigma(w) = B^1_\Sigma(w) + B^2_\Sigma(w) \le C\int_\Sigma \left[|D(\Delta w)|^2 + (\Delta w)^2 \right] d\Sigma. \qquad \square \quad (28)$$

The proof of Corollary 2 will be given in Section 3.3.

Carleman estimates, second version, on (1) without B.C. In order to formulate the Carleman estimates for regular solutions of (1) in a second, definitive version, we need to impose an additional assumption on the dynamics (1) [in addition to (H.1) on F, and (3) on f]:

(H.5) With reference to the energy $E(t)$ defined in (16), for regular solutions of (1), we assume that the following energy inequality holds true: for all $0 \le s \le t \le T$, there exists $C_T > 0$ such that

$$|E(t) - E(s)| \le C_T \left[\int_s^t E(\sigma)d\sigma + G(T) \right], \qquad (29)$$

where

$$G(T) = B_\Sigma(w) + BT_e(w) + \int_0^T \|f\|^2_{H^1(\Omega)}dt + lot(w). \qquad \square \quad (30)$$

In (30), $B_\Sigma(w)$ is defined by (12)–(14); $lot(w)$ is defined in (18), while $BT_e(w)$ includes those 'boundary terms' that are generated by the energy method (the subscript 'e' stands for energy), which is employed in seeking to establish inequality (29). This energy method consists in multiplying Eqn. (1) by Δw_t and integrating by parts. This procedure may or may not succeed in establishing inequality (29) depending on the structure of the lower-order differential operator $F(w)$ in (2), as explained in the next Remark 2.

Remark 2 Assumption (H.5) = (29) is certainly satisfied whenever the differential operator $F(w)$ satisfies the inequality

$$\int_Q |F(w)|^2dQ \le C_T\int_0^T\int_\Omega [|D^2w|^2 + |Dw|^2 + w^2 + w_t^2]dQ, \qquad (31)$$

that is, F contains only terms up to second space derivatives on w and zero space derivatives on w_t, rather than terms up to third space derivatives on w and first space derivatives on w_t as in (2). Indeed, the energy method of multiplying Eqn. (1) by Δw_t and integrating by parts succeeds in establishing inequality (29) not only under the more stringent condition (31) but also in some cases satisfying the more relaxed condition (2), where, however, F is suitably structured. For instance, energy level terms such as $\frac{\partial}{\partial x} w_t$, $\frac{\partial}{\partial y} w_t$ say on $\Omega = \{(x,y) : c \leq y \leq d, h_1(y) \leq x \leq h_2(y)\}$ are acceptable for F to satisfy (H.5) = (29). In fact, with $2\nabla w_t \cdot \nabla \left(\frac{\partial}{\partial x} w_t\right) = \frac{\partial}{\partial x} |\nabla w_t|^2$, we obtain by Green's first theorem

$$\int_\Omega \Delta w_t \frac{\partial w_t}{\partial x} d\Omega = \int_\Gamma \frac{\partial w_t}{\partial \nu} \frac{\partial w_t}{\partial x} d\Gamma - \frac{1}{2} \int_c^d \int_{h_1(y)}^{h_2(y)} \frac{\partial}{\partial x} |\nabla w_t|^2 dx\, dy$$

$$= \int_\Gamma \frac{\partial w_t}{\partial \nu} \frac{\partial w_t}{\partial x} d\Gamma - \frac{1}{2} \int_c^d \{|\nabla w_t|^2\}_{x=h_1(y)}^{x=h_2(y)} dy.$$

In the case of *clamped* B.C., the right-hand side of the above identity vanishes, while it produces the term BT_e in the case of *hinged* B.C. for solutions of the class (10). However, there are also counterexamples of operators F satisfying (2) for which inequality (29) does not hold true. [7].

Theorem 2 Assume (H.1) = (2) on F, $f \in L_2(0,T; H^1(\Omega))$, (H.2) = (4), and (H.5) = (29). Let w be a solution of Eqn. (1) on $(0,T] \times \Omega$ (with no boundary conditions imposed] within the class (2). Let $\phi(x,t)$ be the function defined by (2.1).

Then, for all $\tau > 0$ sufficiently large, there exist constants $k_{\phi,\tau} > 0$, $C_T > 0$, $c > 0$, such that the following one-parameter family of estimates holds true:

$$[1 + C_T(t_1 - t_0)]B_\Sigma(w) + C_T(t_1 - t_0)$$
$$\left\{BT_e(w) + \int_0^T \|f\|_{H^1(\Omega)}^2 dt + lot(w)\right\} + \frac{2}{\tau} \int_Q e^{\tau\phi} f^2 dQ$$
$$+ C_T e^{-\delta\tau} \max_{[0,T]}$$
$$\left\{\int_\Omega [|D^2 w|^2 + |Dw|^2 + w^2 + w_t^2] d\Omega\right\} \geq k_{\phi,\tau}[E(T) + E(0)].$$

(32)

where the boundary terms $B_\Sigma(w)$ and $BT_e(\Sigma)$ are defined by (26)–(28), and by (30), respectively. □

The proof of Theorem 2 is given in Section 4.

3.   Proof of Theorem 1: Carleman estimate (first version). Proof of Corollaries 1 and 2

3.1   Preliminaries

We collect here a few formulas to be invoked in the sequel [[26], Prop. 4.1, p. 138].

Green's first theorem If $f, q \in H^2(\Omega)$, then

$$\int_\Omega q\Delta f \, d\Omega = \int_\Gamma q \frac{\partial f}{\partial n} \, d\Gamma - \int_\Omega \langle Dq, Df \rangle d\Omega. \tag{33}$$

Divergence or Gauss theorem [[26], Thm. 2.1, p. 128]

$$\int_\Omega \text{div } X \, d\Omega = \int_\Gamma \langle X, n \rangle d\Gamma. \tag{34}$$

3.2   Energy methods in the Riemann metric

We will complete the proof of Theorem 1 through several propositions.

Step 1. Proposition 1 (i) Let w be a solution of Eqn. (1) within the class (10). Then, the following identity holds true, where $\Sigma \equiv [0, T] \times \Gamma$; $Q \equiv [0, T] \times \Omega$:

$$\begin{aligned}
B_\Sigma^1(w) &\equiv \int_\Sigma \left\{ [H(w_t) + w_t \text{ div} H] \frac{\partial w_t}{\partial n} + H(\Delta w) \frac{\partial \Delta w}{\partial n} \right\} d\Sigma \\
&\quad - \int_\Sigma \left\{ \left[w_t \Delta w_t + \frac{1}{2}(|Dw_t|^2 + |D(\Delta w)|^2 \right] \langle H, n \rangle \right. \\
&\quad \left. + \frac{1}{2} w_t^2 \langle D(\text{div} H), n \rangle \right\} d\Sigma \\
&= \int_Q [DH(Dw_t, Dw_t) + DH(D(\Delta w), D(\Delta w))] dQ \quad (35) \\
&\quad + \frac{1}{2} \int_Q [|Dw_t|^2 - |D(\Delta w)|^2] \text{div} H \, dQ \\
&\quad + \tau \int_Q w_t \phi_t H(\Delta w) dQ - \frac{1}{2} \int_Q w_t^2 \Delta(\text{div} H) dQ \\
&\quad - \left[\int_\Omega w_t H(\Delta w) d\Omega \right]_0^T - \int_Q [F(w) + f] H(\Delta w) dQ.
\end{aligned}$$

In (35), we have $H = e^{\tau\phi} Dv = e^{\tau\phi} D\phi = e^{\tau\phi} h$, see (15) [(5) and (9a)] while $DH(\cdot, \cdot)$ is the covariant differential of H defined by the bilinear form (A.3).

(ii) Moreover, if X and Y are vector fields on M, we have via (A.3) that DH is given by

$$DH(Y,X) \quad = \quad \tau e^{\tau\phi}X(v)Y(v) + e^{\tau\phi}D^2v(Y,X) \qquad (36\text{a})$$

$$(\text{by (A.2)}) \quad = \quad \tau e^{\tau\phi}\langle Dv, X\rangle\langle Dv, Y\rangle + e^{\tau\phi}D^2v(Y,X), \qquad (36\text{b})$$

see also [18, 19], (4.1.4), (4.1.5), p. 34]; in particular, see (A.2)

$$DH(Dw_t, Dw_t) = \tau e^{\tau\phi}[\langle Dv, Dw_t\rangle]^2 + e^{\tau\phi}D^2v(Dw_t, Dw_t) \qquad (37)$$

$$DH(D(\Delta w), D(\Delta w)) = \tau e^{\tau\phi}[\langle Dv, D(\Delta w)\rangle]^2 + e^{\tau\phi}D^2v(D(\Delta w), D(\Delta w)). \qquad (38)$$

In (3.2.2)–(38), D^2v is the Hessian (defined by (A.6)) of the assumed function v. [Of the two forms in (A.2), we have chosen the one that explicitly defines the number of derivatives involved.]

Proof of Proposition 1 (i) We multiply both sides of Eqn. (1) by the following (main) multiplier (see (15)):

$$H(\Delta w) = e^{\tau\phi}Dv(\Delta w) = e^{\tau\phi}h(\Delta w) = e^{\tau\phi}\langle Dv, D(\Delta w)\rangle, \qquad (39)$$

counterpart of the multiplier in [15] in the Euclidean case in the notation of (A.2), (15), and integrate over $Q = [0,T] \times \Omega$ by parts. See also [12, 13, 14], [20], in the Euclidean case with $F \equiv 0$.

Term w_{tt} in (1) With reference to the first term w_{tt} in (1), we shall show below that the following identity holds true:

$$\int_Q w_{tt}H(\Delta w)dQ = [(w_t, H(\Delta w))]_0^T + \int_\Sigma [H(w_t) + w_t \text{ div}H]\frac{\partial w_t}{\partial n}\,d\Sigma$$

$$- \int_\Sigma \left\{\left[w_t\Delta w_t + \frac{1}{2}|Dw_t|^2\right]\langle H, n\rangle + \frac{1}{2}w_t^2\langle D(\text{div}H), n\rangle\right\}d\Sigma$$

$$- \int_Q DH(Dw_t, Dw_t)dQ - \tau\int_Q w_t\phi_t H(\Delta w)dQ$$

$$+ \frac{1}{2}\int_Q [w_t^2\Delta(\text{div}H) - |Dw_t|^2 \text{ div}H]dQ. \qquad (40)$$

Proof of (40) From (39), we find, preliminarily via (15),

$$\frac{d}{dt}H(\Delta w) = \tau e^{\tau\phi}\phi_t Dv(\Delta w) + e^{\tau\phi}Dv(\Delta w_t) = \tau\phi_t H(\Delta w) + H(\Delta w_t). \qquad (41)$$

Thus, integrating by parts in t, we obtain by (41),

$$\int_\Omega \int_0^T w_{tt}H(\Delta w)dQ =$$

$$[(w_t, H(\Delta w))]_0^T - \tau \int_Q \phi_t w_t H(\Delta w)dQ - \int_Q w_t H(\Delta w_t)dQ. \quad (42)$$

Next, using the following direct identities ([[26], Eqn. (2.19), p. 128]) for the derivative (2.1),

$$w_t H(\Delta w_t) = H(w_t \Delta w_t) - \Delta w_t H(w_t)$$

$$= \operatorname{div}(w_t \Delta w_t H) - w_t \Delta w_t \operatorname{div}H - \Delta w_t H(w_t) \quad (43)$$

in the last term of (42), we obtain

$$\int_\Omega \int_0^T w_{tt}H(\Delta w)dQ = [(w_t, H(\Delta w))]_0^T$$

$$- \tau \int_Q \phi_t w_t H(\Delta w)dQ + \int_Q w_t \Delta w_t \operatorname{div}H \, dQ$$

$$+ \int_Q \Delta w_t H(w_t)dQ - \int_\Sigma w_t \Delta w_t \langle H, n \rangle d\Sigma, \quad (44)$$

after using the divergence (or Gauss) theorem (34) or the term $\operatorname{div}(w_t \Delta w_t H)$. Regarding the third term on the right side of (44), we set $p = \operatorname{div} H$, and compute by using Green's first identity (33) twice, the first time with $D(pw_t) = pDw_t + w_t Dp$,

$$\int_\Omega pw_t \Delta w_t d\Omega = \int_\Gamma pw_t \frac{\partial w_t}{\partial n} d\Gamma - \int_\Omega p \langle Dw_t, Dw_t \rangle d\Omega$$

$$- \frac{1}{2} \int_\Omega \langle Dp, D(w_t^2) \rangle d\Omega \quad (45)$$

$$= \int_\Gamma \left[pw_t \frac{\partial w_t}{\partial n} - \frac{1}{2} w_t^2 \langle Dp, n \rangle \right] d\Gamma$$

$$- \int_\Omega \left[p|Dw_t|^2 - \frac{1}{2} w_t^2 \Delta p \right] d\Omega, \quad (46)$$

and the second time on the term $\int_\Omega w_t^2 \Delta p d\Omega$ in (46).

Regarding the fourth term on the right side of (44), we first invoke again Green's first identity (33), next we invoke identity (A.4) with $f = w_t$, and finally use the Gauss theorem (34) on the term $\operatorname{div}(|Dw_t|^2 H)$.

We obtain

$$\int_\Omega \Delta w_t H(w_t) d\Omega = \int_\Gamma H(w_t) \frac{\partial w_t}{\partial n} d\Gamma - \int_\Omega \langle Dw_t, D(H(w_t)) \rangle d\Omega \quad (47)$$

$$\text{(by (A.4))} = \int_\Gamma H(w_t) \frac{\partial w_t}{\partial n} d\Gamma$$

$$- \int_\Omega \left\{ DH(Dw_t, Dw_t) - \frac{1}{2} |Dw_t|^2 \text{ div} H \right\} d\Omega$$

$$- \frac{1}{2} \int_\Gamma |Dw_t|^2 \langle H, n \rangle d\Gamma. \quad (48)$$

Finally, we substitute (46) with $p = \text{div } H$, and (48) for the third and fourth term on the right side of (44), and readily obtain (40), as desired.

Term $\Delta^2 w$ in (1)   With reference to the second term $\Delta^2 w$ in (1), we shall show below that the following identity holds true:

$$\int_Q \Delta^2 w H(\Delta w) dQ = \int_\Sigma \left[H(\Delta w) \frac{\partial \Delta w}{\partial n} - \frac{1}{2} |D(\Delta w)|^2 \langle H, n \rangle \right] d\Sigma$$

$$- \int_Q DH(D(\Delta w), D(\Delta w)) dQ + \frac{1}{2} \int_Q |D(\Delta w)|^2 \text{div} H \, dQ. \quad (49)$$

This identity was already observed in [[31], Eqn. (2.2.23)]. We provide here a proof for completeness.

Proof of (49)   By Green's first identity (33), followed by use of identity (A.4) with $f = \Delta w$ once more, we compute

$$\int_Q \Delta^2 w H(\Delta w) d\Omega \, dt =$$

$$\int_\Sigma H(\Delta w) \frac{\partial \Delta w}{\partial n} d\Sigma - \int_Q \langle D(\Delta w), D(H(\Delta w)) \rangle dQ$$

$$\text{(by (A.4))} = \int_\Sigma H(\Delta w) \frac{\partial \Delta w}{\partial n} d\Sigma - \int_Q DH(D(\Delta w), D(\Delta w)) dQ$$

$$- \frac{1}{2} \int_Q \text{div}(|D(\Delta w)|^2 H) dQ + \frac{1}{2} \int_Q |D(\Delta w)|^2 \text{div} H \, dQ. \quad (50)$$

Next, using the divergence (Gauss) theorem (34) on the term $\text{div}(D(\Delta w)|^2 H)$ carries (50) into (49), as desired.

Finally, we return to Eqn. (1), apply the multiplier $H(\Delta w)$ in (39) throughout and obtain

$$\int_Q w_{tt} H(\Delta w) dQ + \int_Q \Delta^2 w\, H(\Delta w) dQ + \int_Q [F(w) + f] H(\Delta w) dQ = 0.$$
(51)

We now substitute identities (40) and (49) for the first and second term in (51), and obtain identity (35), as desired. Part (i) of Proposition 1 is proved.

(ii) Identity (36) was already verified in [19], Eqn. (4.1.5), p. 34]. We re-prove it here for completeness. From the definition of $H = e^{\tau\phi} Dv$ in (15), we obtain preliminarily, with X a vector field on M:

$$D_X H = D_X(e^{\tau\phi} Dv) = e^{\tau\phi} D_X(Dv) + X(e^{\tau\phi}) Dv.$$
(52)

Thus, the definition (A.3) of DH yields via (52)

$$
\begin{aligned}
DH(Y, X) &= \langle D_X H, Y \rangle = X(e^{\tau\phi})\langle Dv, Y \rangle + e^{\tau\phi}\langle D_X(Dv), Y \rangle \\
&= X(e^{\tau\phi})\langle Dv, Y \rangle + e^{\tau\phi} D^2 v(Y, X),
\end{aligned}
$$
(53)

by recalling the definition of $D^2 v(\cdot, \cdot)$ in (A.6) in the last step. Finally, (53) yields (36) since, plainly, via (A.2):

$$X(e^{\tau\phi}) = \tau e^{\tau\phi} X(\phi) = \tau e^{\tau\phi} X(v) = \tau e^{\tau\phi}\langle Dv, X \rangle.$$
(54)

Proposition 1 is fully proved.   □

Step 2 The following lemma will be invoked below for a suitable choice of the function $m(x, t)$:

Lemma 1 Let w be a solution of (1) within the class (10). Let m be a smooth function (C^2 in space, C^1 in time) defined on Q. Then the following identity ("kinetic energy minus potential energy") holds true:

$$\int_Q [|Dw_t|^2 - |D(\Delta w)|^2]\, m\, dQ = \frac{1}{2}\int_Q [w_t^2 - (\Delta w)^2]\Delta m\, dQ$$

$$+ \int_\Sigma \left\{ \left[w_t \frac{\partial w_t}{\partial n} - \Delta w \frac{\partial \Delta w}{\partial n} \right] m + \frac{1}{2}[(\Delta w)^2 - w_t^2]\langle Dm, n \rangle \right\} d\Sigma$$

$$+ \int_Q m_t w_t \Delta w\, dQ - \left[\int_\Omega m\, w_t\, \Delta w\, d\Omega \right]_0^T$$

$$- \int_Q [F(w) + f] m\, \Delta w\, dQ.$$
(55)

[Note: The boundary term on $\int_\Sigma$ becomes $(-B_\Sigma^2(w))$ given by (14), if $m = \frac{1}{2} \operatorname{div} H$.] $\square$

Proof We shall use the multiplier $m\Delta w$. Indeed, we shall show below that

$$
\begin{aligned}
- \int_Q \Delta^2 w (m\Delta w) dQ &= \int_Q \left[|D(\Delta w)|^2 m - \frac{1}{2}(\Delta w)^2 \Delta m \right] dQ \\
&+ \int_\Sigma \left[\frac{1}{2}(\Delta w)^2 \frac{\partial m}{\partial n} - m\Delta w \frac{\partial \Delta w}{\partial n} \right] d\Sigma.
\end{aligned} \tag{56}
$$

In fact, by Green's first identity (33), we first obtain

$$
\begin{aligned}
- \int_Q (\Delta w)^2 (m\Delta w) d\Omega\, dt &= \int_Q \langle D(\Delta w), D(m\Delta w) \rangle dQ \\
&\quad - \int_\Sigma m\Delta w \frac{\partial \Delta w}{\partial n} d\Sigma \\
&= \int_Q [|D(\Delta w)|^2 m + \frac{1}{2}\langle D((\Delta w)^2), Dm \rangle] dQ \\
&\quad - \int_\Sigma m\Delta w \frac{\partial \Delta w}{\partial n} d\Sigma,
\end{aligned} \tag{57}
$$

using also, in the last step, both $D(m\Delta w) = mD(\Delta w) + \Delta w Dm$, as well as the term $\langle D((\Delta w)^2), Dm \rangle = 2\langle \Delta w D(\Delta w), Dm \rangle$. Next, another application of Green's first identity (33) gives

$$
\frac{1}{2}\int_Q (\Delta w)^2 \Delta m\, dQ = \frac{1}{2}\int_\Sigma (\Delta w)^2 \frac{\partial m}{\partial n} d\Sigma - \frac{1}{2}\int_Q \langle D((\Delta w)^2), Dm \rangle dQ. \tag{58}
$$

Substituting the last term on the right of (58) for the second term on the right of (57) yields (56), as claimed.

Similarly, we shall show that

$$
\begin{aligned}
\int_Q m w_t \Delta w_t\, dQ &= \int_Q \left[\frac{1}{2} w_t^2 \Delta m - |Dw_t|^2 m \right] dQ \\
&+ \int_\Sigma \left[m w_t \frac{\partial w_t}{\partial n} - \frac{1}{2} w_t^2 \frac{\partial m}{\partial n} \right] d\Sigma.
\end{aligned} \tag{59}
$$

Indeed, a first application of Green's first identity (33) yields likewise

$$
\begin{aligned}
\int_Q m w_t \Delta w_t\, dQ &= \int_\Sigma m w_t \frac{\partial w_t}{\partial n} d\Sigma - \int_Q m |Dw_t|^2 dQ \\
&\quad - \frac{1}{2}\int_Q \langle D(w_t^2), Dm \rangle dQ,
\end{aligned} \tag{60}
$$

using both $D(mw_t) = mDw_t + w_t Dm$ and $\langle D(w_t^2), Dm \rangle = 2\langle w_t Dw_t, Dm \rangle$.
Next, a second application of Green's first identity (33) gives

$$\frac{1}{2} \int_Q w_t^2 \Delta m \, dQ = \frac{1}{2} \int_\Sigma w_t^2 \frac{\partial m}{\partial n} \, d\Sigma - \frac{1}{2} \int_Q \langle D(w_t^2), Dm \rangle dQ. \qquad (61)$$

Subisituting the last term on the right of (61) for the last term on the right of (60) yields (59), as claimed.

Finally, we return to Eqn. (1), apply the multiplier $m\Delta w$ throughout and obtain first

$$\int_Q w_{tt} m\Delta w \, dQ + \int_Q \Delta^2 w (m\Delta w) dQ + \int_Q [F(w) + f] m\Delta w \, dQ = 0, \qquad (62)$$

and then, upon integrating by parts in t the first term in (62):

$$\left[\int_\Omega w_t m\Delta w \, d\Omega \right]_0^T - \int_Q w_t m_t \Delta w \, dQ - \int_Q m w_t \Delta w_t \, dQ$$

$$+ \int_Q \Delta^2 w (m\Delta w) dQ + \int_Q [F(w) + f] m\Delta w \, dQ = 0. \qquad (63)$$

We now substitute identities (59) and (56) for the third and fourth term in (63), and readily obtain identity (55), as desired. $\qquad \square$

Proposition 2 (Final identity) Let w be a solution of Eqn. (1) within the class (10). Then, the following identity holds true:

$$B_\Sigma^1(w) + B_\Sigma^2(w) = \int_Q [DH(Dw_t, Dw_t) + DH(D(\Delta w), D(\Delta w))] dQ$$

$$- \int_Q [F(w) + f] H(\Delta w) dQ - \frac{1}{2} \int_Q [F(w) + f] \Delta w (\text{div} H) dQ$$

$$+ \tau \int_Q w_t \phi_t H(\Delta w) dQ + \frac{1}{4} \int_Q [w_t^2 - (\Delta w)^2] \Delta(\text{div} H) dQ$$

$$+ \frac{1}{2} \int_Q (\text{div} H)_t \, w_t \Delta w \, dQ - [(w_t, H(\Delta w))]_0^T$$

$$- \frac{1}{2} [(w_t \, \text{div} H, \Delta w)]_0^T, \qquad (64)$$

where $B_\Sigma^1(w)$ and $B_\Sigma^2(w)$ are the boundary terms defined in (13), (14).

Proof We substitute identity (55) of Lemma 1 with the choice $m = \frac{1}{2} \text{div} H$ into the right-hand side of identity (35) of Proposition 1. We notice that with $m = \frac{1}{2}$ div H, the boundary terms on the right-hand side of (55) sum up to, precisely, $-B_\Sigma^2(w)$, as noted just below (55). □

Step 4 Beginning with this step, we concentrate our 3analysis on the right-hand side (RHS) of the fundamental identity (64), where we have listed the terms in an appropriate order, as it will be clear from the next lemma.

Lemma 2 Let w be a solution of Eqn. (1) within the class (10). Assume (H.2) = (4). Then, with reference to the (driving) term on the right-hand side of identity (64), we have

$$DH(Dw_t, Dw_t) + DH(D(\Delta w), D(\Delta w))$$

$$\equiv e^{\tau\phi}D^2v(Dw_t, Dw_t) + e^{\tau\phi}D^2v(D(\Delta w), D(\Delta w))$$

$$+\tau\, e^{\tau\phi}\left[|\langle Dv, Dw_t\rangle|^2 + |\langle Dv, D(\Delta w)\rangle|^2\right] \tag{65}$$

$$\geq a\, e^{\tau\phi}[|Dw_t|^2 + |D(\Delta w)|^2] + \tau\, e^{\tau\phi}\left[|\langle Dv, Dw_t\rangle|^2 + |\langle Dv, D(\Delta w)\rangle|^2\right], \tag{66}$$

where $a > 0$ is the constant in assumption (H.2) = (4).

Proof Identity (65) was already established in (37), (38). The passage from (65) to (66) is *precisely* the spot where the main assumption (H.2) = (4) on the existence of a strictly convex function v with coercive vector field Dv is used.   □

Lemma 3 Let w be a solution of Eqn. (1) within the class (10). Assume (H.2) = (4). Then, the following estimate holds true:

$$B_\Sigma^1(w) + B_\Sigma^2(w) \geq a \int_Q e^{\tau\phi}[|Dw_t|^2 + |D(\Delta w)|^2]dQ$$

$$+\tau \int_Q e^{\tau\phi}[|\langle Dv, Dw_t\rangle|^2 + |\langle Dv, D(\Delta w)\rangle|^2]dQ$$

$$- \int_Q e^{\tau\phi}[F(w) + f]\langle Dv, D(\Delta w)\rangle dQ - \frac{1}{2}\int_Q [F(w) + f]\Delta w(\mathrm{div}H)dQ$$

$$+\tau \int_Q e^{\tau\phi}\phi_t w_t \langle Dv, D(\Delta w)\rangle dQ + \frac{1}{4}\int_Q [w_t^2 - (\Delta w)^2]\Delta(\mathrm{div}H)dQ$$

$$+\frac{1}{2}\int_Q (\mathrm{div}H)_t\, w_t \Delta w\, dQ - [(w_t, H(\Delta w))]_0^T - \frac{1}{2}[(w_t\, \mathrm{div}H, \Delta w)]_0^T. \tag{67}$$

Proof We start from identity (64): we then use inequality (66) of Lemma 2 for the first integral term of (64); and, moreover, definition (39) for $H(\Delta w)$ for the second and fourth integral term of (64). This way we obtain (67). □

Step 6 We now invoke assumption (H.1) and examine its consequences.

Lemma 4 (i) There exists a constant $C > 0$ such that

$$\int_\Omega |D^3 u|^2 dx \le C \int_\Omega [|D(\Delta u)|^2 + |D^2 u|^2 + |Du|^2 + u^2] dx, \quad \forall u \in H^3(\Omega);$$
(68)

(ii) these are constants $C > 0$, $C_{\tau,T} > 0$ such that

$$\int_\Omega |D^3 w|^2 e^{\tau\phi} dx \le C \int_\Omega |D(\Delta w)|^2 e^{\tau\phi} dx + C_{\tau,T} \int_\Omega [|D^2 w|^2 + |Dw|^2 + w^2] dx.$$
(69)

(iii) Assume (H.1) = (2). Then, as to the second and third terms on the RHS of (67), for any $\epsilon > 0$, we have the following inequality for any smooth solution of (1)

$$\tau \int_Q e^{\tau\phi}[|\langle Dv, Dw_t\rangle|^2 + |\langle Dv, D(\Delta w)\rangle|^2] dQ$$

$$- \int_Q e^{\tau\phi}[F(w) + f]\langle Dv, D(\Delta w)\rangle dQ \ge \tau \int_Q e^{\tau\phi}|\langle Dv, Dw_t\rangle|^2 dQ$$

$$+ \left[\tau - \frac{1}{2\epsilon}\right] \int_Q e^{\tau\phi}|\langle Dv, D(\Delta w)\rangle|^2 dQ$$

$$-\epsilon C_T \int_Q e^{\tau\phi}[|Dw_t|^2 + |D(\Delta w)|^2] dQ$$

$$-\epsilon C_{\tau,T} \int_Q [|D^2 w|^2 + |Dw|^2 + w^2 + w_t^2] dQ - \epsilon \int_Q e^{\tau\phi} f^2 dQ. \quad (70)$$

Proof (i) Inequality (68) is a consequence of ellipticity.
(ii) A direct computation gives

$$D^3\left(e^{\tau\frac{\phi}{2}} w\right) = e^{\tau\frac{\phi}{2}} D^3 w + \psi_1(w); \quad (71)$$

$$D\left(\Delta\left(e^{\tau\frac{\phi}{2}} w\right)\right) = e^{\tau\frac{\phi}{2}} D(\Delta w) + \psi_2(w), \quad (72)$$

where $\psi_1(w)$ and $\psi_2(w)$ are tensors that satisfy the following inequality

$$|\psi_1(w)|^2 + |\psi_2(w)|^2 \leq C_{\tau,T}[|D^2w|^2 + |Dw|^2 + w^2] \quad \text{for } x \in \Omega. \quad (73)$$

Next, we substitute relations (71) and (72) into inequality (68) with $u = e^{\tau\frac{\phi}{2}}w$ and obtain

$$\int_\Omega |D^3w|^2 e^{\tau\phi}dx \leq C\int_\Omega |D(\Delta w)|^2 e^{\tau\phi}dx + C_{\tau,T}\int_\Omega [|D^2w|^2 + |Dw|^2 + w^2]dx,$$
$$(74)$$

and (69) is established.

(iii) For any $\epsilon > 0$, we estimate the third integral term on the RHS of (67),

$$-\int_\Omega e^{\tau\frac{\phi}{2}}[F(w) + f]e^{\tau\frac{\phi}{2}}\langle Dv, D(\Delta w)\rangle d\Omega$$

$$\geq -\frac{\epsilon}{2}\int_\Omega e^{\tau\phi}[F(w) + f]^2 d\Omega - \frac{1}{2\epsilon}\int_\Omega e^{\tau\phi}|\langle Dv, D(\Delta w)\rangle|^2 d\Omega, \quad (75)$$

where, by assumption (H.1) = (2), we have

$$-\int_\Omega e^{\tau\phi}[F(w) + f]^2 d\Omega \geq -2\int_\Omega e^{\tau\phi}[|F(w)|^2 + f^2]d\Omega$$

$$\geq -2C_T\int_\Omega e^{\tau\phi}\{|D^3w|^2 + |Dw_t|^2 + |D^2w|^2 + |Dw|^2$$

$$+ w^2 + w_t^2\}d\Omega - 2\int_\Omega e^{\tau\phi}f^2 d\Omega. \quad (76)$$

Hence, by using (69) in (76), we obtain

$$-\int_\Omega e^{\tau\phi}[F(w) + f]^2 d\Omega \geq -2C_T\int_\Omega e^{\tau\phi}[|D(\Delta w)|^2 + |Dw_t|^2]d\Omega$$

$$-2C_{\tau,T}\int_\Omega [|D^2w|^2 + |Dw|^2 + w^2 + w_t^2]d\Omega - 2\int_\Omega e^{\tau\phi}f^2 d\Omega. \quad (77)$$

Inserting (77) into the right side of (75) yields

$$-\int_\Omega e^{\tau\phi}[F(w) + f]\langle Dv, D(\Delta w)\rangle d\Omega$$

$$\geq -\epsilon C_T\int_\Omega e^{\tau\phi}[|D(\Delta w)|^2 + |Dw_t|^2]d\Omega$$

$$-\epsilon C_{\tau,T}\int_\Omega [|D^2w|^2 + |Dw|^2 + w^2 + w_t^2]d\Omega$$

$$- \epsilon \int_\Omega e^{\tau\phi} f^2 d\Omega - \frac{1}{2\epsilon} \int_\Omega e^{\tau\phi} |\langle Dv, D(\Delta w)\rangle|^2 d\Omega. \qquad (78)$$

Then inequality (78) yields readily inequality (70). Lemma 3.2.6 is proved. □

Step 6. Lemma 5 Assume hypothesis (H.1) = (2). Let w be a smooth solution of Eqn. (1). Then, the following estimate holds true:

(i) For any $\epsilon_1 > 0$, and $H(\Delta w) = e^{\tau\phi}\langle Dv, D(Aw)\rangle$ by (39), we have

$$\int_Q w_t \phi_t H(\Delta w) dQ \geq -\epsilon_1 \int_Q e^{\tau\phi} |D(\Delta w)|^2 dQ - C_{\epsilon_1, T, \phi} lot(w), \qquad (79)$$

where $lot(w)$ is defined in (18).

(ii) For any $\epsilon_2 > 0$, and recalling again (18)

$$\int_Q [F(w) + f]\Delta w \, \text{div} H \, dQ \geq -\epsilon_2 \int_Q e^{\tau\phi} [|Dw_t|^2$$

$$+ |D(\Delta w)|^2] dQ - \epsilon_2 \int_Q e^{\tau\phi} f^2 dQ - C_{\epsilon_2, T} lot(w). \qquad (80)$$

Proof (i) Since $H(\Delta w) = e^{\tau\phi}\langle Dv, D(\Delta w)\rangle$ by (39), then, given $\epsilon_1 > 0$ we estimate

$$\int_Q w_t \phi_t H(\Delta w) dQ = \int_Q \left[w_t \phi_t e^{\tau\frac{\phi}{2}} \right] \left[e^{\tau\frac{\phi}{2}} \langle Dv, D(\Delta w)\rangle \right] dQ$$

$$\geq - \int_Q \left[|w_t| \, |\phi_t| e^{\tau\frac{\phi}{2}} \right] |Dv| e^{\tau\frac{\phi}{2}} |D(\Delta w)| dQ \quad (81)$$

$$\geq - \epsilon_1 \int_Q e^{\tau\phi} |D(\Delta w)|^2 dQ - C_{\epsilon_1, \phi, T} \, lot(w), (82)$$

recalling $lot(w)$ in (18), and (59) is established.

(ii) Recalling $H = e^{\tau\phi} Dv$, $Dv = D\phi$ from (15), we obtain first

$$p \equiv \text{div} H = \text{div}\left(e^{\tau\phi} Dv \right) = e^{\tau\phi} \text{div} \, Dv + \left\langle D\left(e^{\tau\phi} \right), Dv \right\rangle$$

$$= e^{\tau\phi} \Delta v + \tau e^{\tau\phi} |Dv|^2. \qquad (83)$$

Then, with $\epsilon_1 > 0$ given, we use (83) and proceed as in the proof of Lemma 4, via (2), (18), (69),

$$\int_Q F(w)\Delta w \, \text{div} H \, dQ = \int_Q \left[e^{\tau\frac{\phi}{2}} F(w) \right] e^{\tau\frac{\phi}{2}} \Delta w [\Delta v + \tau |Dv|^2] dQ \quad (84)$$

(by (2) and (18)) $\geq -\epsilon_1 \int_Q e^{\tau\phi}[|D^3 w|^2 + |Dw_t|^2]dQ - C_{\epsilon_1,T,\tau}lot(w)$

(by (69)) $\geq -\epsilon_1 C \int_Q e^{\tau\phi}[|D(\Delta w)|^2 + |Dw_t|^2]dQ$

$$-C_{\epsilon_1,T,\tau}lot(w). \tag{85}$$

Thus, (85) plus a similar argument for $\int_Q f\Delta w \ \text{div}H \ dQ$ yield (80).
$\square$

Step 8. Lemma 6 Assume (H.1) = (2) and (H.2) = (4). Let w be a solution of Eqn. (1) within the class (10). Then the following estimate holds true.

$$B_\Sigma^1(w) + B_\Sigma^2(w)$$

$$+ \tau C_{\epsilon,T} \int_Q [|D^2 w|^2 + |Dw|^2 + w^2 + w_t^2]dQ + (\epsilon + \epsilon_2)\int_Q e^{\tau\phi}f^2 dQ$$

$$\geq [a - \epsilon C_T - \epsilon_2 - \tau\epsilon_1]\int_Q e^{\tau\phi}[|D(\Delta w)|^2 + |Dw_t|^2]dQ$$

$$+ \tau \int_Q e^{\tau\phi}|\langle Dv, Dw_t\rangle|^2 dQ$$

$$+ \left[\tau - \frac{1}{2\epsilon}\right]\int_Q e^{\tau\phi}|\langle Dv, D(\Delta w)\rangle|^2 dQ - \beta_{0,T}, \tag{86}$$

where

$$\beta_{0,T} = [(w_t, H(\Delta w))]_0^T + \frac{1}{2}[(w_t \ \text{div}H, \Delta w)]_0^T. \tag{87}$$

(ii) Moreover, the time boundary term $\beta_{0,T}$ satisfies the estimate

$$|\beta_{0,T}| = |[((\text{div}H)w_t, \Delta w)]_0^T| + |[(w_t, H(\Delta w))]_0^T|$$

$$\leq C_T e^{-\tau\delta}[P_w(0) + P_w(T)], \quad \tau \geq 1, \tag{88}$$

where $\delta > 0$ is the constant in property (9c) and $P_w(\cdot)$ is defined in (17).

Proof (i) We return to the estimate (67) and invoke here estimate (70) for the second and third integral term; estimates (80) and (79) for the fourth and fifth integral terms, while the sixth and seventh integral terms involve only lower-order terms. This way estimate (86) is obtained.

(ii) To prove (88) we return to (83) for $p = \text{div}H$: thus, recalling property (9c) for $\phi(x, 0)$ and $\phi(x, T)$, we obtain by (83),

$$|p(T)| + |p(0)| \leq C_v \tau e^{-\delta\tau}. \tag{89}$$

Thus, by using (89) and recalling (43), we estimate

$$|[((\text{div}H)w_t, \Delta w)]_0^T| = |[(pw_t, \Delta w)]_0^T| \leq |p(T)| \, |w_t(T)| \, |\Delta w(T)|$$

$$+ \ |p(0)| \, |w_t(0)| \, |\Delta w(0)| \tag{90}$$

$$(\text{by (89)}) \quad \leq \quad C_v \tau e^{-\delta\tau} [|w_t(T)|^2 + |\Delta w(T)|^2$$

$$+ \ |w_t(0)|^2 + |\Delta w(0)|^2] \tag{91}$$

$$(\text{by (17)}) \quad \leq \quad C_v \tau e^{-\delta\tau} [P_w(0) + P_w(T)]. \tag{92}$$

Similarly,

$$|[(w_t, H(\Delta w))]_0^T| = |[(w_t, e^{\tau\phi}\langle Dv, D(\Delta w)\rangle)]_0^T| \leq C_v e^{-\delta\tau}[P_w(0) + P_w(T)]. \tag{93}$$

Then, combining (92) and (93) in definition (87) yields (88).

Step 9. Remark 3 So far, the parameter $\tau > 1$ has been arbitrary. The next result, however, which leads to the desired Theorem 1, Eqn. (11a), shows the key virtue of the free parameter τ entering the present multiplier (39) in dealing with the final estimate (86) of Lemma 6. More precisely, in the second integral term: $\int_Q e^{\tau\phi} F(w)\langle Dv, D(\Delta w)\rangle dQ$ on the left-hand side of (70), both factors $F(w)$ and $\langle Dv, D(\Delta w)\rangle$ are "energy level," when F is a general operator, subject only to condition (2). Our analysis has then led from estimate (70) to the final estimate (86) of Lemma 6. The virtue of the free parameter τ is then seen in the right-hand side of estimate (86): by choosing, say, $\epsilon = \epsilon_2 = \frac{1}{\tau}$, and $\epsilon_1 = \frac{1}{\tau^2}$, once ϵ, ϵ_2, ϵ_1 have been chosen, we then obtain:

(i) the coefficient $\left[\tau - \frac{1}{2\epsilon}\right] = \frac{\tau}{2} > 0$ in front of the 'bad' energy-level term

$$\int_Q e^{\tau\phi} |\langle Dv, D(\Delta w)\rangle|^2 dQ,$$

which therefore can be dropped, along with the other 'bad' energy term

$$\tau \int_Q e^{\tau\phi} |\langle Dv, Dw_t\rangle|^2 dQ,$$

both on the RHS of (86);

(ii) the coefficient in front on the 'driving' term (the first on the RHS of (86) becomes $[a - \epsilon C_T - \epsilon_2/2 - \tau\epsilon_1] = \left[a - \frac{C_T}{\tau}\right] > 0$ for $\tau > 0$ sufficiently large. This way, estimate (86), along with (88), yields Eqn. (11a) of the sought-after Theorem 1, which we repeat here as

Theorem 3 Assume (H.1) = (2) on F, (3) on f, and (H.2) = (4). Let w be a solution of Eqn. (1) within the class (10). Then, for all $\tau > 0$ sufficiently large, there exists constants $C_T > 0$, $C > 0$, $C_{\tau,T} > 0$ such that the following one-parameter family of estimates holds true, where $a > 0$ is the constant in (4):

$$B_\Sigma(w) + \frac{2}{\tau}\int_Q e^{\tau\phi}f^2 dQ + C_{T,\tau}\int_Q [|D^2 w|^2 + |Dw|^2 + w^2 + w_t^2] dQ$$

$$\geq \left(a - \frac{C_T}{\tau}\right)\int_Q e^{\tau\phi}[|Dw_t|^2 + |D(\Delta w)|^2] dQ$$

$$- C_\tau e^{-\delta\tau}[P_w(T) + P_w(0)]. \tag{94}$$

Proof As explained in Remark 3, we take $\epsilon = \epsilon_2 = \frac{1}{\tau}$, $\epsilon_1 = \frac{1}{\tau^2}$ in (86) and drop the last two integral terms in (86); moreover, we use (12) for $B_\Sigma(w)$; (88) for $\beta_{0,T}$. This way, estimate (86) becomes (93), as desired. $\square$

Thus, Theorem 3 is a repetition of Theorem 1, Eqn. (11a). Next, the passage from (11a) to (11b) uses property (9d) for the function ϕ in the ('driving') first integral term on the RHS of (11a), as well as identity (17b) relating $P_w(t)$ and $E(t)$. Thus, estimate (11b) is obtained. Theorem 1 is fully proved. $\square$

3.3 Proof of Corollary 1 (Case 1) and Corollary 2 (Case 2)

Case 1: Proof of Corollary 1 Here, under the additional assumptions (H.3) = (20) on Γ_0, our task is to show (22)–(24) for B_{Σ_0}.

Let η denote the unit tangent vector along the boundary $\Gamma = \partial\Omega$ of Ω on M. Then, for any $x \in \Gamma$, $[n, \eta]$ is an orthonormal basis of the tangent space M_x, where n is the outward normal field of the boundary Γ, defined below (20). Hypothesis (21) of Case 1 implies

$$\langle Dw_t, \eta\rangle = \frac{\partial w_t}{\partial\eta} = \frac{\partial\Delta w}{\partial\eta} = \langle D(\Delta w), \eta\rangle \equiv 0, \quad \forall\, x \in \Gamma_0. \tag{95}$$

For the vector field X of M_x, we can write, for each $x \in \Gamma_0$:

$$X = \langle X, n \rangle n + \langle X, \eta \rangle \eta, \tag{96}$$

hence, for $X = H$ defined in (15), we have by (A.2), via (95),

$$H(w_t) = \langle Dw_t, H \rangle = \langle Dw_t, n \rangle \langle H, n \rangle + \langle Dw_t, \eta \rangle \langle H, \eta \rangle; \tag{97}$$

$$H(\Delta w) = \langle D(\Delta w), H \rangle = \langle D(\Delta w), n \rangle \langle H, n \rangle + \langle D(\Delta w), \eta \rangle \langle H, \eta \rangle; \tag{98}$$

$$\begin{cases} H(w_t) = \langle Dw_t, n \rangle \langle H, n \rangle = \dfrac{\partial w_t}{\partial n} \langle H, n \rangle \quad \text{on } \Gamma_0; & (99) \\[2mm] H(\Delta w) = \langle D(\Delta w), n \rangle \langle H, n \rangle = \dfrac{\partial \Delta w}{\partial n} \langle H, n \rangle \quad \text{on } \Gamma_0; & (100) \end{cases}$$

since $\langle Dw_t, \eta \rangle = \frac{\partial w_t}{\partial \eta} \equiv 0$, $\langle D\Delta w, \eta \rangle \equiv \frac{\partial \Delta w}{\partial \eta} \equiv 0$ on Γ_0 by (95). Moreover, again by (95),

$$\begin{cases} Dw_t = \langle Dw_t, n \rangle n + \langle Dw_t, \eta \rangle \eta = \dfrac{\partial w_t}{\partial n} n \quad \text{on } \Gamma_0; & (101) \\[2mm] D(\Delta w) = \langle D(\Delta w), n \rangle + \langle D(\Delta w), \eta \rangle \eta = \dfrac{\partial \Delta w}{\partial n} n \quad \text{on } \Gamma_0. & (102) \end{cases}$$

Then, using $w_t \equiv \Delta w \equiv 0$ on Γ_0 from assumption (21) readily yields $B_{\Sigma_0}^2(w) \equiv 0$ by using (14) with Σ replaced by Σ_0, and (23) is verified. Similarly, using (99)–(102), as well as $w_t \equiv 0$ on Γ_0 in (13) for $B^1(w)$ with Σ replaced by Σ_0, readily yields

$$\begin{aligned} B_{\Sigma_0}^1(w) &= \int_{\Sigma_0} \left[\left(\frac{\partial w_t}{\partial n} \right)^2 + \left(\frac{\partial \Delta w}{\partial n} \right)^2 \right] \langle H, n \rangle d\Sigma_0 \\ &\quad - \frac{1}{2} \int_{\Sigma_0} \left[\left(\frac{\partial w_t}{\partial n} \right)^2 + \left(\frac{\partial \Delta w}{\partial n} \right)^2 \right] \langle H, n \rangle d\Sigma_0, \end{aligned} \tag{103}$$

and (103) proves (22). The negative sign for $B_{\Sigma_0}(w)$ in (22) is due via (15) to $\langle H, n \rangle = e^{\tau \phi} \langle Dv, n \rangle \le 0$ on Γ_0 by hypothesis (20). This yields the first inequality in (24). It remains to show the second inequality in (24). Returning to (13) and (14) for B^1 and B^2 with Σ replaced by Σ_1, we estimate with $p = \text{div} H$ and (A.2),

$$B_{\Sigma_1}^1(w)$$

$$= \int_{\Sigma_1} [\langle Dw_t, H \rangle + pw_t] \langle Dw_t, n \rangle + \langle D(\Delta w), H \rangle \langle D(\Delta w), n \rangle] d\Sigma_1$$

$$-\int_{\Sigma_1}\left\{\left[w_t\Delta w_t+\frac{1}{2}(|Dw_t|^2+|D(\Delta w)|^2)\right]\langle H,n\rangle\right.$$

$$\left.+\frac{1}{2}w_t^2\langle Dp,n\rangle\right\}d\Sigma_1 \tag{104}$$

$$\leq C\int_{\Sigma_1}\left[|Dw_t|^2+w_t^2+|D(\Delta w)|^2+|w_t\Delta w_t|\right]d\Sigma_1 \tag{105}$$

$$B_{\Sigma_1}^2(w)$$

$$=\frac{1}{2}\int_{\Sigma_1}\left\{\left[\Delta w\frac{\partial\Delta w}{\partial n}-w_t\frac{\partial w_t}{\partial n}\right]p+\frac{1}{2}[w_t^2-(\Delta w)^2]\langle Dp,n\rangle\right\}d\Sigma_1 \tag{106}$$

$$\leq C\int_{\Sigma_1}\left[\left|\Delta w\frac{\partial\Delta w}{\partial n}\right|+w_t^2+\left(\frac{\partial w_t}{\partial n}\right)^2+(\Delta w)^2\right]d\Sigma_1. \tag{107}$$

Then (105), (107) and $B_{\Sigma_1}=B_{\Sigma_1}^1+B_{\Sigma_1}^2$ by (12) yield the second inequality in (24), as desired, as $(\partial w_t/\partial n)$ is absorbed by Dw_t. Corollary 1 is proved.   □

Proof of Corollary 2 First, we note that the B.C.'s assumed in (25) imply

$$w_t|_\Sigma\equiv 0,\ \left.\frac{\partial w_t}{\partial n}\right|_\Sigma\equiv 0,\ Dw_t=\frac{\partial w_t}{\partial n}n+\frac{\partial w_t}{\partial\eta}\eta\equiv 0,\ x\in\Gamma, \tag{108}$$

using a decomposition of Dw_t on Γ as in (96). Next, using $w_t|_\Gamma=\left.\frac{\partial w_t}{\partial n}\right|_\Gamma\equiv 0$, $Dw_t\equiv 0$ on Γ by (25), (108), we return to (13) for $B_\Sigma^1(w)$ and (14) for $B_\Sigma^2(w)$, and readily obtain (26), (27). Then, (28) follows at once, since $\frac{\partial\Delta w}{\partial n}=\langle D(\Delta w),n\rangle$.   □

4.   Proof of Theorem 2: Carleman estimates, second version

Our starting point is Theorem 1, Eqn. (11b).

Step 1 Using the additional assumption (H.5) = (29), we shall show that

$$E(t)\geq\frac{E(0)+E(T)}{2}e^{-C_TT}-C_TG(T),\ 0\leq t\leq T. \tag{109}$$

In fact, by assumption (H.5) = (29), we have both inequalities

$$E(t)\ \leq\ E(s)+C_TG(T)+C_T\int_s^t E(\sigma)d\sigma,\ 0\leq s\leq t\leq T, \tag{110}$$

$$E(s) \;\leq\; E(t) + C_T G(T) + C_T \int_s^t E(\sigma)d\sigma, \; 0 \leq s \leq t \leq T. \quad (111)$$

Applying Gronwall's inequality on each (110) and (111), we obtain for $0 \leq s \leq t \leq T$:

$$E(t) \leq [E(s) + C_T G(T)]e^{C_T(t-s)}; \quad E(s) \leq [E(t) + C_T G(T)]e^{C_T(t-s)}. \quad (112)$$

Then the inequality on the right of (112) with $s = 0$, and that on the left with $t = T$ and $s = t$ yield then

$$E(0) \leq [E(t) + C_T G(T)]e^{C_T T}; \quad E(T) \leq [E(t) + C_T G(T)]e^{C_T T}. \quad (113)$$

Summing up these two inequalities in (113), we arrive at (109), as desired.

Step 2 We insert (109) for $E(t)$ into the integral $\int_{t_0}^{t_1} E(t)dt$ on the right-hand side of (11b) and obtain for its left-hand side:

$$\text{LHS of (11b)} \;\geq\; \left\{ \left(a - \frac{c}{\tau}\right) e^{-\tau\frac{\delta}{2}} \frac{(t_1 - t_0)e^{-C_T T}}{2} - C_T e^{-\delta\tau} \right\}$$

$$[E(T) + E(0)] - C_T G(T)(t_1 - t_2). \quad (114)$$

Then, recalling $G(T)$ from (2.2.2), we see that (114) readily yields (32), by taking τ sufficiently large as to make positive the coefficient $\{ \quad \}$ in front of $[E(T) + E(0)]$ in (114). Theorem 2.4 is proved. $\qquad \square$

5. Implications on observability/stabilization inequalities

The Carleman estimates at the $H^3(\Omega) \times H^1(\Omega)$-level of Section 2—in particular, estimate (32) of Theroem 2.4—form the key basic ground for obtaining continuous observability inequalities (hence, by duality, exact controllability results)/stabilization inequalities for Eqn. (1), when this equation is accompanied by suitable boundary conditions (B.C.)

The $H^3(\Omega) \times H^1(\Omega)$-level of the estimates in Section 2 are most directly of use when equation (1) is supplemented by *hinged* B.C. However, a full, sharp account even in the case of hinged B.C.—and surely more so for other B.C. such as *clamped* B.C.—requires additional tools [15, 16], [23, 24, 25]to deal with a few remaining issues. These include: (i) the issue of reducing the number of traces needed in the continuous observability estimates (that is, the issue of reducing the number of controls needed to obtain the corresponding (dual) exact controllability results); (ii) the issue of shifting (in particular, shifting down) the

$H^3(\Omega) \times H^1(\Omega)$—level of topologies of the original estimate (32) [which is good, as we said, for *hinged* B.C.]. This shifting is required in order to obtain continuous observability estimates (exact controllability results) also in the case of *clamped* B.C. While lack of space induces us to leave this program to a subsequent paper, we here derive the correponding continuous observability/stabilization estimate for *hinged* B.C., in the case of two traces/two controls), since this follows readily from (32). More general consequences will be derived in a subsequent paper.

Uniqueness assumption Below we shall need the following *uniqueness property for an over-determined problem* associated with (1): Let w be a sufficiently smooth solution of (1) as in (10) which satisfies the following B.C.

$$w|_\Sigma = \Delta w|_\Sigma \equiv 0 \text{ on } (0,T] \times \Gamma \equiv \Sigma, \tag{115}$$

$$\left.\frac{\partial w}{\partial n}\right|_{\Sigma_1} = \left.\frac{\partial \Delta w}{\partial n}\right|_{\Sigma_1} \equiv 0 \text{ on } (0,T] \times \Gamma_1 \equiv \Sigma_1, \tag{116}$$

where the geometrical condition (20):

$$\frac{\partial v}{\partial n} = \langle Dv, n \rangle \leq 0, \qquad x \in \Gamma_0 \tag{117}$$

holds true on $\Gamma_0 = \Gamma/\Gamma_1$. Then, in fact: $w \equiv 0$ on $Q = (0,T] \times \Omega$.

Remark 4 The following are a few known cases, where the unique continuation property required above for problem (1), (115)–(117) holds true, say, in the setting of Appendix B, for Eqn. (B.6) with variable coefficients, on a bounded domain $\Omega \subset \mathbb{R}^n$:

(1) The case where the coefficients $a_{ij}(x)$ of the elliptic operator $\mathcal{A}$ in (B.1) satisfy $a_{ij}(x) \in C^1(\Omega)$, while the coefficients q_α and r_α of the energy level term F in (B.5) are *time-independent* and in $L_\infty(\Omega)$ in the space variable.

Indeed, in this case, the unique continuation property for the dynamical problem (1), (115)–(117) can be converted (via Laplace transform) into a unique continuation property for the corresponding fourth-order elliptic problem with space-variable coefficients. This latter problem has all four boundary conditions (Cauchy data) zero on the portion Γ_1 of the boundary. As a consequence, the solution of the corresponding elliptic problem has to vanish in a suitable neighborhood of Γ_1, interior to Ω. At this point, we can apply Hörmander's 1959-theorem [H.2] (of which a new proof has been recently given in [[1], Theorem 1.1]) and conclude that, then, the solution of such elliptic problem must vanish on all of

Ω: $w \equiv 0$ in Ω, as desired. [A stronger version of the unique continuation result in [8] for certain fourth-order elliptic equations, which are iterations of two second-order elliptic equations, is given in [22], under "weaker vanishing requirements," beyond our present needs. Papers [8] and [22] improve upon a prior result of Peterson, quoted in [1], where the fourth-order equation has principal part restricted to Δ^2, followed by *all* third-order terms. In turn, Peterson's result improves upon [21], where not all third-order terms were allowed.]

(2) The case—given in [[11], Theorem 5.1, p. 137]—of the equation (B.6) in a Euclidean setting as in Appendix B, where, however, the differential operator $\mathcal{A} = \Delta$ in (B.1) $[a_{ij} = \delta_{ij}]$, but with coefficients of F allowed to vary in both *time* and *space*:

$$F(w) = \sum_{|\beta| \leq 1} a_\beta \partial^\beta \Delta w + \sum_{|\alpha| \leq 2} a_\alpha \partial^\alpha w,$$

$a_\alpha \in L(Q)$, $Q = (0,T) \times \Omega$, plus some additional conditions. This result in [I.1] requires, however, all four zero B.C. on the *entire* boundary Γ (so, in (116), we must take $\Gamma_1 = \Gamma$). We remark that a readjustment (*mutatis mutandis*) of the proof for unique continuation result given in [2], Section 10], [3] for the corresponding Kirchhoff plate with all four B.C. zero, but this time only on an arbitrary common portion of the boundary, of positive measure, is likely to admit a counterpart version to the Euler-Bernoulli plate. [The constant $\gamma > 0$ for the Kirchhoff plate, which accounts for rotational inertia, becomes $\gamma = 0$ on the Euler-Bernoulli plate.] Should this be the case, the required $\Gamma_1 \subset \Gamma$ in (116) subject to (117) would be allowed. The unique continuation result in [2], Section 10, [3] for the Kirchhoff equation with all four zero B.C. on an arbitrary portion of the boundary extended the prior result [10], Theorem 1.2, p. 136], where vanishing of all four B.C. on the entire boundary was required. $\square$

Theorem 4 With reference to Eqn. (1) with $f = 0$, assume (H.1) on F and (H.2) = (4). In addition, assume: (i) the hinged boundary condition (115) on all of $\Sigma = (0,T] \times \Gamma$, as well as (ii) the geometrical condition (20) = (117) on Γ_0.

(a) Then, given $T > 0$, there exists $C_T > 0$, such that the following estimate holds true:

$$\int_0^T \int_{\Gamma_1} \left[\left(\frac{\partial w_t}{\partial n} \right)^2 + \left(\frac{\partial \Delta w}{\partial n} \right)^2 \right] d\Gamma_1 \, dt + \ell ot(w) \geq C_T[E(0) + E(T)].$$

$$(118)$$

(b) Assume further the uniqueness property stated in Remark 5.1: (1), (115), (116), (117) imply $w \equiv 0$ on Q. Then, estimate (118) simplifies to the following *Continuous Observability Inequality*:

$$\int_0^T \int_{\Gamma_1} \left[\left(\frac{\partial w_t}{\partial n} \right)^2 + \left(\frac{\partial \Delta w}{\partial n} \right)^2 \right] d\Gamma_1 \, dt \geq c_T[E(0) + E(T)]. \quad (119)$$

Proof (a) Under the hinged B.C. assumption (115) [that is, assumption (21), this time on all of $\Sigma = (0, T] \times \Gamma$], then conclusions (22) and (23) hold true on all of Σ:

$$
\left\{
\begin{aligned}
B_\Sigma(w) &= B_\Sigma^1(w) + B_\Sigma^2(w) = B_\Sigma^1(w) \\
&= \frac{1}{2} \int_0^T \int_\Gamma \left[\left(\frac{\partial w_t}{\partial n} \right)^2 + \left(\frac{\partial \Delta w}{\partial n} \right)^2 \right] \langle H, n \rangle d\Sigma \qquad (120) \\
\text{(by (117))} \quad &\leq \frac{1}{2} \int_0^T \int_{\Gamma_1} \left[\left(\frac{\partial w_t}{\partial n} \right)^2 + \left(\frac{\partial \Delta w}{\partial n} \right)^2 \right] \langle H, n \rangle d\Sigma_1, \qquad (121) \\
B_\Sigma^2(w) &\equiv 0. \qquad (122)
\end{aligned}
\right.
$$

In going from (120) to (121) we have used (20) = (117), along with (15). Moreover (122) follows from (27). Finally, in the case of hinged B.C. we have $BT_e(w) \equiv 0$ (see the paragraph below (30) and Remark 2). Hence, estimate (32) yields then estimate (118), as desired.
(b) The usual compactness/uniqueness argument under the uniqueness assumption of Remark 4 permits one to absorb the lower-order terms $\ell ot(w)$ in (118) and obtain (119). □

Remark 5 As is well-known [12, 13, 14], by duality, the continuous observability inequality (119) is equivalent to the property of exact controllability of the corresponding non-homogeneous boundary control problem, with controls in the hinged B.C., in the space of 'optimal regularity' [12, 13, 14], [20].

Appendix A

(i) Throughout this paper, M is an n-dimensional, $n \geq 2$, Riemann manifold with metric $g(\cdot, \cdot) = \langle \cdot, \cdot \rangle$ and norm $| \ |$:

$$
\left\{
\begin{aligned}
& g(X, Y) = \langle X, Y \rangle = \sum_{i,j=1}^n g_{ij}(x) \alpha_i \beta_j, \ X, Y \in M_x; \\
& |X| = \{ g(X, X) \}^{\frac{1}{2}}, \ X = \sum_{i=1}^n \alpha_i \frac{\partial}{\partial x_i}, \qquad Y = \sum_{i=1}^n \beta_i \frac{\partial}{\partial x_i} \in M_x,
\end{aligned}
\right.
\tag{123}
$$

where, for each $x \in M$, we let M_x denote the tangent space of M at x. We denote by $\mathcal{X}(M)$ the set of all vector fields on M.

(ii) Throughout this paper, D denotes the Levi-Civita connection, so that if f is a scalar C^1-function on M and X a vector field on M, then the continuous linear functional $X(f)$ [= derivative of f in the direction of X] is given by

$$X(f) = \langle Df, X \rangle = \langle \nabla f, X \rangle, \tag{124}$$

∇f being the gradient of f. Furthermore, if H, X, Y are vector fields on M, then DH denotes the covariant differential of H, a 2-covariant tensor: it determines a bilinear form on $M_x \times M_x$, for each $x \in M$, defined by

$$DH(X,Y) = \langle D_Y H, X \rangle, \quad \forall\, X, Y \in M_x,\ x \in M, \tag{125}$$

where $D_Y H$ is the covariant derivative of H with respect to Y.

(iii) For any function f and vector field H on M, the following identity holds true on each $x \in M$ [30, 31], [19], p. 22],

$$\langle Df, D(H(f)) \rangle = DH(Df, Df) + \frac{1}{2}[\operatorname{div}(|Df|^2 H) - |Df|^2 \operatorname{div} H], \tag{126}$$

where all the symbols have been defined above, save for div X, which may be defined in terms of the Levi-Civita connection as [G-P-V.1, p. 153]

$$\operatorname{div} X = \sum_{i=1}^{n} \left[D_{\frac{\partial}{\partial x_i}} X \right]_i. \tag{127}$$

(iv) Finally, if f is a C^2-scalar function on M, then, by definition, its Hessian $D^2 f(\cdot, \cdot)$ with respect to the metric g is a 2-covariant tensor defined by

$$D^2 f(X,Y) = \langle D_Y(Df), X \rangle. \tag{128}$$

(v) The Laplace-Beltrami operator Δ is then

$$\Delta f = \operatorname{div} Df \tag{129}$$

with div $\cdot$ defined in (A.5). Moreover, if n and τ are respectively the normal and tangential unit vectors along the boundary Γ of $\Omega \subset M$, then $\frac{\partial \cdot}{\partial n} = \langle D\cdot, n \rangle$ and $\frac{\partial \cdot}{\partial \tau} = \langle D\cdot, \tau \rangle$.

(vi) We have denoted the set of all vector fields on M by $\mathcal{X}(M)$. Likewise, we denote the set of all k-order tensor fields by $T^k(M)$, and the set of all k-forms (alternating k-tensors) by $\Lambda^k(M)$, where k is a non-negative integer. Then: $\Lambda^k(M) \subset T^k(M)$. In particular, $\Lambda^0(M) = T^0(M) = C^\infty(M) =$ the set of all smooth functions on M; and $T^1(M) = T(M) =$

$\Lambda(M) = \mathcal{X}(M)$, where $\Lambda(M) = \mathcal{X}(M)$ is the following isomorphism: for a given $X \in \mathcal{X}(M)$, then $U(Y) = \langle Y, X \rangle$, $\forall\, Y \in \mathcal{X}(M)$, determines a unique $U \in \Lambda(M)$.

(vii) For each $x \in M$, the k-order tensor space T_x^k on M_x is an inner product space, defined as follows. Let $e_1, \ldots, e_n$ be an orthonormal basis of M_x. For any $\alpha, \beta \in T_x^k$, $x \in M$, the inner product is given by

$$\langle \alpha, \eta \rangle_{T_x^k} = \sum_{i_1, \ldots, i_k = 1}^{n} \alpha(e_{i_1}, \ldots, e_{i_k}) \beta(e_{i_1}, \ldots, e_{i_k}) \text{ at } x. \qquad (130)$$

In particular, for $k = 1$, definition (15) becomes $g(\alpha, \beta) = \langle \alpha, \beta \rangle_{T_x} = \langle \alpha, \beta \rangle$, $\forall\, \alpha, \beta \in M_x$; that is, the inner product of M_x.

(viii) Let Ω be a bounded region of M with regular boundary Γ or without boundary (Γ is empty). By (A.8), $T^k(\Omega)$ are inner product spaces in the following inner product

$$\langle T_1, T_2 \rangle_{T^k(\Omega)} = \int_{\Omega} \langle T_1, T_2 \rangle_{T_x^k} dx, \quad T_1, T_2 \in T^k(\Omega), \qquad (131)$$

where dx is the volume element of M in the metric g. The completions of $T^k(\Omega)$ in the inner product (A.9) are denoted by $L^2(\Omega, T^k)$. In particular, $L^2(\Omega, \Lambda) = L^2(\Omega, T)$. Furthermore, $L^2(\Omega)$ is the completion of $C^\infty(\Omega)$ in the usual inner product

$$(f_1, f_2)_{L^2(\Omega)} = \int_{\Omega} f_1(x) f_2(x) dx, \quad f_i \in C^\infty(\Omega). \qquad (132)$$

(ix) With D the Levi-Civita connection on M, the Sobolev space $H^k(\Omega)$ is the completion of $C^\infty(\Omega)$ with respect to the norm $\|\ \ \|$,

$$\|f\|_{H^k(\Omega)}^2 = \sum_{i=1}^{k} \|D^i f\|_{L^2(\Omega, T^i)}^2 + \|f\|_{L^2(\Omega)}^2, \quad f \in C^\infty(\Omega), \qquad (133)$$

where $D^i f$ is the ith covariant differential of f in the metric g of M, which is an i-order tensor field on Ω, and $\|\ \ \|_{L^2(\Omega, T^k)}$ are the corresponding norms, induced by the inner products (A.8), (A.9). For details on Sobolev spaces on Riemann manifolds, we refer to Hebey [6] or Taylor [26].

Remark A.1 Throughout this paper, in order to streamline the notation, we shall use two vertical bars $|\ \ |$ to denote the L^2-norm, without distinguishing whether the argument of the norm is a function or a k-tensor (k-form). Thus, the norms of $L^2(\Omega)$, $L^2(\Omega, \Lambda)$, $L^2(\Omega, T)$, etc., are denoted by the same symbol. This already occurs in (2), where

$|w|$, $|w_t|$ are the $L^2(\Omega)$-norms of functions, but $|Dw|$, $|Dw_t|$ are in the $L^2(\Omega, \Lambda = T)$-norms of 1-forms (vector fields); finally, $|D^2w|$, $|D^3w|$ are the norms of D^2w, D^3w as 2-order tensors and 3-order tensors defined by (A.8), (A.9).

Appendix B: The case arising from an E-B plate equation with variable coefficients defined on an open bounded domain $\tilde{\Omega}$ of $\mathbb{R}^n$: The Riemann manifold $(\mathbb{R}^n, g)$

The abstract geometric setting of the present paper includes, in particular, the case of an Euler-Bernoulli plate-like equation with space variable coefficients, defined on an open bounded domain $\tilde{\Omega}$ of $\mathbb{R}^n$, say $n \geq 2$, with boundary $\tilde{\Gamma}$ of class C^2. In this case, let

$$\mathcal{A}w = \sum_{i,j=1}^n \frac{\partial}{\partial x_i} \left(a_{ij}(x) \frac{\partial w}{\partial x_j} \right) \tag{134}$$

be a second-order differential operator with real coefficients $a_{ij} = a_{ji}$ of class C^1, see Remark B.2 below, satisfying the uniform ellipticity condition

$$\sum_{i,j=1}^n a_{ij}(x)\xi_i\xi_j \geq C \sum_{i=1}^n \xi_i^2, \quad x \in \tilde{\Omega} \tag{135}$$

for some positive $C > 0$. Assume further that

$$\sum_{i,j=1}^n a_{ij}(x)\xi_i\xi_j > 0, \quad \forall\, x \in \mathbb{R}^n,\ \xi = (\xi_1,\ldots,\xi_n) \in \mathbb{R}^n,\ \xi \neq 0. \tag{136}$$

Denote by $A(x)$ and $G(x)$, respectively, the coefficient matrix and its inverse

$$A(x) = (a_{ij}(x)),\ G(x) = [A(x)]^{-1} = (g_{ij}(x)). \tag{137}$$

Both $A(x)$ and $G(x)$ are $n \times n$ positive definite matrices for any $x \in \mathbb{R}^n$ by assumption (B.3). Moreover, we can take with $\tilde{Q} = (0, T] \times \tilde{\Omega}$:

$$F(w) = \sum_{|\alpha|=0}^3 q_\alpha(t, x)\partial_x^\alpha w + \sum_{|\alpha|=0}^1 r_\alpha(t, x)\partial_x^\alpha w_t, \quad q_\alpha, r_\alpha \in L_\infty(\tilde{Q}). \tag{138}$$

Problem (1) then covers the following Euclidean plate-like equation

$$w_{tt} + \mathcal{A}^2 w + F(w) = f \text{ in } (0, T] \times \tilde{\Omega} \equiv \tilde{Q}, \tag{139}$$

with $\mathcal{A}$ given by (B.1) and F given by (B.5). The main assumption (H.2) = (4) for problem (B.5) is discussed in Remark B.1 below.

Riemann metric Let $\mathbb{R}^n$ have the usual topology and $x = [x_1, \ldots, x_n]$ be the natural coordinate system. For each $x \in \mathbb{R}^n$, define the inner product and the norm on the tangent space $\mathbb{R}_x^n = \mathbb{R}^n$ by

$$g(X, Y) = \langle X, Y \rangle_g = \sum_{i,j=1}^{n} g_{ij}(x)\alpha_i\beta_j; \tag{140}$$

$$|X|_g = [\langle X, X \rangle]^{\frac{1}{2}}, \quad \forall \; X = \sum_{i=1}^{n} \alpha_i \frac{\partial}{\partial x_i}, \quad Y = \sum_{i=1}^{n} \beta_i \frac{\partial}{\partial x_i}. \tag{141}$$

It is easily checked from (B.3) that $(\mathbb{R}^n, g)$ is a Riemann manifold. We may denote g by $g = \sum_{i,j=1}^{n} g_{ij} dx_i dx_j$ (if $A(x) = I$, i.e., $\mathcal{A} = \Delta$, then $G(x) = I$, and g is the Euclidean $\mathbb{R}^n$-metric). If X is any vector field on the manifold $(\mathbb{R}^n, g)$, and if $f \in C^1(\tilde{\Omega})$, then $X(f)$ is a continuous linear functional which via the Riesz representation theorem is given by (A.2),

$$X(f) = \langle \nabla_g f, X \rangle_g = \langle Df, X \rangle_g, \tag{142}$$

where D is the Levi-Civita connection in the Riemann metric g. The following relationships are useful.

Lemma B.1 Let $x = [x_1, x_2, \ldots, x_n]$ be the natural coordinate system in $\mathbb{R}^n$. Let $f, h \in C^1(\tilde{\Omega})$. Finally, let H, X be vector fields. Then:
(a)

$$\langle X(x), A(x)X(x) \rangle_g = H(x) \cdot X(x), \quad x \in R^n; \tag{143}$$

$$A(x)H(x) = \sum_{i=1}^{n} \left(\sum_{j=1}^{n} a_{ij}(x)\alpha_j(x) \right) \frac{\partial}{\partial x_j}; \tag{144}$$

(b)

$$\nabla_g f(x) = \sum_{i=1}^{n} \left(\sum_{j=1}^{n} a_{ij}(x) \frac{\partial f}{\partial x_j} \right) \frac{\partial}{\partial x_i} = A(x)\nabla_0 f; \tag{145}$$

(c) if $X = \sum_{i=1}^{n} \xi_i \frac{\partial}{\partial x_i}$, then by (B.9) and (B.12)

$$X(f) = \langle \nabla_g f, X \rangle_g = \langle A\nabla_0 f, X \rangle_g = \nabla_0 f \cdot X = \sum_{i=1}^{n} \xi_i \frac{\partial f}{\partial x_i}; \tag{146}$$

(d) by (B.11) and (B.12),

$$\frac{\partial w}{\partial \nu_{\mathcal{A}}} = \sum_{i=1}^{n} \left(\sum_{j=1}^{n} a_{ij}(x) \frac{\partial w}{\partial x_j} \right) \nu_i = (A(x)\nabla_0 w) \cdot \nu = \nabla_g w \cdot \nu; \tag{147}$$

234

(e) by (B.9), (B.12), (B.10) and $x \in \mathbb{R}^n$,

$$\langle \nabla_g f, \nabla_g h \rangle_g = \langle A(x) \nabla_0 f, \nabla_g h \rangle_g = \nabla_0 f \cdot \nabla_g h = \nabla_0 f \cdot A(x) \nabla_0 h; \quad (148)$$

(f) recalling (B.1) and using (B.12), (A.7),

$$\mathcal{A}w \equiv \sum_{i=1}^{n} \frac{\partial}{\partial x_i} \left(\sum_{j=1}^{n} a_{ij}(x) \frac{\partial w}{\partial x_j} \right)$$

$$= \text{div}_0(A(x)\nabla_0 w) = \text{div}_0(\nabla_g w)$$

$$= \text{div}_g(\nabla_g w) + \nabla_g w = \Delta_g w + \nabla_g w. \quad (149)$$

Remark B.1 Several classes of examples where the main assumption (H.2) = (4) holds true in $\{\mathbb{R}^n, g\}$ can be given [18], [30, 31]. These exploit general criteria [4], [29], [30], [4]. Indeed, in many cases, the conditions on the coefficients a_{ij} guarantee the existence of a strictly convex function in $\{\mathbb{R}^n, g\}$ *globally* so that *any* smooth bounded domain $\tilde{\Omega}$ in $\mathbb{R}^n$ can then be taken. A very recent result in dimension 2, giving the existence of the required strictly convex function v in (H.2) = (4) was given in [4]: it requires that the boundary $\partial\Omega$ of Ω be strictly convex in the metric g. □

Remark B.2 Let the coefficients a_{ij} in (B.1) be of class C^1, as assumed. Then, the entries g_{ij} in (B.7) are of class C^1 as well. Thus, the connection coefficients Γ_{ik}^{ℓ} (Christoffel symbols) are of class C^0. The geodesic-solutions to a corresponding second-order nonlinear ordinary differential equation are then of class C^2. Thus, the square of the distance function $d_g^2(x, x_0)$ is in C^2. In many caes the square of the distance function $d_g^2(x, x_0)$ provides the requires strictly convex function v in assumption (H.2) = (4). In our case, where the manifolds are complete, the geodesics exist globally. □

Acknowledgements

I. Lasiecka's and R. Triggiani's Research partially supported by the National Science Foundation under grant DMS-9804056 and by the ARO under Grant DAAW-96-1-0059.

P. F. Yao's Research performed while visiting the Mathematics Department, University of Virginia, Charlottesville, VA 22904 partially supported by the NSF grant DMS-9804056 and by the National Science Foundation of China.

References

[1] Eller, M.:A remark on a theorem by Hörmander. Proceedings of 3rd International ISAAC Congress, Berlin, August 2001. World Scientific, Singapore, to appear.

[2] Eller, M., Lasiecka, I., Triggiani, R.: Simultaneous exact/approximate boundary controllability of thermo-elastic plates with variable transmission coefficients. Lecture Notes in Pure and Applied Mathematics 216, Marcel Dekker, New York, 2001, 109–230.

[3] Eller, M., Lasiecka, I., Triggiani, R.: Unique continuation for over-determined Kirchhoff plate equations and related thermo-elastic systems. J. Inv. & Ill-Posed Problems 9(2) (2001), 103–148.

[4] Gulliver, R., Littman, W.: The use of geometric tools in the boundary control of partial differential equations. Proceedings of 3rd International ISAAC Congress, Berlin, August 2001. World Scientific, Singapore, to appear.

[5] Greene, R.-E., Wu, H.: C^∞ convex functions and manifolds of positive curvature. Acta Math. 137 (1976), 209–245.

[6] Hebey, E.: Sobolev spaces on Riemannian manifolds. Lecture Notes in Mathematics, Springer, Berlin, 1996.

[7] Hörmander, L.: On the characteristic Cauchy problem. Ann. Math. (2) 88 (1968), 341–370.

[8] Hörmander, L.: On the uniqueness of the Cauchy problem, II, Math. Scand. 7 (1959), 177–190.

[9] Hörmander, L.: Linear Partial Differential Operators. Springer, Berlin etc. 1963.

[10] Isakov, V.: On uniqueness on a lateral Cauchy problem with multiple characteristics. J. Diff. Eqns. 134 (1997), 134–147.

[11] Isakov, V.: Inverse problems for partial differential equations. Springer, Berlin etc. 1998.

[12] Lasiecka, I., Triggiani, R.: Exact controllability of the Euler-Bernoulli equation with $L_2(\Sigma)$-control only in the Dirichlet B.C. Atti Accad. Naz. Lincei Rend. Cl. Sc. Fis. Mat. Nat. Vol. LXXXII, No. 1 (1988), Rome.

[13] Lasiecka, I., Triggiani, R.: Exact controllability of the Euler-Bernoulli equations with controls in the Dirichlet and Neumann B.C.: A non-conservative case. SIAM J. Control Optim. 27 (1989), 330–373.

[14] Lasiecka, I., Triggiani, R.: Exact controllability of the Euler-Bernoulli equation with boundary controls for displacement and moment. JMAA 146 (1990), 1–33.

[15] Lasiecka, I., Triggiani, R.: Uniform stabilization of the wave equation with Dirichlet or Neumann feedback control without geometric conditions. Appl. Math. Optim. 25 (1992), 189–224.

[16] Lasiecka, I., Triggiani, R.: A-priori observability inequalities. Chapter 1, Lecture Notes, University of Virginia, May 1995.

[17] Lasiecka, I., Triggiani, R.: Carleman estimates and exact boundary controllability for a system of coupled, non-conservative second-order hyperbolic equations. Lecture Notes in Pure and Applied Mathematics 188, eds. J.P. Zolesio and G. Da Prato. Marcel Dekker, New York, 1997, 125–243.

236

[18] Lasiecka, I., Triggiani, R., Yao, P.F.: Exact controllability for second-order hyperbolic equations with variable coefficient-principal part and first-order terms. Nonlinear Analysis: Theory, Methods, Applications 30(1) (1997), 111–122.

[19] Lasiecka, I., Triggiani, R., Yao, P.F.: Inverse/observability estimates for second order hyperbolic equations with variable coefficients. JMAA 235 (1999), 13–57.

[20] Lions, J.L.: Controllabilite exacte, stabilization et perturbation des systemes distribues. 1, Masson, Paris, 1988.

[21] Nirenberg, L.: Uniqueness in Cauchy problems for differential equations with constant leading coefficients. Comm. Pure Appl. Math. 10 (1957), 89–105.

[22] Shirota, T.: A remark on the unique continuation theorem for certain fourth order elliptic equations. Proc. Japan Acad. 36 (1960), 571–573.

[23] Tataru, D.: A-priori pseudoconvexity energy estimates in domains with boundary and applications to exact boundary controllability for conservative PDEs. Ph.D. Thesis, University of Virginia, May 1992.

[24] Tataru, D.: Boundary controllability for conservative PDEs. Appl. Math. & Optimiz. 31 (1995), 257–295.

[25] Tataru, D.: A-priori estimates of Carleman's type in domains with boundaries. J. Math. Pures et Appl. 73 (1994), 355–387.

[26] Taylor, M.: Partial Differential Equations. 1, 2, Springer, Berlin etc. 1991.

[27] Triggiani, R.: Carleman estimates and exact boundary controllability for a system of coupled non-conservative Schrödinger equations. Rend. Istit. Math. U-niv. Trieste (Italy), Suppl. Vol. XXVIII (1997), 453–504; volume in memory of P. Grisvard.

[28] Triggiani, R., Yao, P.F.: Inverse/observability estimates for Schrödinger equations with variable coefficients. Special volume on control of PDEs. Control & Cybernetics 28 (1999), 627–664.

[29] Wu, H., Shen, C.L., Yu, Y.L.: An introduction to Riemann geometry. University of Beijing, 1989 (in Chinese).

[30] Yao, P.F.: On the observability inequalities for the exact controllability of wave equations with variable coefficients, SIAM Control and Optimiz. 37(5) (1999), 1568–1599.

[31] Yao, P.F.: Observability inequalities for the Euler-Bernoulli plate with variable coefficients. Contemporary Mathematics, Amer. Math. Soc. 268 (2000), 383–406.

HYPERBOLICITY FOR SYSTEMS

Tatsuo Nishitani

Department of Mathematics

Osaka University

Machikaneyama 1-16

Toyonaka Osaka 560-0043, Japan

tatsuo@math.wani.osaka-u.ac.jp

Abstract   We study the Cauchy problem for (mainly) first order systems. Our main concern is to investigate for which systems the Cauchy problem is C^∞ well posed for any lower order terms (strong hyperbolicity), or for which systems the Cauchy problem is C^∞ well posed (hyperbolicity). We here present a survey of the subject, in particular focussing the interests on the necessary conditions for strong hyperbolicity or just hyperbolicity.

1.   Introduction

Let us consider

$$Lu = D_0 u + \sum_{j=1}^{n} A_j(x) D_j u + B(x) u = L_1(x, D) u + L_0(x) u,$$

where $u = (u_1, ..., u_m)$, $x = (x_0, x_1, ..., x_n) = (x_0, x')$ and $A_j(x)$, $B(x)$ are $m \times m$ matrix valued smooth functions defined in a neighborhood Ω of the origin of $\mathbf{R}^{n+1}$. We study the Cauchy problem for $L(x, D)$

$$(\text{C.P}) \qquad \begin{cases} L(x, D) u = f, & f = 0 \text{ in } x_0 < \tau, \\ u = 0 & \text{in } x_0 < \tau. \end{cases}$$

We say that the Cauchy problem (C.P) is C^∞ well posed if for any given smooth $f(x)$ vanishing in $x_0 < \tau$ there exists a unique smooth $u(x)$ vanishing in $x_0 < \tau$ and satisfying $L(x, D)u = 0$ in a neighborhood of the origin. We say that $L(x, D)$ is hyperbolic (near the origin) if the Cauchy problem for $L(x, D)$ is C^∞ well posed. We say that $L_1(x, D)$ is strongly hyperbolic if the Cauchy problem for $L(x, D)$ is C^∞ well posed for arbitrary lower order terms $B(x)$. Our main concern is to investigate

237

H.G.W. Begehr et al. (eds.), Analysis and Applications - ISAAC 2001, 237–252.

QUESTION A: for which systems the Cauchy problem is C^∞ well posed (hyperbolicity),

QUESTION B: for which systems the Cauchy problem is C^∞ well posed for arbitrary lower order terms $L_0(x) = B(x)$ (strong hyperbolicity).

For symmetric (or symmetrizable) systems the Cauchy problem is C^∞ well posed for any lower order terms as is well known and hence strongly hyperbolic as we call. On the other hand there is a class of strongly hyperbolic systems which is not symmetrizable. This gives some difficulties when we try to characterize the strong hyperbolicity in a unified way including both symmetrizable and non symmetrizable systems. For triangular systems the Cauchy problem is C^∞ well posed of course if so is for every scalar operator on the diagonal. This causes some troubles when we try to get necessary conditions on lower order terms $B(x)$ for $L(x, D)$ to be hyperbolic which is available without assumptions on the matrix structure.

2. Strong hyperbolicity

Recall that the principal symbol of $L(x, D)$ is defined as

$$L_1(x, \xi) = \xi_0 I + \sum_{j=1}^{n} A_j(x)\xi_j = \xi_0 I - A(x, \xi')$$

and the full symbol $L(x, D)$ by

$$L(x, \xi) = L_1(x, \xi) + L_0(x).$$

We denote by $h(x, \xi)$ the determinant of $L_1(x, \xi)$, where $h(x, \xi)$ is a homogeneous polynomial in ξ of degree m.

Definition *We say that $\bar{z} = (\bar{x}, \bar{\xi})$ is a characteristic of $L_1(x, \xi)$ (or of $h(x, \xi)$) of order r if*

$$\partial_x^\beta \partial_\xi^\alpha h(\bar{x}, \bar{\xi}) = 0, \quad \forall |\alpha + \beta| < r, \quad \partial_x^\beta \partial_\xi^\alpha h(\bar{x}, \bar{\xi}) \neq 0, \quad \exists |\alpha + \beta| = r.$$

We first recall

Theorem 1 (Lax [9], Mizohata [10]) *Let $0 \in \Omega$ and the Cauchy problem for $L(x, D)$ is C^∞ well posed near the origin. Then the roots of $h(x, \xi) = 0$ with respect to ξ_0 are all real when x is close to the origin and $\xi' \in \mathbf{R}^n$.*

This means that every root τ of $h(x, \xi + \tau\theta) = 0$, $\theta = (1, 0, ..., 0)$ is real. In what follows we assume that Theorem 1 holds.

Theorem 2 ([17]) *Assume that $A_j(x)$ are real analytic in Ω and $\bar{z} = (0, \bar{\xi})$ is a characteristic of $L_1(x, \xi)$ of order r. If $L_1(x, D)$ is strongly hyperbolic near the origin, then every $(m-1)$-th minor of $L_1(x, \xi)$ vanishes of order $r - 2$ at $\bar{z}$.*

Remark If $\bar{z}$ is a zero of h of order r, then $\lambda = 0$ is a zero of $h(\bar{z} + \lambda\theta)$ of order r and vice versa. Since $\det(\lambda I + L_1(\bar{z})) = h(\bar{z} + \lambda\theta)$ then $\bar{z}$ is a characteristic of L_1 of order r is equivalent to say that $\lambda = 0$ is an eigenvalue of $L_1(\bar{z})$ of multiplicity r.

Corollary 1 ([17]) *Assume that $A_j(x)$ are real analytic in Ω and $\bar{z} = (0, \bar{\xi})$ is a multiple characteristic of L_1. If $L_1(x, D)$ is strongly hyperbolic near the origin, then*

$$(L_1(\bar{z})|_V)^2 = 0,$$

where $L_1(\bar{z})|_V$ is the restriction of $L_1(\bar{z})$ on the generalized zero eigenspace V of $L_1(\bar{z})$.

On $T_{\bar{z}}(T^*\Omega) \cong \mathbf{R}^{n+1} \times \mathbf{R}^{n+1}$ we have the following bilinear form σ

$$\sigma(z, w) = \langle \xi, y \rangle - \langle x, \eta \rangle, \quad z = (x, \xi), \quad w = (y, \eta) \in \mathbf{R}^{n+1} \times \mathbf{R}^{n+1},$$

where $\langle \xi, x \rangle$ stands for the usual scalar product in $\mathbf{R}^{n+1}$. Recall that a linear subspace $V \subset \mathbf{R}^{2n+2}$ is called involutive if

$$V^\sigma = \{z \in \mathbf{R}^{2n+2} \mid \sigma(z, w) = 0, \ \forall w \in V\} \subset V$$

holds. Let $\bar{z} = (\bar{x}, \bar{\xi})$ be a characteristic of $L_1(z)$ of order r. Then the Taylor expansion of $h(z)$ starts with the term

$$h(\bar{z} + \mu(x, \xi)) = \mu^r \{ \sum_{|\alpha+\beta|=r} \tfrac{1}{\alpha!\beta!} \partial_\xi^\beta \partial_x^\alpha h(\bar{z}) \xi^\alpha x^\beta + O(\mu) \}.$$

Definition *Let $\bar{z}$ be a characteristic of L_1 of order r. We define the localization $h_{\bar{z}}(x, \xi)$ of h at $\bar{z}$ by*

$$h(\bar{z} + \mu z) = \mu^r \{h_{\bar{z}}(z) + O(\mu)\}, \quad \mu \to 0.$$

This is given explicitly by

$$h_{\bar{z}}(x, \xi) = \sum_{|\alpha+\beta|=r} \tfrac{1}{\alpha!\beta!} \partial_\xi^\beta \partial_x^\alpha h(\bar{z}) \xi^\alpha x^\beta$$

which is a homogeneous polynomial in (x, ξ) of degree r and has only real roots with respect to ξ_0 for any (x, ξ').

Definition *Let $q(z)$ be a (homogeneous) polynomial. We define a linear subspace $\Lambda(q) \subset \mathbf{R}^{2n+2}$ as*

$$\Lambda(q) = \{z \in \mathbf{R}^{2n+2} \mid q(tz + w) = q(w), \forall t \in \mathbf{R}, \forall w \in \mathbf{R}^{2n+2}\}$$

and we call $\Lambda(q)$ the linearity space of q.

Choose linear forms $\ell_j(z)$, $1 \leq j \leq p$ which are linearly independent so that $\Lambda(q)$ is given by $\{z \mid \ell_j(z) = 0, 1 \leq j \leq p\}$. Then $q(z)$ is a polynomial in $\ell_j(z)$, that is we can write

$$q(z) = \sum c_\alpha \ell_1(z)^{\alpha_1} \cdots \ell_p^{\alpha_p}(z).$$

Theorem 3 ([17]) *Assume that $A_j(x)$ are real analytic in Ω and $\bar{z} = (0, \bar{\xi})$ is a characteristic of order r. Assume that $\Lambda(h_{\bar{z}})$ is involutive. Then in order that $L_1(x, D)$ is strongly hyperbolic it is necessary that every $(m - 1)$-th minor of $L_1(x, \xi)$ vanishes of order $r - 1$ at $\bar{z}$.*

Corollary 2 ([17]) *Assume that $A_j(x)$ are real analytic in Ω and $\bar{z}$ is a multiple characteristic such that $\Lambda(h_{\bar{z}})$ is involutive. Then if $L_1(x, D)$ is strongly hyperbolic we have*

$$\mathrm{Ker} L_1(\bar{z}) \cap \mathrm{Im} L_1(\bar{z}) = \{0\}. \tag{1}$$

Corollary 3 (Kajitani [13]) *Assume that the multiplicity of characteristics are constant. If $L_1(x, D)$ is strongly hyperbolic, then $A(x, \xi')$ is smoothly diagonalizable.*

As we have remarked in Introducton, there is a class of strongly hyperbolic systems which is not symmetrizable. We now give this class. Recall that the hyperbolic cone $\Gamma(h_{\bar{z}})$ of $h_{\bar{z}}$ is defined by

$\Gamma(h_{\bar{z}})$ = the connected component of $\{z \in \mathbf{R}^{2n+2} \mid h_{\bar{z}}(z) \neq 0\}$ containing $(0, e_n)$.

Definition *We define the propagation cone $C(h_{\bar{z}})$ of $h_{\bar{z}}$ by*

$$C(h_{\bar{z}}) = \{z \in \mathbf{R}^{2n+2} \mid \sigma(z, w) \leq 0, \forall w \in \Gamma(h_{\bar{z}})\}.$$

Note that $\Lambda(h_{\bar{z}})$ is involutive if and only if $C(h_{\bar{z}}) \subset \Lambda(h_{\bar{z}})$. If $\bar{z}$ is a simple characteristic, then $h_{\bar{z}}(z) = dh(\bar{z}; z)$, the differential at $\bar{z}$, and clearly $\Lambda(h_{\bar{z}})$ is a hyperplane defined by $dh(\bar{z}; z) = 0$. The hyperbolic cone is the half space given by $\{z \mid dh(\bar{z}; z) > 0\}$ and the propagation cone $\Gamma(h_{\bar{z}})$ is just $H_h(\bar{z}) \cdot \mathbf{R}_+$, where H_h is the Hamilton vector field of h.

Recall that a scalar operator P is strongly hyperbolic if and only if every multiple characteristic is at most double and at every double characteristic z the following holds

$$C(P_z) \cap \Lambda(P_z) = \{0\}. \tag{2}$$

From Corollary 2 it follows, that if L_1 is strongly hyperbolic and z is a characteristic of order m then

$$0 \leq \operatorname{rank} L_1(z) \leq [m/2]. \tag{3}$$

Let us denote by Σ the set of characteristics of order m and assume that Σ is a smooth manifold.

Theorem 4 ([20]) *Assume that* $\operatorname{rank} L_1 = [m/2]$ *on* Σ. *Assume that every* $(m-1)$-*th minor vanishes of order* $m-2$ *on* Σ *and* $C(h_{\bar{z}}) \cap \Lambda(h_{\bar{z}}) = \{0\}$. *We also assume that* $h_{\bar{z}}$ *is strictly hyperbolic on* $T_{\bar{z}}(T^*\Omega)/T_{\bar{z}}\Sigma$. *Then* $L_1(x, D)$ *is (microlocally near* $\bar{z}$) *strongly hyperbolic.*

Remark Since $z \in \Sigma$ is a characteristic of order m then $\operatorname{rank} L_1(z) > 0$ means that $L_1(z)$ is not symmetrizable.

Remark When $m = 2$ then $h_{\bar{z}}$ is always strictly hyperbolic on $T_{\bar{z}}(T^*\Omega)/T_{\bar{z}}\Sigma$.

Question Let $\bar{z}$ be a multiple characteristic. Then if L_1 is strongly hyperbolic can we conclude

$$\text{either} \quad C(h_{\bar{z}}) \cap \Lambda(h_{\bar{z}}) = \{0\}$$

$$\text{or} \quad \operatorname{Ker} L_1(\bar{z}) \cap \operatorname{Im} L_1(\bar{z}) = \{0\}.$$

When the characteristic is double or $n = 1$ this is true.

3. Non degenerate characteristics

We now study the case $\operatorname{Ker} L_1(\bar{z}) \cap \operatorname{Im} L_1(\bar{z}) = \{0\}$. To simplify notations and to treat higher order systems we change a little bit the previous notations. Let $P(x)$ be a $m \times m$ matrix valued smooth function defined near $\bar{x} \in \mathbf{R}^{n+1}$. We assume that $P(x)$ is a polynomial in x_0 so that

$$P(x) = \sum_{j=0}^{q} A_j(x')x_0^{q-j}, \tag{4}$$

where $x' = (x_1, ..., x_n)$. We always assume that $\det A_0(x') \neq 0$ near $x' = \bar{x}'$ and

$$\det P(x + \lambda\theta) = 0 \implies \lambda \text{ is real} \tag{5}$$

with $\theta = (1, 0, ..., 0)$. Recall that $\bar{x}$ is a characteristic of order r if

$$\partial_x^\alpha(\det P)(\bar{x}) = 0, \ \forall|\alpha| < r, \quad \partial_x^\alpha(\det P)(\bar{x}) \neq 0, \ \exists|\alpha| = r.$$

We first define the localization of $P(x)$ at a multiple characteristic $\bar{x}$ verifying

$$\mathrm{Ker}P(\bar{x}) \cap \mathrm{Im}P(\bar{x}) = \{0\}. \tag{6}$$

Let $\dim\mathrm{Ker}P(\bar{x}) = r$ and let $v_1, ..., v_r$ be a basis for $\mathrm{Ker}P(\bar{x})$. Taking (6) into account we can choose linear forms $\ell_1, ..., \ell_r$ so that

$$\ell_i(\mathrm{Im}P(\bar{x})) = 0, \quad \ell_i(v_j) = \delta_{ij},$$

where δ_{ij} is the Kronecker's delta. Then we define the $r \times r$ matrix $P_{\bar{x}}(x)$, which we call the localization of $P(x)$ at $\bar{x}$, by

$$(\ell_i(P(\bar{x} + \mu x)v_j))_{1 \leq i,j \leq r} = \mu[P_{\bar{x}}(x) + O(\mu)], \quad \mu \to 0.$$

It is easy to see that $P_{\bar{x}}(x)$ is a well defined map from $\mathrm{Ker}P(\bar{x})$ to $\mathrm{Ker}P(\bar{x})$. We can write

$$P_{\bar{x}}(x) = \sum_{j=0}^{n} P_j x_j$$

where P_j are $r \times r$ matrices.

Definition *We call the dimension of the linear subspace spanned by P_0, $P_1, ..., P_n$ the reduced dimension of $P_{\bar{x}}$ and we denote it by $d(P_{\bar{x}})$.*

We introduce the non degenerate characteristics.

Definition *We say that $\bar{x}$ is a non degenerate characteristic (of order r) of $P(x)$ if the following conditions are verified:*

$$\mathrm{Ker}P(\bar{x}) \cap \mathrm{Im}P(\bar{x}) = \{0\}, \tag{7}$$

$$d(P_{\bar{x}}) = r(r+1)/2, \quad r = \dim\mathrm{Ker}P(\bar{x}), \tag{8}$$

$$\det P_{\bar{x}}(\theta) \neq 0, \ \mathrm{Ker}(P_{\bar{x}}(\theta)^{-1}P_{\bar{x}}(x)) \cap \mathrm{Im}(P_{\bar{x}}(\theta)^{-1}P_{\bar{x}}(x)) \tag{9}$$

$$= \{0\}, \forall x \in \mathbf{R}^{n+1}.$$

Simple characteristics verify (7)–(9) with $r = 1$ and hence are non degenerate. Non degenerate double characteristics have a special feature:

Lemma 1 ([18]) *Assume that* $\dim \mathrm{Ker} P(\bar{x}) = 2$ *and* $\mathrm{Ker} P(\bar{x}) \cap \mathrm{Im} P(\bar{x})$ $= \{0\}$. *Then a double characteristic* $\bar{x}$ *is non degenerate if and only if the rank of the Hessian of* $P(x)$ *at* $\bar{x}$ *is maximal, that is 3.*

We prove that non degenerate characteristics are stable under hyperbolic perturbations.

Theorem 5 *Assume that* $P(x)$ *is a* $m \times m$ *real matrix valued smooth function of the form* (4) *verifying* (5) *in a neighborhood of* $\bar{x}$ *and let* $\bar{x}$ *be a non degenerate characteristic of order* r *of* P. *Let* $\tilde{P}(x)$ *be another* $m \times m$ *real matrix valued smooth function of the form* (4) *verifying* (5) *which is sufficiently close to* $P(x)$ *in* C^{q+2}, *then* $\tilde{P}(x)$ *has a non degenerate characteristic of the same order close to* $\bar{x}$. *Moreover, near* $\bar{x}$, *the characteristics of order* r *are non degenerate and they form a smooth manifold of codimension* $r(r+1)/2$. *In particular, near* $\bar{x}$ *the set of characteristics of order* r *of* $P(x)$ *itself consists of non degenerate ones which form a smooth manifold of codimension* $r(r+1)/2$.

We turn to the study of the Cauchy problem. Let us study a system of general order

$$P(x, D) = \sum_{|\alpha| \leq q} A_\alpha(x) D^\alpha, \quad D_j = \frac{1}{i} \frac{\partial}{\partial x_j} \tag{10}$$

where $A_\alpha(x)$ are $m \times m$ matrix valued smooth functions defined in Ω. We assume that

$$A_{(q,0,\dots,0)}(x) = I.$$

Let $P_q(x, \xi)$ be the principal symbol of $P(x, D)$:

$$P_q(x, \xi) = \sum_{|\alpha|=q} A_\alpha(x) \xi^\alpha,$$

and we assume that

$$\det P_q(x, \xi) = 0 \implies \xi_0 \in \mathbf{R}, \ \forall x \in \Omega, \ \forall \xi' = (\xi_1, \dots, \xi_n) \in \mathbf{R}^n. \tag{11}$$

We have

Theorem 6 *Assume that every characteristic over* $(0, \xi')$, $|\xi'| = 1$, *of* $P_q(x, \xi)$ *is at most double and non degenerate. Then the Cauchy problem for* $P(x, D)$ *is* C^∞ *well posed near the origin for arbitrary lower order terms. Moreover, if* $\tilde{P}(x, D)$ *is another system of the form* (10) *verifying* (11) *with the principal symbol* $\tilde{P}_q(x, \xi) = \sum_{|\alpha|=q} \tilde{A}_\alpha(x) \xi^\alpha$ *of* *which $\tilde{A}_\alpha(x)$ are sufficiently close to $A_\alpha(x)$ in $C^2(\Omega)$ for $|\alpha| = q$, then the Cauchy problem for $\tilde{P}(x,D)$ is C^∞ well posed near the origin for arbitrary lower order terms.*

Assuming the analyticity of the coefficients we have

Theorem 7 ([18]) *Assume that $A_\alpha(x)$, $|\alpha| = q$, are real analytic in Ω and every characteristic of $P_q(x,\xi)$ over $(0,\xi')$, $|\xi'| = 1$, is non degenerate. Then the Cauchy problem for $P(x,D)$ is C^∞ well posed near the origin for arbitrary lower order terms.*

Remark From Theorem 5 if

$$P(x,\xi) = \sum_{j=0}^{q} A_j(x,\xi')\xi_0^{q-j}, \quad A_q = I \qquad (12)$$

has a multiple non degenerate characteristic, then near P there is no strictly hyperbolic system of the form (12). Let P be a first order ($q = 1$) system with constant coefficients:

$$P(\xi) = \xi_0 - \sum_{j=1}^{n} A_j\xi_j = \xi_0 - A(\xi') \qquad (13)$$

where A_j are $m \times m$ constant coefficients. We always assume that all eigenvalues of $A(\xi')$ are real. From [8], $P(\xi)$ can not be strictly hyperbolic if $n > 2$ and $m \equiv 2$ modulo 4. Contrary to this if $q = 2$ it is clear that for any n and any m there exist strictly hyperbolic systems.

Proposition 1 *Let $m = 2$ and all eigenvalues of $A(\xi')$ are real. If $n \leq 2$, then $P(\xi)$ can be approximated by strictly hyperbolic systems. If $n > 2$ there is no strictly hyperbolic system.*

We now consider first order systems $L(x,D)$ which has a multiple characteristic $\bar{z}$ such that

$$\mathrm{Ker}L_1(\bar{z}) \cap \mathrm{Im}L_1(\bar{z}) = \{0\}.$$

Then we have defined the localization $L_{\bar{z}}(z)$ of L_1 at $\bar{z}$. We will be concerned with the question when we have

$$\mathrm{Ker}L_{\bar{z}}(z) \cap \mathrm{Im}L_{\bar{z}}(z) = \{0\} \qquad (14)$$

(this is the condition (9)) for multiple characteristics z of $L_{\bar{z}}(z)$.

Theorem 8 ([19]) *Let z_0 be a characteristic of L_1 of order m such that $\Lambda(h_{z_0})$ is involutive. Let z_1 be a characteristic of L_{z_0} of order r. If $L_1(x, D)$ is strongly hyperbolic, then every $(m - 1)$-th minor of L_{z_0} vanishes of order $r - 2$ at z_1.*

Let Σ be the set of characteristics of order m of L_1 and assume that Σ is an involutive manifold. Let $w \in \Sigma$, $z \in N_w\Sigma = T_w(T^*\Omega)/T_w\Sigma$. Then we can define the localization h_Σ of h along Σ by

$$h_\Sigma(w, z) = \lim_{\mu \to 0} h(w + \mu z).$$

Let $\tilde{\sigma}$ be the relative symplectic form on $N\Sigma$. Let Λ be a linear subspace in $T_X N\Sigma$. Then we say that Λ is 2-involutive if

$$\Lambda^{\tilde{\sigma}} = \{X \mid \tilde{\sigma}(X, Y) = 0, \ \forall Y \in \Lambda\} \subset \Lambda.$$

Theorem 9 ([19]) *Assume that Σ is an involutive manifold and $z_1 \in N_{z_0}\Sigma$ is a characteristic of order r of h_{z_0}. Assume that $\Lambda(h_{\Sigma,X})$, $X = (z_0, z_1)$ is 2-involutive. Then if $L_1(x, D)$ is strongly hyperbolic, we have*

$$\mathrm{Ker}L_{z_0}(z_1) \cap \mathrm{Im}L_{z_0}(z_1) = \{0\}.$$

In the definition of the non degenerate characteristics, the diagonal-izability of the localization is required. When the localization can be symmetrized then we could say much more for the original systems. Let $z_0 = (0, e_n)$ be a characteristic of order m and let $L_1(z_0) = 0$. Then one can write

$$
\begin{aligned}
L_1(x, \xi) &= \xi_n L_{z_0}(x, \xi'/\xi_n) + \xi_n R(x, \xi'/\xi_n), \ \ R(x, \xi'/\xi_n) \\
&= O(|x| + |\xi'/\xi_n|)^2,
\end{aligned}
\tag{15}
$$

where we assume that $L_{z_0}(x, \xi')$ is symmetric. We can regard (15) as "hyperbolic perturbation" of symmetric system $L_{z_0}(x, \xi')$ with perturbation $R(x, \xi')$. To simplify notations we write

$$L(x) = \sum_{j=0}^{n} A_j x_j, \ \ P(x) = L(x) + R(x), \ \ R(x) = O(|x|^2), \ \ x \to 0$$

where A_j are real symmetric and $R(x)$ is real analytic near $x = 0$. We assume that the perturbation is hyperbolic, that is

$$\det P(x + \lambda\theta) = 0 \implies \lambda \in \mathbf{R}.$$

We say that the hyperbolic perturbation is trivial if there exist real analytic $A(x)$, $B(x)$ defined near $x = 0$ such that $A(0)B(0) = I$ and

$A(x)P(x)B(x)$ is symmetric near $x = 0$. Let us denote by $M^s(m; \mathbf{R})$ the space of all real $m \times m$ symmetric matrices.

Theorem 10 ([21]) *Assume $d_m - m + 1 \leq p = d(L) \leq d_m$, where $d_m = m(m + 1)/2$. Then, generically, every hyperbolic perturbation is trivial, that is, in the $(d_m - p)(p - 1)$ dimensional Grassmannian of p dimensional subspaces of $M^s(m; \mathbf{R})$ containing the identity, the subset for which every hyperbolic perturbation is trivial is an open dense subset. In particular, when $p = d_m$, then every hyperbolic perturbation is trivial.*

Theorem 11 ([21]) *Assume that $m = 3$ and $4 \leq p = d(L) \leq 6 = d_3$. Then in the $(6 - p)(p - 1)$ dimensional Grassmannian of p dimensional subspaces of $M^s(3; \mathbf{R})$ containing the identity, the subset for which every hyperbolic perturbation is trivial is an open and dense subset.*

We return to $L_1(x, \xi)$.

Corollary 4 *Assume that $L_{z_0}(z)$ is symmetric and $d_m - m + 1 \leq d(L_{z_0}) \leq d_m$. Then, generically, $L_1(x, \xi)$ is smoothly symmetrizable near z_0.*

4. Pseudosymmetric systems

As noted in Introduction, the Cauchy problem for triangular systems (the lower order term is triangular at the same time) is always C^∞ well posed. The class of pseudosymmetric systems was introduced by D'Ancona and Spagnolo [4] which includes symmetric systems and triangular systems. Let us again consider the first order system $L(x, D)$ with symbol

$$L(x, \xi) = \xi_0 + \sum_{j=1}^{n} A_j(x)\xi_j = \xi_0 - A(x, \xi'). \tag{16}$$

We recall the definition:

Definition *The matrix $A(x, \xi') = (a_{ij}(x, \xi'))_{1 \leq i,j \leq m}$ is called pseudosymmetric when the following conditions are fulfilled for all choices of the indices h, k, $h_1,...,h_\nu \in \{1, ..., m\}$:*

$$a_{hk}(x, \xi') \cdot a_{kh}(x, \xi') \geq 0, \tag{17}$$

$$a_{h_1 h_2} a_{h_2 h_3} \cdots a_{h_\nu h_1} = \overline{a_{h_1 h_\nu}} \cdots \overline{a_{h_3 h_2}} \, \overline{a_{h_2 h_1}}. \tag{18}$$

It is clear that both symmetric systems and triangular systems are pseudosymmetric. For pseudosymmetric systems with coefficients depending only on x_0 we have

Theorem 12 (D'Ancona and Spagnolo [4]) *Assume that $A_j(x)$ depends only on x_0 real analytically and $A(x_0, \xi')$ is pseudosymmetric. Then the Cauchy problem is C^∞ well posed.*

For another special case that $n = 1$ and $A_1(x)$ depends only on x_1 real analytically we have

Theorem 13 ([22]) *Assume that $n = 1$ and $A_1(x)$ depends only on x_1 real analytically. Assume that $A(x_1)$ is pseudosymmetric and*

$$a_{ij}(0)a_{ji}(0) = 0, \quad \forall i, j = 1, ..., m.$$

Then the Cauchy problem is C^∞ well posed near the origin.

We can expect the C^∞ well posedness for pseudosymmetric systems under less restrictive conditions on the coefficient matrices, improving the results in Theorems 12 and 13.

5. Uniformly diagonalizable systems

We first recall the definition:

Definition *We say that $A(x, \xi')$ is uniformly diagonalizable if for every $(x, \xi') \in \Omega \times \mathbf{R}^n$, $|\xi'| = 1$, there exists a $m \times m$ matrix $S(x, \xi')$ verifying $\|S(x, \xi')^{-1}\|$, $\|S(x, \xi')\| \leq C$ with C independent of (x, ξ') such that*

$$S(x, \xi')^{-1} A(x, \xi') S(x, \xi')$$

is diagonal.

In the case of first order systems with constant coefficients it is known ([12]) that $L_1(\xi)$ is strongly hyperbolic if and only if $A(\xi')$ is uniformly diagonalizable. This fact motivates the study of uniformly diagonalizable systems with variable coefficients. We denote by $\gamma^{(s)}(\Omega)$ the class of Gevrey functions of order s defined in Ω, $1 \leq s$.

Theorem 14 (Kajitani [14]) *Assume that $A_j(x)$, $B(x) \in \gamma^{(s)}(\Omega)$ and $A(x, \xi')$ is uniformly diagonalizable. Then the Cauchy problem is $\gamma^{(s)}$ well posed near the origin if $1 \leq s < 2$.*

We present here some special results which show the complications for uniformly diagonalizable systems with variable coefficients. In what follows we study 2×2 systems which depend only on x_0.

Proposition 2 ([3]) *Assume that $A_j(x_0)$ depends on x_0 real analytically and $A(x_0, \xi')$ is uniformly diagonalizable. Then the Cauchy problem is C^∞ well posed for arbitrary smooth $B(x_0, x_1)$.*

Proposition 3 ([3]) *Assume that $A_j(x_0) \in C^k(I)$ and $A(x_0, \xi')$ is uniformly diagonalizable. Then the Cauchy problem is $\gamma^{(s(k))}$ well posed for every $B(x_0, x_1) \in C^1(I; \gamma^{(s(k))})$, where*

$$s(1) = 1, \quad s(2) = \frac{3}{2}, \quad s(k) = \frac{\sqrt{2k+1}+1}{2} \quad (k \geq 3).$$

Proposition 4 ([3], [24]) *Let $n = 1$. Then we can find $A_1(x_0) \in \cap_{s>2} \gamma^{(s)}$ for which $A(x_0, \xi_1)$ is uniformly diagonalizable but the Cauchy problem for L (with $B = 0$) is not C^∞ well posed.*

It is obvious, from Proposition 4 for instance, the diagonalizablity (or symmetrizability) at every fixed (x, ξ'), or fixed x and every ξ' does not imply the existence of a smooth diagonalizer (symmetrizer). But under some conditions it may happen. We make some observations on this subject.

Let us consider $L(x, \xi)$ in (16).

Theorem 15 *Assume that for every x there is $S(x)$ such that $S(x)^{-1} L(x, \xi) S(x)$ is symmetric for every ξ and the reduced dimension of $L(\bar{x}, \cdot) \geq m(m+1)/2 - [m/2]$ and $m \geq 3$. Then $L(x, \xi)$ is smoothly symmetrizable near $\bar{x}$, that is, there is a smooth matrix $T(x)$ defined near $\bar{x}$ such that*

$$T(x)^{-1} L(x, \xi) T(x)$$

is symmetric for any ξ and x near $\bar{x}$.

In some cases we can replace the symmetrizability by the diagonalizability in the assumption.

Theorem 16 ([23]) *Assume that $d(L(\bar{x}, \cdot)) \geq m(m+1)/2 - 1$ and $m \geq 3$ and $L(x, \xi)$ is diagonalizable for every (x, ξ), x near $\bar{x}$. Then $L(x, \xi)$ is smoothly symmetrizable.*

Corollary 5 *Assume that for every frozen x, $L(x,\xi)$ is strongly hyperbolic and $d(L(x,\cdot)) \geq m(m+1)/2 - 1$. Then $L(x,\xi)$ is strongly hyperbolic.*

6. Levi conditions

We now study the conditions required for lower order terms in order that the Cauchy problem is C^∞ well posed. We first study 2×2 systems.

Definition *Let $K(x,\xi)$ be a matrix with entries which are meromorphic functions. Let p, $q \in \mathbf{Q}_+$ such that $q < 1+p$. Denote $\delta = (1+p-q)^{-1}$, $\mu = 1 + \delta q$ and $\sigma = (\sigma_0, ..., \sigma_n)$ with $\sigma_j = \delta p$ for $0 \leq j \leq n-1$, $\sigma_n = \delta q$. We define*

$$K_\lambda(y,\eta) = K(\lambda^{-\sigma}y, \lambda^\sigma\eta + \lambda^\mu e_n),$$

where $\lambda^{-\sigma}y = (\lambda^{-\sigma_0}y_0, ..., \lambda^{-\sigma_n}y_n)$. We can write

$$K_\lambda(y,\eta) = \sum_{j=s}^\infty K_j(y,\eta)\lambda^{-\epsilon j}, \quad \epsilon \in \mathbf{Q}_+,$$

where $K_s(y,\eta)$ is not identically zero and define

$$\mathrm{Ord}K_\lambda = -\epsilon s.$$

We introduce the symbols

$$\mathcal{L}(x,\xi) = \left(L_0(x) + \frac{i}{2}\sum_{j=0}^n \frac{\partial^2 L_1}{\partial x_j \partial \xi_j}(x,\xi) \right) {}^{co}L_1(x,\xi) - \frac{i}{2}\{L_1, {}^{co}L_1\}(x,\xi)$$

and

$$\mathcal{L}^\#(x,\xi) = \frac{i}{2}\sum_{j=0}^n \frac{\partial^2 L_1}{\partial x_j \partial \xi_j}(x,\xi){}^{co}L_1(x,\xi) - \frac{i}{2}\{L_1, {}^{co}L_1\}(x,\xi),$$

where ${}^{co}L_1$ stands for the cofactor matrix of L_1.

Theorem 17 ([1]) *Assume that $A_j(x)$, $B(x)$ are real analytic and that there exist p, $q \in \mathbf{Q}_+$, $q < p+1$ such that*

$$\mathrm{Ord}h_\lambda = 2\delta p.$$

Then if the Cauchy problem is C^∞ well posed, then there is a meromorphic C such that $\mathrm{Ord}C_\lambda \leq 2\delta p$ and

$$\mathrm{Ord}(\mathcal{L} + L_1 C)_\lambda \leq 2\delta p.$$

Theorem 18 ([1]) *Assume that $A_j(x)$ are real analytic and there exist $p, q \in \mathbf{Q}_+$, $q < p + 1$ such that*

$$\mathrm{Ordh}_\lambda = 2\delta p.$$

Then if L_1 is strongly hyperbolic we have

$$\mathrm{Ord}L_{1\lambda} \leq 2\delta p$$

and there exists a meromorphic C such that $\mathrm{Ord}C_\lambda \leq \delta p$ and

$$\mathrm{Ord}(\mathcal{L}^\# + L_1 C)_\lambda \leq 2\delta p.$$

We turn to $m \times m$ systems. Assume that $z_0 = (0, e_n)$ be a characteristic of order r. We study the simplest case among truly vectorial cases, that is we assume

$$\mathrm{rank}L_1(0, e_n) = r - 2. \tag{19}$$

Let us denote

$$K(\theta) = \{f(x, \xi; \lambda)$$

$$= \sum_{j=n_f}^{\infty} \lambda^{-\theta j} f_j(x, \xi) \mid f_j(x, \xi) \text{ is meromorphic at } (0,0)\}.$$

We introduce in $K(\theta)$ a product $\#$, depending on s, defined by

$$f \# g = \sum \frac{1}{\alpha!} \lambda^{-\theta(i+j+s|\alpha|)} f_i^{(\alpha)}(x, \xi) g_{j(\alpha)}(x, \xi)$$

$$= \sum_k \left[\sum_{i+j+s|\alpha|=k} \frac{1}{\alpha!} f_i^{(\alpha)}(x, \xi) g_{j(\alpha)}(x, \xi) \right] \lambda^{-\theta k},$$

where $f_{(\beta)}^{(\alpha)}(x, \xi) = \partial_\xi^\alpha D_x^\beta f(x, \xi)$. Then $K(\theta)$ becomes a non commutative field. According to Dieudonné [5] we can introduce, in a unique way, the determinant "$\mathrm{Det}_{(s)}$" on $M(m; K(\theta))$ with values in

$$\overline{K}(\theta) = \{f(x, \xi; \lambda)$$

$$= \sum_{j=n_f}^{n_f+s-1} \lambda^{-\theta j} f_j(x, \xi) \mid f_j(x, \xi) \text{ is meromorphic at } (0,0)\}$$

with the product given by

$$f \cdot g = \sum_{k=n_f+n_g}^{n_f+n_g+s-1} \lambda^{-\theta k} \sum_{i+j=k} f_i(x, \xi) g_j(x, \xi).$$

Recall the localization h_{z_0} of h at z_0 is given by

$$h(\lambda^{-\theta}x, e_n + \lambda^{-\theta}\xi) = \lambda^{-r\theta}[h_{z_0}(x,\xi) + O(\lambda^{-\theta})].$$

Theorem 19 ([2]) *Let $2s + 2 \geq r$ and put*

$$G = L_1(\lambda^{-\theta}x, e_n + \lambda^{-\theta}\xi) + \lambda^{-(s+2)\theta}L_0(\lambda^{-\theta}x).$$

In order that the Cauchy problem for $L_1(x, D) + L_0(x)$ is C^{∞} well posed it is necessary that

$$\mathrm{Det}_{(s)}G = O(\lambda^{-r\theta}), \quad \sigma(\mathrm{Det}_{(s)}G) = h_{z_0},$$

where $\sigma(A)$ denotes the principal part of the λ-expansion of A.

It seems to be natural that in Theorem 19 we can remove the restriction $\mathrm{rank} L_1(0, e_n) = r - 2$.

References

[1] Benvenuti, S., Nishitani, T.: Necessary conditions for the hyperbolicity of 2 × 2 systems, Japanese J. Math. 25 (1999), 377–408.

[2] Bove, A., Nishitani, T.: Necessary conditions for hyperbolic systems, in preparation.

[3] Colombini, F., Nishitani, T.: Two by two strongly hyperbolic systems and the Gevrey classes. Ann. Univ. Ferrara Sc. Mat. Suppl. XLV (1999), 291–312.

[4] D'Ancona, P., Spagnolo, S.: On pseudosymmetric hyperbolic systems. Ann. Scuola Norm. Sup. Pisa 25 (1997), 397–417.

[5] Dieudonné, J.: Les déterminants sur un corps non commutatif. Bull. Soc. Math. France 71 (1943), 27–45.

[6] John, F.: Algebraic conditions for hyperbolicity of systems of partial differential equations. Comm. Pure Appl. Math. 31 (1978), 787–793.

[7] John, F.: Addendum to: Algebraic conditions for hyperbolicity of systems of partial differential equations. Comm. Pure Appl. Math. 31 (1978), 787–793.

[8] Lax, P.D.: The multiplicity of eigenvalues. Bull. Amer. Math. Soc. 6 (1982), 213–214.

[9] Lax, P.D.: Asymptotic solutions of oscillatory initial value problems. Duke Math. J. 24 (1957), 627–646.

[10] Mizohata, S.: Some remarks on the Cauchy problem. J. Math. Kyoto Univ. 1 (1961), 109–127.

[11] Ivrii, V.Ja., Petkov, V.M.: Necessary conditions for the Cauchy problem for non strictly hyperbolic equations to be well posed. Uspehi Mat. Nauk. 29 (1974), 3–70.

[12] Kasahara, K., Yamaguti, M.: Strongly hyperbolic systems of linear partial differential equations with constant coefficients. Mem. Coll. Sci. Univ. Kyoto Ser A. 33 (1960), 1–23.

[13] Kajitani, K.: Strongly hyperbolic systems with variable coefficients. Publ. RIMS. Kyoto Univ. 9 (1974), 597–612.

[14] Kajitani, K.: The Cauchy problem for uniformly diagonalizable hyperbolic systems in Gevrey classes. Hyperbolic Equations and Related Topics, Ed., S.Mizohata 101-123, Kinokuniya, 1986.

[15] Kajitani, K., Nishitani, T., Wakabayashi, S.: The Cauchy problem for hyperbolic operators of strong type. Duke Math. J. 75 (1994), 353–408.

[16] Hörmander, L.: Hyperbolic systems with double characteristics. Comm. Pure Appl. Math. 46 (1993), 261–301.

[17] Nishitani, T.: Necessary conditions for strong hyperbolicity of first order systems. J. Analyse Math. 61 (1993), 181–229.

[18] Nishitani, T.: Symmetrization of hyperbolic systems with non degenerate characteristics. J. Func. Anal. 132 (1995), 251–272.

[19] Nishitani, T.: On localization of a class of strongly hyperbolic systems. Osaka J. Math. 32 (1995), 41–69.

[20] Nishitani, T.: Strongly hyperbolic systems of maximal rank. Publ. RIMS. 33 (1997), 765–773.

[21] Nishitani, T.: Stability of symmetric systems under hyperbolic perturbations. Hokkaido Math. J. 26 (1997), 509–527.

[22] Nishitani, T., Spagnolo, S.: On pseudosymmetric systems with one space variable, to appear in Ann. Scuola Norm. Sup. Pisa.

[23] Nishitani, T., Vaillant, J.: Smoothly symmetrizable systems and the reduced dimensions. Tsukuba J. Math. 25 (2001), 165–177.

[24] Tarama, S.: Une note sur les systèmes hyperboliques uniformément disgonalisable. Mem. Fac. Eng. Kyoto Univ. 56 (1993), 9–18.

STRICTLY HYPERBOLIC OPERATORS AND APPROXIMATE ENERGIES

Ferruccio Colombini
Dipartimento di Matematica
Università di Pisa
Via F. Buonarroti 2
56127 Pisa, Italy
colombini@dm.unipi.it

Daniele Del Santo
Dipartimento di Scienze Matematiche
Università di Trieste
Via A. Valerio 12/1
34127 Trieste, Italy
delsanto@univ.trieste.it

Dedicated to Kunihiko Kajitani on the occasion of his sixtieth birthday

Abstract   In this note we collect some results on Sobolev-, C^∞- and Gevrey-well-posedness of the Cauchy problem for linear strictly hyperbolic operators having non Lipschitz-continuous coefficients. These results are obtained modifying the classical method of the energy estimates by the introduction of the so-called approximate energies, i.e. a family of energies which depend on a small parameter.

1.   Introduction

This note is devoted to the Cauchy problem for second order strictly hyperbolic operators. For sake of simplicity we will restrict ourselves to consider homogeneous operators, in view of the fact that in the strictly hyperbolic case the lower order terms are always "dominated" by the homogeneous part of maximal order.

Let us say before going on that the results for weakly hyperbolic operators, i.e. operators with multiple characteristics with constant or variable multiplicity, are very different from those ones we will present

H.G.W. Begehr et al. (eds.), Analysis and Applications - ISAAC 2001, 253–277.

here, as was pointed out for the first time by E. E. Levi [75] by the well known example

$$\partial_t^2 - \partial_x$$

for which the Cauchy problem is well-posed neither in C^∞ nor in the Gevrey classes γ^s for $s > 2$ (see also [51]).

Let us then consider

$$L = \partial_t^2 - \sum_{i,j=1}^n \partial_{x_i}(a_{ij}(t,x)\partial_{x_j}), \qquad (1)$$

where $a_{ij} = a_{ji}$ and

$$0 < \lambda_0 \le a(t,x,\xi) = \sum_{i,j=1}^n a_{ij}(t,x)\xi_i\xi_j/|\xi|^2 \le \lambda_0^{-1} \quad \text{for all } \xi \in \mathbf{R}^n \setminus \{0\} \tag{2}$$

with $a_{ij} \in C([0,T] \times \mathbf{R}_x^n)$. Actually in this section we will suppose that the coefficients are smooth functions.

Consider the Cauchy problem

$$\begin{cases} L(u) = 0 \\ u(0,x) = u_0(x), \quad \partial_t u(0,x) = u_1(x). \end{cases} \tag{3}$$

We will say that (3) is well-posed in a space $\mathcal{B}$ of indefinitely differentiable functions if, for all u_0, $u_1 \in \mathcal{B}$, there exists a unique solution u to (3) such that

$$u \in C^1([0,T]; \mathcal{B}). \tag{4}$$

This concept has been introduced by J. Hadamard: "problème correctement posé" ([48, p. 4 and pp. 41-66]).

The importance of the equations considered here above, in the study of propagation phenomena, is well known since the 18'th century. For studying the behavior of the vibrating string, J. d'Alembert considered the simplest operator of type (1):

$$L = \partial_t^2 - \partial_x^2, \tag{5}$$

and he gave the general solution $u(t,x) = f(t+x) + g(t-x)$, where f and g are arbitrary functions. We are at the beginning of the theory of Partial Differential Equations. We quote some passages from the interesting historical studies of S. Demidov [35] and of M. Kline [63] on this subject: "L'origine de cette théorie est souvent attribuée (voir [100] [101]) au mémoire de L. Euler *Additamentum ad dissertationem de infinitis curvis ejusdem generis* [37] publié en 1740. Comme l'indique son titre, ce mémoire se rapporte à la géometrie et, notamment, au problème des courbes dites isogones. [...] Dans le mémoire [37], L. Euler obtient des expressions que l'on pourrait aujourd'hui considérer comme des équations aux dérivées partielles du premier, second et troisième ordre. Mais ce n'est qu'à une seule d'entre elles, $\frac{\partial z}{\partial x} = P(x, a)$, qu'il attribue une telle signification, les autres expressions, bien qu'ayant pour nous une forme d'équations aux dérivées partielles, ont pour lui un sens tout différent. [...] On peut donc en conclure que L. Euler, tout en ayant affaire à des équations aux dérivées partielles, ne leur a pas à cette époque accordé une grande attention. La signification des résultats obtenus ne semble avoir été totalement comprise ni de l'auteur lui-même, ni, à plus forte raison, de la plupart de ses contemporains. Ce n'est que bien plus tard que ses continuateurs s'en sont servis pour tenter de démontrer sa priorité.

Si les résultats obtenus par Euler [37] préfigurent la naissance de la théorie des équations aux dérivées partielles, les travaux de d'Alembert des années 1740 marquent véritablement celle-ci. D'Alembert, en effet, y obtient plusieurs équations de physique mathématique de ce type et en effectue l'intégration.

D'après C. Truesdell [99, p.192], la première équation aux dérivées partielles apparaît en 1743 dans le *Traité de dynamique* de d'Alembert, dans l'étude du problème de la vibration d'une corde pesante suspendue à une extremité [1, pag.117]. " ([35, pp. 4-6]).

"Dans les mémoires [2] [3] publiés en 1749, d'Alembert écrit pour la première fois l'équation aux dérivées partielles des vibrations d'une corde, mais sous la forme suivante, qui diffère de celle d'aujourd'hui:

$$dp = \alpha \, dt + \nu \, ds,$$
$$dq = \nu \, dt + \frac{\alpha}{c^2} \, ds,$$

où $p = \frac{\partial y}{\partial t}$ et $q = \frac{\partial y}{\partial s}$ ($y(t, s)$ est l'élongation verticale du point s de la corde au temps t). Au debut ([2]), il traite le cas $c^2 = 1$.

En intégrant ce système suivant la méthode proposée dans [4], il écrit la solution sous la forme

$$y = \Psi(t + s) + \Gamma(t - s),$$

où $\Psi(u)$ et $\Gamma(u)$ sont des fonctions arbitraires de leurs arguments. Ensuite, dans [3], il étudia le cas de c^2 non égal à l'unité, dont l'intégration par la même méthode fournit

$$y = \Psi(ct + s) + \Gamma(ct - s).$$

[...] Dans le mémoire [3], on trouve également, comme le note C. Truesdell [99, pag.241], les origines de la méthode de séparation des variables.

[...] Dans le mémoire [6], contenu dans le premier tome de ses *Opuscules* (1761), d'Alembert présente, entre autres, l'équation de vibration de la corde d'épaisseur variable:

$$\frac{1}{S(s)} \cdot \frac{\partial^2 y}{\partial s^2} = \frac{\partial^2 y}{\partial t^2}$$

(we observe that this is perhaps the first example of an operator like (1) with variable coefficients). [...] Les études [2] [3] [4] ont immédiatement attiré l'attention des grands mathématiciens de cette époque, ont suscité l'intérêt d'Euler lui-même pour un nouveau domaine d'analyse (et lui ont rappelé ses propres recherches antérieures). Peu après la parution des travaux [2] [3] [...] dans un mémoire écrit en 1748 et publié en 1750: *De vibratione chordarum exercitatio* [38], Euler donna, d'après d'Alembert, sa méthode de résolution du problème, peu différente en fait de celle de ce dernier. Toutefois, à la différence de d'Alembert, il supposa que la fonction primitive pouvait être une courbe mécanique quelconque (cet arbitraire, nota-t-il plus tard [40], n'est limité que par l'exigence de contiguïté de la courbe). [...] La solution obtenue par Euler n'est pas classique: non seulement les dérivées secondes peuvent être discontinues, mais aussi, les dérivées premières. En fait, il introduisit, comme le note A. P. Juškevič [62], les solutions faibles de l'équation (5)." ([35, pp. 16-34]).

As M. Kline [63] says: "Though in method of solution he followed d'Alembert, Euler by this time had a totally different idea as to what functions could be admitted as initial curves and therefore as solutions of partial differential equations. Even before the debate on the vibrating-string problem, in fact in a work of 1734, he allowed functions formed from parts of different well-known curves and even formed by drawing curves freehand. [...] In his basic paper [38] Euler points out that all possible motions of the vibrating string are periodic in time whatever the shape of the string; that is, the period is (usually) the period of what we now call the fundamental. He also realized that individual modes whose periods are one half, one third, and so on of the basic (fundamental) period can occur as the vibrating figure. He gives such special solutions as

$$y(t, x) = \sum A_n \sin \frac{n\pi x}{l} \cos \frac{n\pi ct}{l}$$

when the initial shape is

$$y(0, x) = \sum A_n \sin \frac{n\pi x}{l},$$

[where l denotes the length of the string] but does not say whether the summation covers a finite or infinite number of terms. Neverthless he has the idea of superposition of modes. Thus Euler's main point of disagreement with d'Alembert is that he would admit all kinds of initial curves, and therefore non-analytic solutions, whereas d'Alembert accepted only analytic initial curves and solutions.

In introducing his «discontinuous» functions, Euler appreciated that he had taken a big step forward. He wrote to d'Alembert on December 20, 1763, that «considering such functions as are subject to no law of continuity [analyticity] opens to us a wholly new range of analysis» [40].

The solution of the vibrating-string problem was given in entirely different form by Daniel Bernoulli; this work stirred up another ground for controversy about the allowable solutions. [...] When he read d'Alembert first paper of 1746 and Euler's paper of 1749 on the vibrating string, he hastened to publish the ideas he had had for many years [8] [9]. [...] He reasserts that many modes of a vibrating string can exist simultaneously (the string then responds to the sum of superposition of all modes) and claims that this is all that Euler and d'Alembert have shown. Then comes a major point. He insists that *all* possible initial curves are representable as

$$f(x) = \sum_{n=1}^{\infty} a_n \sin \frac{n\pi x}{l} \tag{6}$$

because there are enough constants to make the series fit any curve. Hence, he asserts, *all* subsequent motions would be

$$y(t, x) = \sum_{n=1}^{\infty} a_n \sin \frac{n\pi x}{l} \cos \frac{n\pi ct}{l}.$$

Thus every motion corresponding to any initial curve is no more than a sum of sinusoidal periodic modes, and the combination has the frequency of the fundamental. However, he gives no mathematical arguments to back up his contentions; he relies on the physics. [...] Euler objected to Bernoulli's last assertion. In fact Euler's 1753 paper presented to the Berlin Academy [39] was in part a reply to Bernoulli's two papers. [...] D'Alembert, in his article «Fondamental» in Volume 7 (1757) of the *Encyclopédie*, also attacked Bernoulli. He did not believe that all odd and periodic functions could be represented by a series such as (6), because the series is twice differentiable but all odd and periodic functions need not be so. [...] Bernoulli did not retreat from his position. [...] The argument between d'Alembert, Euler, and Bernoulli continued for a decade with no agreement reached. The essence of the problem was the extent of the class of functions that could be represented by the sine series, or, more generally, by Fourier series.

In 1759, Lagrange, then young and unknown, entered the controversy. In his paper, which dealt with the nature and propagation of sound [68] (see also [67]) he gave some results on that subject and then applied his method to the vibrating string " ([63, pp. 505-510]).

But only during the following century these questions have been solved by J. Fourier [43]: "He points out, finally, that his work settles the arguments on solutions of the vibrating-string problem in favor of Daniel Bernoulli " ([63, p. 677]).

Let us now consider the operator corresponding to (5) but in more than one spatial variable. "In three spatial dimensions the basic form is

$$\frac{\partial^2 u}{\partial x^2} + \frac{\partial^2 u}{\partial y^2} + \frac{\partial^2 u}{\partial z^2} = \frac{\partial^2 u}{\partial t^2}. \tag{7}$$

As we know, this equation had already been introduced in the 18'th century and had also been expressed in spherical coordinates. During the 19'th century new uses of the wave equation were found, especially in the burgeoning field of elasticity. The vibrations of solid bodies of a variety of shapes with different initial and boundary conditions and the propagation of waves in elastic bodies produced a host of problems. Further work in the propagation of sound and light raised hundreds of additional problems.

Where separation of variables is possible, the technique of solving (7) is no different from what Fourier did with the heat equation or Lamé did after expressing the potential equation in some system of curvilinear coordinates. Mathieu's use of curvilinear coordinates to solve the wave equation by separation of variables is typical of hundreds of papers.

Quite another and important class of results dealing with the wave equation was obtained by treating the equation as an entirety. The first of such major results deals with initial-value problems and goes back to Poisson. [...] His principal achievement [90] [91] was a formula (P) for the propagation of a wave $u(x, y, z, t)$ whose initial state is described by the initial conditions " ([63, p. 690]).

"Progress in the solution of the wave equation by other methods is intimately connected with what are called steady-state problems, which lead to the reduced wave equation. The wave equation, by its very form, involved the time variable. In many physical problems, where one is interested in simple harmonic waves, one assumes that $u = w(x, y, z)e^{ikt}$, and by substituting this into the wave equation one obtains

$$\Delta w + k^2 w = \frac{\partial^2 w}{\partial x^2} + \frac{\partial^2 w}{\partial y^2} + \frac{\partial^2 w}{\partial z^2} + k^2 w = 0.$$

This is the reduced wave equation or the Helmholtz equation. The e-quation $\Delta w + k^2 w = 0$ represents all harmonic, acoustic, elastic, and electromagnetic waves. While the older authors were satisfied to find particular integrals, Hermann von Helmholtz (1821-94), in his work on the oscillations of air in a tube (organ pipe) with an open end, gave the first general investigation of its solutions [49]. [...] The work of Helmholtz was used by Gustav R. Kirchhoff (1824-87), one of the great German nineteenth-century mathematical physicists, to obtain another solution of the initial-value problem for the wave equation [57] ([63, pp. 693-694]).

The resolving formula to the problem (3) with respect to the operator

$$L = \partial_t^2 - \Delta_x \qquad (8)$$

for the case $n = 2$ can be deduced from that one of the case $n = 3$ using the so-called method of the descent, proposed by J. Hadamard for the solution of (8): "il suffit de supposer que, dans la formule (P), les functions u_0 et u_1 sont indépendantes de z. On a ainsi un premier exemple de ce que nous appellerons la "méthode de descente". Il peut paraître superflu de créer un mot spécial pour une remarque, en somme, puérile et qui a été employée dès les premiers stades de la théorie; mais comme nous aurons souvent à la faire intervenir, il nous sera commode de disposer d'un terme pour la désigner. Elle consiste à remarquer que qui peut le plus peut le moins: si on peut intégrer les équations a m variables, on peut en faire autant pour celles où n'interviennent que $(m - 1)$ variables ([48, pp. 69-70]).

Many other techniques in the study of hyperbolic problems are by now classical. We will only recall some of them: the successive approximations technique, developed and widely utilized one century ago (see [88], [83], [77]).

The method of Riemann (see [93]) is typical for the case of two variables: "The modern theory of hyperbolic partial differential equations was initiated by Bernhard Riemann's representation of the solution of the initial value problem for an equation of second order. His paper gives neither a general existence proof, nor a construction for a solution, except in explicit solvable examples. Assuming the existence of a solution u, it merely offers an elegant explicit integral representation of u in a form analogous to the representation of solutions of boundary value problems for elliptic equations by Green's function, and in fact, typical of a great variety of other such formulas expressing "linear functionals". The elementary presentation of Riemann's theory will subsequently be greatly generalized ([31, p. 449]). Riemann's representation formula was first generalized to higher order linear equations in two independent variables by P. Burgatti [15] and F. Rellich [92], and by E. Holmgren [52] to systems of first order equations with two independent variables. The method of Riemann's invariants has become very useful in the study of the formation of singularities for nonlinear hyperbolic equations in one space variable (see e.g. [64]; see also [7] for a wide treatment of this subject).

Next we may cite the technique of "Decomposition into plane waves". "This method is based on the expansion of the desired solution as a sum of plane waves, more precisely on a representation of it in the form of an integral

$$\int_{|\alpha|=1} f(t, \alpha \cdot x) dS_\alpha,$$

where dS_α is the standard measure on the unit sphere. As with Fourier's method, the plane wave method is applicable to the solution of the Cauchy problem for a general hyperbolic equation $P(D_t, D_x)u = f$ of order m with constant coefficients ([36, p. 145]) (see [32]; for a comprehensive monograph see [60]; see also [31, pp. 699-718] and [61, pp. 71-74]).

More recently the so-called method of parametrix has been developed; this method consists in searching for "asymptotic solutions " to the problem considered. The original idea of this technique, usually ascribed to P. D. Lax [69], can be found, even if only for the elliptic case, in [76] and [50]: "Le résultat [...] pour l'équation générale du type *elliptique*, a été obtenu indépendamment, par E. Elia Levi - un des jeunes géomètres italiens qui donnaient les plus belles espérances (il a sacrifié sa vie dans la dernière guerre), - et, sous une forme finalement équivalente, par M. Hilbert. Leur méthode à tous deux consiste à former une première approximation (la "parametrix" d'Hilbert), la quelle *ne satisfait pas* à l'équation donnée, mais, substituée dans cette équation, donne simplement un résultat qui, au point singulier, n'est que du premier ordre d'infinitude. Grâce à l'introduction de cette "parametrix", E. Elia Levi réussit à former la solution élémentaire; de son côté, M. Hilbert n'introduit pas cette dernière: il fait jouer à la parametrix le rôle qui appartient généralement à la solution élémentaire elle-même. Les deux questions n'en font réellement qu'une, et l'analyse de E. Elia Levi est, en somme, identique à celle de M. Hilbert ([48, pp. 377-378]). For a first application of this technique to hyperbolic equations see [48, pp. 377-415] for the case of analytic coefficients and [48, pp. 415-438] for non analytic coefficients; see finally [54] for a modern treatment of this subject.

The method of the Fourier transform has been always widely used in the study of hyperbolic problems, starting from the works of Cauchy ([17, Vol. 1, pp. 275-357]). See [44] for a modern utilization of this technique in the scalar case and [70] in the case of systems, both of them with constant coefficients. Fourier transform is usually applied with respect to the spatial variables and it is usually, but not necessarily, utilized in the case of coefficients depending only on time.

The method which may be considered the most powerful and fruitful is that one of the "energy estimates". In this technique an essential role is played by the L^2 norms of the solution or of its derivatives, with or without a previous application of the Fourier transform. "The method of energy integrals was first used by S. Zaremba [102]. It was rediscovered and extended by A. Rubinowicz [94], [95] and by K. O. Friedrichs and H. Lewy [42] and was used for the treatment of symmetric hyperbolic systems. Various later publications by K. O. Friedrichs [41] as well as J. Schauder [96] have clearly established the power and versatility of the method ([31, p. 642]). Fundamental in this direction are the works of Petrowski [85] and [86], of Leray [72] and [73] (see also the version due to Gårding [45] of Leray's beautiful argument) and of Gårding [46]. A different approach has been used in the work of Calderón and Zygmund [16]: "Their investigations on singular integral equations have provided a flexible tool [...] for the study [...] of linear partial differential equations. The key is the fact that by applying suitable integral operators one can symmetrise partial differential operators and thus provide other methods than that by Leray for the derivation of energy integrals ([31, p. 666]) (see also the papers by S. Mizohata [80] and [81]). For a wide and by now classical treatment of this subject see e.g. [31, pp. 625-667].

Other results on hyperbolic Cauchy problems and in particular some other applications of the method of energy estimates can be found in the books [71], [87], [61], [82] and [53]; let us recall finally the recent and very interesting review of results in [58] and the fundamental work [54, Vol. III].

We observe that, in the C^∞-well-posedness results obtained by using the energy estimates technique, the coefficients are usually supposed, for simplicity, to be C^∞. Actually to obtain energy estimates it is enough to suppose the Lipschitz-continuity of the coefficients of the principal part of the operator: see e.g. [53, pp. 234-241] and [82, pp. 311-316]. Moreover, it is important to remark that in the strictly hyperbolic case the energy estimates are obtained without "loss of derivatives (see e.g. [58, pp. 156-158]).

Considering operators of the type (1) which verify (2), if one would obtain an energy estimate of the type

$$\sup_{0 \leq t \leq T} (\|\nabla_x u(t, \cdot)\|_{L^2(\mathbf{R}^n)} + \|\partial_t u(t, \cdot)\|_{L^2(\mathbf{R}^n)})$$
$$\leq C(\|\nabla_x u(0, \cdot)\|_{L^2(\mathbf{R}^n)} + \|\partial_t u(0, \cdot)\|_{L^2(\mathbf{R}^n)}),$$

where C is a constant depending on λ_0, T, $\|\partial_t a(t, x, \xi)\|_{L^\infty}$, the usual regularity condition is that the coefficients of the leading term are Lipschitz-continuous in the variable t. If moreover we want solutions in the sense of (4) with $\mathcal{B} = C^\infty$, the regularity condition becomes the following: for all multi-indices α, the functions $\partial_x^\alpha a_{ij}(t, x)$ are Lipschitz-continuous with respect to $t \in [0, T]$, uniformly on the compact sets of $\mathbf{R}_x^n$, i.e.

$$\sum_{i,j=1}^n \|\partial_x^\alpha \partial_t a_{ij}(t, x)\|_{L^\infty([0,T] \times K)} \le C_{\alpha, K} \tag{9}$$

for all $\alpha \in \mathbf{N}^n$ and for all compact sets $K \subset \mathbf{R}_x^n$.

2. Nonregular coefficients: the case of the uniform nonregularity

One may ask if the condition (9) should be possibly weakened, and in which term: this question was posed by I. M. Gelfand in 1959 (see [47]). The simple example

$$\partial_t^2 - \partial_x(\alpha(t - x)\partial_x)$$

with $\alpha(s) = 1/2$ for $s < 0$ and $\alpha(s) = 2$ for $s > 0$, for which the Cauchy problem is not well-posed in C^∞ and not even in a weaker sense (see [56]), shows that one cannot hope to find solutions to the problem (1)–(3) without any regularity assumption on the coefficients of the operator. On the other hand the interest in weakening the condition (9) arise naturally starting from several problems. One of this is the so called "G-convergence" introduced by S. Spagnolo in [97] (see also [34]) in the elliptic case, for sequences $\{L^k\}$ of operators of type (1) verifying (2) uniformly (obviously the L^k are independent of the variable t). Let us recall the definition of G-convergence for the elliptic operators

$$L^k = -\sum_{i,j=1}^n \partial_{x_i}(a_{ij,k}(x)\partial_{x_j}) \qquad (k = 1, 2, 3, \ldots),$$

where $a_{ij,k}$ are real measurable functions on the bounded open set Ω, verifying condition (2) uniformly in k.

Definition 1 *We say that* $[a_{ij,k}] \xrightarrow{G} [a_{ij}]$ *on* Ω, *as* $k \to \infty$, *if* $u^k(f) \to u(f)$ *in* $L^2(\Omega)$ *for all* $f \in H^{-1}(\Omega)$, *where* $u^k(f)$ *and* $u(f)$ *denote the solutions of the Dirichlet problems*

$$\sum_{i,j=1}^n \partial_{x_i}(a_{ij,k}(x)\partial_{x_j}u^k) = f \quad on \ \Omega, \ u^k \in H_0^1(\Omega),$$

and

$$\sum_{i,j=1}^{n} \partial_{x_i}(a_{ij}(x)\partial_{x_j}u) = f \quad on \ \Omega, \ u \in H_0^1(\Omega).$$

See [98] and [10] for a complete picture of the property of the G-convergence in the elliptic case; we want only notice here that the G-limit does not depend on the open set Ω, in the sense that if $L^k \xrightarrow{G} L$ in Ω and $\Omega' \subset \Omega$, then $L^k \xrightarrow{G} L$ in Ω'.

In [26], for a sequence of parabolic operators

$$\partial_t - \sum_{i,j=1}^{n} \partial_{x_i}(a_{ij,k}(t,x)\partial_{x_j})$$

verifying (2) uniformly, it has been studied the link between the G-convergence for every fixed t and a sort of "parabolic G-convergence".

The hyperbolic case has been considered in [27], showing that if the sequence of the coefficients $a_{ij,k}$ is such that condition (9) for $\alpha = 0$ holds uniformly, then all the properties of the elliptic G-convergence transfer to the hyperbolic case (see [27, Th. 1 and Th. 2]). Obviously in these results an essential role is played by some energy estimates uniform with respect to the parameter k. Such kind of estimates require that conditions (2) and (9) are satisfied uniformly in k. Unfortunately the, may be, most interesting case of G-convergence is that one of the so-called homogenization: in the elliptic case the coefficients $a_{ij,k}(x)$ are given by $\alpha_{ij}(kx)$, where the α_{ij} are periodic. The similar situation in the hyperbolic case should be that in which the coefficients $a_{ij,k}(t,x)$ are given in the form $\alpha_{ij}(kt, k^\gamma x)$, with α_{ij} periodic, γ a positive constant. It is immediate to see that for such sequences of coefficients the condition (9) is never uniformly verified, apart the trivial case of α_{ij} independent of t.

The interesting fact is that in the case of the hyperbolic homogenization there is not convergence of the solutions of the problem (3) in L^2 and not even in $\mathcal{D}'$ or in $(\mathcal{D}^s)'$, the space of the Gevrey ultradistributions, for any $s > 1$. This phenomenon has been pointed out for the first time in [27, Ex. 2] with the following example

$$\begin{cases} \partial_t^2 u^k - a_k(t)\partial_x^2 u^k = 0, \\ u^k(0,x) = 0, \qquad \partial_t u^k(0,x) = u_1(x), \end{cases}$$

where u_1 is a suitable function and $a_k(t) = \alpha_\varepsilon(kt)$ with

$$\alpha_\varepsilon(\tau) = 1/4 - 2\varepsilon \sin \tau - \varepsilon^2(1 - \cos \tau), \qquad (10)$$

where ε is a fixed constant in $]0, 1/10]$.

Using the theory of Hill (see [78]), it is possible to see that this result holds *always* in the case of the hyperbolic homogenization, at least for coefficients depending only on t and not all constant; this is shown in the following theorem ([29, Par. 3 and 4]).

Theorem 1 *Let us consider a sequence of Cauchy problems as (1)-(3) with $a_{ij,k}(t) = \alpha_{ij}(kt)$, α_{ij} periodic functions not all constant. Let us assume that the initial data u_0 and u_1 belong to $L^2(\mathbf{R}_x^n)$ and let $u^k \in C^1([0,T]; L^2(\mathbf{R}_x^n))$ the solutions to (3)$_k$.*

Then the sequence $u^k(t,\cdot)$ is necessarily unbounded in $\mathcal{D}'$, however u_0 and u_1 have been taken, unless they are analytic on some complex strip $\{|\mathrm{Im}\, z_j| < \delta\}$ in which case $\{u^k(t,\cdot)\}$ must be unbounded for $t > C\delta$.

The situation described above leads to search for existence of the solution to (3) under the hypothesis (2) with a weaker form for the condition (9). To obtain some results in this direction, it was introduced in [20] a variant of the energy estimates method, using a sort of *approximate energy*. In [20] operators of type (1) with coefficients not depending on x are considered. Bounding ourselves to the case of initial data and solutions with compact support in $\mathbf{R}_x^n$ (notice that the well-known finite speed propagation property shows that this choice is not restrictive with respect to the general situation) we can use the Fourier transform with respect to x. The problem (3) becomes

$$\begin{cases} v'' + a(t,\xi)|\xi|^2 v = 0, \\ v(0,\xi) = \hat{u}_0(\xi), \qquad v'(0,\xi) = \hat{u}_1(\xi), \end{cases} \tag{11}$$

where $v(t,\xi)$ is the Fourier transform of u with respect to x and v' is the derivative of v with respect to t. Using the classical energy

$$E(t,\xi) = |v'|^2 + |\xi|^2 a(t,\xi)|v|^2 \tag{12}$$

we obtain, by differentiating E, from (11) and Gronwall's lemma,

$$E(t,\xi) \le E(0,\xi) \exp(\frac{1}{\lambda_0} \int_0^t |a'(s,\xi)| ds). \tag{13}$$

It is easily seen, looking at this formula, that, on one side it gives immediately the C^∞-well-posedness to the problem (3) under the hypotheses (2) and (9), but, on the other side, nothing can be deduced from it if one of the cited hypotheses does not hold. To overcome the difficulties deriving from the lack of regularity of the coefficients, in [20] it was for the fist time introduced an "approximate energy", i.e. an energy depending on a parameter ε, which shall be linked to the dual variable ξ

and which tends to 0 for ξ tending to ∞. We define, at the place of (12), the following energy:

$$E_\varepsilon(t,\xi) = |v'|^2 + |\xi|^2 a_\varepsilon(t,\xi)|v|^2, \qquad (14)$$

where $a_\varepsilon(t,\xi)$ is a suitable "approximation" of the function $a(t,\xi)$; $a_\varepsilon(t,\xi)$ is, by the way, Lipschitz-continuous in t and strictly positive. Arguing as before we obtain

$$E(t,\xi) \le E(0,\xi)\exp\left(\int_0^t \frac{|a'_\varepsilon(s,\xi)|}{a_\varepsilon(s,\xi)}ds + |\xi|\int_0^t \frac{|a(s,\xi) - a_\varepsilon(s,\xi)|}{(a_\varepsilon(s,\xi))^{1/2}}ds\right). \qquad (15)$$

In [20] the choice is the following: $a_\varepsilon(t,\xi) = (a(\cdot,\xi) * \rho_\varepsilon)(t)$, where ρ_ε is the usual Friedrichs' mollifier and the function a has been extended with continuity to the whole $\mathbf{R}$. The problem reduces to the estimate of the two integrals in (15) in term of ε (and consequently in term of ξ). We obtain three different cases

i) suppose that the coefficients a_{ij} belong to $C([0,T])$, then

$$\lim_{\varepsilon \to 0}\int_0^T \frac{|a'_\varepsilon(s,\xi)|}{a_\varepsilon(s,\xi)}ds = +\infty, \qquad \lim_{\varepsilon \to 0}\int_0^T \frac{|a(s,\xi) - a_\varepsilon(s,\xi)|}{(a_\varepsilon(s,\xi))^{1/2}}ds = 0;$$
$$(16)$$

ii) let $a_{ij} \in C^{0,\alpha}([0,T])$, then

$$\int_0^T \frac{|a'_\varepsilon(s,\xi)|}{a_\varepsilon(s,\xi)}ds \le C\varepsilon^{\alpha-1}, \qquad \int_0^T \frac{|a(s,\xi) - a_\varepsilon(s,\xi)|}{(a_\varepsilon(s,\xi))^{1/2}}ds \le C\varepsilon^\alpha;$$
$$(17)$$

iii) suppose finally $a_{ij} \in LL([0,T])$, i.e. there exists a positive constant K such that for all $|t - t'| \le 1/2$ it holds

$$|a_{ij}(t) - a_{ij}(t')| \le K|t - t'|\log(|t - t'|^{-1}), \qquad (18)$$

then

$$\int_0^T \frac{|a'_\varepsilon(s,\xi)|}{a_\varepsilon(s,\xi)}ds \le CK\log\frac{1}{\varepsilon},$$
$$\int_0^T \frac{|a(s,\xi) - a_\varepsilon(s,\xi)|}{(a_\varepsilon(s,\xi))^{1/2}}ds \le CK\varepsilon\log\frac{1}{\varepsilon}. \qquad (19)$$

From these inequalities it is easy to prove the next theorem ([20, Th. 3 and 4]).

Theorem 2 *In the situation described above, in the case i) the Cauchy problem (1)–(3) is well-posed in the space of real analytic functions; in the case ii) it is well-posed in the Gevrey space γ^s for all $s < 1/(1 - \alpha)$; in the case iii) it is well-posed in C^∞ with a loss of a finite number of derivatives which is proportional to the constant K of (18).*

To show this theorem it is enough to take $\varepsilon = |\xi|^{-1}$ in the estimates (16), (17) and (19) and to use the Paley-Wiener theorem, for the case in which the initial data have compact support. To avoid this last hypothesis, by using the finite speed of propagation property, it will be sufficient to consider some suitable regular cut-off functions in the cases *ii)* and *iii)*. In the case *i)*, on the contrary, some arguments dealing with analytic functionals (see [79]) and a duality procedure will be necessary.

These results have been extended to the case of coefficients depending also on the space variables, for the case *ii)* by Nishitani [84] and by Jannelli [59], using, in some sense, energies of infinite order. The case *i)* has been considered in [28], where the case of initial data in the space of the analytic functional is first studied applying the Fourier transform with respect to x. In this situation it is necessary to deal with some differential operators in the dual variable $\zeta \in \mathbf{C}^n$ of infinite order, to which again some approximate energy techniques are applied; the result for analytic initial data follows also in this case by a duality procedure. Similar results obtained in a totally different way can be found in [12], [13] and [14].

The case *iii)* is studied in [25]. First of all the case of the "isotropically" LL coefficients (i.e. for the coefficients the condition (18) is verified with the variable t replaced by $y = (t, x) \in [0, T] \times \mathbf{R}_x^n$) is considered and the following energy estimate holds ([25, Th. 2.1]).

Theorem 3 *Let $0 < \theta \le 1/4$ be a given constant. Then there exist $\beta > 0$, $T^* > 0$ and $C > 0$ such that for all $0 \le t \le T^*$, if $u \in C^\infty([0, T]) \times \mathbf{R}_x^n)$ solves the Cauchy problem (1)–(3), we have*

$$\sup_{0 \le s \le t} \|\partial_t u(s, \cdot)\|_{H^{-\theta-\beta s}} + \sup_{0 \le s \le t} \|u(s, \cdot)\|_{H^{1-\theta-\beta s}} \le C(\|u_1\|_{H^{-\theta}} + \|u_0\|_{H^{1-\theta}}).$$

$$(20)$$

Here $\beta = \lambda_0^{-1} M$, where λ_0 is defined in (2), M is a positive constant depending only on K, the LL-norm of the a_{ij}'s given in (18), and $T^* = 1/\beta$.

A second situation is then faced in [25]: the coefficients a_{ij} are smooth functions of the space variables, verifying (18), with K uniform with respect to x, and such that $\partial_x^\alpha a_{ij}(t, x) \in L^\infty([0, T] \times \mathbf{R}_x^n)$ for $\alpha \le 2$. The following estimate holds true ([25, Th. 2.3]).

Theorem 4 *There exists β and, for any $m \in \mathbf{N}$, there exists C_m such that, if $u \in C^\infty([0,T] \times \mathbf{R}_x^n)$ solves the problem (1)–(3), we get, for $T^* = 1/\beta$,*

$$\sup_{0 \le t \le T^*} \|\partial_t u(t, \cdot)\|_{H^{m-\beta t}} + \sup_{0 \le t \le T^*} \|u(t, \cdot)\|_{H^{1+m-\beta t}}$$

$$\le C_m(\|u_1\|_{H^m} + \|u_0\|_{H^{1+m}}). \tag{21}$$

From the estimates (20) and (21) the well-posedness follows immediately. The proofs of (20) and (21) are based on the Littlewood-Paley decomposition technique (see [11]) and again on the method of the approximate energies. However we wish to provide some informal arguments and a sketch of proof, in a model case, for our energy estimate. Let $u(x,t)$ $(t \in \mathbf{R}, x \in \mathbf{R})$ be a solution to

$$\begin{cases} \partial_t^2 u - \partial_x(a(t,x)\partial_x u) = 0 \\ u(0,x) = u_0(x), \quad \partial_t u(0,x) = u_1(x), \end{cases}$$

with $a \ge 1$. We consider

$$\varphi \in C_0^\infty(\tfrac{1}{2} < \xi < 2) \tag{22}$$

and we set

$$u_\nu(t,x) = (\varphi(\tfrac{D_x}{2^\nu})u)(t,x) = \int e^{2i\pi x \xi} \varphi(\tfrac{\xi}{2^\nu}) \hat{u}(t,\xi) d\xi, \tag{23}$$

where $\hat{u}$ stands for the Fourier transform in the x variable and $D_x = i^{-1}\partial/\partial_x$. We set also $\varphi_\nu = \varphi(\tfrac{D_x}{2^\nu})$. We get from (23)

$$\begin{cases} \partial_t^2 u_\nu - \partial_x(a(t,x)\partial_x u_\nu) = \partial_x[\varphi_\nu, a]\partial_x u, \\ u_\nu(0,x) = u_{0,\nu}(x), \quad \partial_t u_\nu(0,x) = u_{1,\nu}(x). \end{cases}$$

For simplicity of our exposition, let us assume $a(t,x)$ is LL in the t variable and C^∞ in the x variable. In this case, the commutator $[\varphi_\nu, a]$ does not give rise to any difficulty since φ_ν acts only in the x variables. The reader must be warned that the handling of this commutator is nontrivial when a is isotropically LL. We define

$$\Omega_\nu = 2\operatorname{Re} \int_0^T (\partial_t^2 u_\nu + D_x a D_x u_\nu, \partial_t u_\nu)_{L^2(\mathbf{R}_x)} e^{-\lambda_\nu t} dt, \tag{24}$$

where $\lambda_\nu > 0$ is to be chosen later. We have to deal with the low regularity of a in the t variable: this leads us to introduce a mollified version of a, a_ε defined as in the Theorem 2. We obtain some estimates similar to (19). We get from (24), with a certain amount of computations, where $\varepsilon = 2^{-\nu}$, which corresponds to the choice $\varepsilon = |\xi|^{-1}$ in the proof of Theorem 2, and $\lambda_\nu = \beta\nu$, with $\beta > 0$ sufficiently large,

$$|u_{1,\nu}|^2 + |D_x u_{0,\nu}|^2 + \Omega_\nu \geq$$

$$e^{-\lambda_\nu T}(|\partial_t u_\nu(T)|^2 + |D_x u_\nu(T)|^2)$$

$$+\frac{\beta}{2}\int_0^T e^{-\lambda_\nu t}(\nu|\partial_t u_\nu(t)|^2 + 2^{2\nu-2}\nu|u_\nu(t)|^2)dt.$$

This energy estimate gives a control of $e^{-\lambda_\nu t}|\partial_t u_\nu(t)|^2 + e^{-\lambda_\nu t}2^{2\nu}|u_\nu(t)|^2$, and since $\lambda_\nu = \beta\nu$, this amounts to the control of

$$\||D_x|^{-\alpha t}\partial_t u(t,\cdot)\|_{L^2(\mathbf{R}_x)} + \||D_x|^{1-\alpha t}u(t,\cdot)\|_{L^2(\mathbf{R}_x)},$$

which explains the loss of derivatives we referred to earlier. Essential tools in proving Theorems 3 and 4 are also the continuity from H^s to H^s of the multiplication by a for $|s| < 1$ when $a \in LL(\mathbf{R}^n)$, and the estimate of the commutator $[\varphi_\nu, a]$ (see [25, Prop. 3.5 and 3.6]).

3. Nonregular coefficients: the case of a singularity in a point

In the previous section we showed as, starting from the homogenization problem, we was led to study the Cauchy problem (1)–(3) for operators having coefficients with a regularity which is on one hand weaker than the Lipschitz-continuity, but on the other hand uniformly distributed on the interval $[0, T]$. On the contrary there is a large class of problems which brings naturally to consider linear Cauchy problem of the type (1)–(3) such that the coefficients a_{ij} have a singularity in a point $\bar{t} \in [0, T]$.

More precisely suppose to be interested to the lifespan $\bar{t}$ of the solutions to nonlinear hyperbolic equations with smooth (C^∞) coefficients a_{ij} which depend on its variables under the form $a_{ij}(u)$ or $a_{ij}(\nabla u)$. The possible formation of singularity (blow up) in the solution or in its gradient for t tending to $\bar{t}$ are strictly linked to the solvability of the linear problem (1)–(3) when $\partial_t a$ and $\nabla_x a$ are unbounded for t tending to $\bar{t}$ (for a wide and exhaustive survey on nonlinear hyperbolic problems see [7] and [55]).

This last problem, in the case of coefficients not depending on the space variables, have been discussed in [22]. The results of [22] have been extended to some classes of operators depending on all variables by Cicognani [18] and by Kubo and Reissig [65], [66].

In [22] we consider operators of the type (1), (2), with coefficients depending only on t, under the hypothesis that there exists $\bar{t} \in [0, T]$ such that $a_{ij} \in C^1([0, T] \setminus \{\bar{t}\})$ and

$$|\partial_t a_{ij}(t)| \leq C|t - \bar{t}|^q \tag{25}$$

for some $C > 0$ and $q \geq 1$. The result is the following ([22, Th. 1 and 2]):

Theorem 5 *i) Suppose that (25) holds with $q = 1$, then the problem (1)–(3) is C^∞-well-posed;*

ii) suppose that (25) holds with $q > 1$, then (1)–(3) is γ^s-well-posed for all $s < q/(q-1)$.

Adding to the condition (25) a Hölder-regularity of the coefficients, we obtain a result which is intermediate between the second part of the previous theorem and Theorem 2, *ii)*; we can prove the next theorem ([22, Th. 3]):

Theorem 6 *Suppose that (25) holds with $q > 1$ and $a_{ij} \in C^{0,\alpha}([0, T])$. Then (1)–(3) is γ^s-well-posed for all $s < q/((1 - \alpha)(q - 1))$.*

We give now a sketch of the proof of Theorems 5 and 6. Since the other situations are similar, we will only consider the case $\bar{t} = 0$.

We use again an approximate energy E_ε defined as in (14) from a suitable function $a_\varepsilon(t, \xi)$ and we obtain the estimate (15). In the proof of Theorem 5 we define

$$a_\varepsilon(t, \xi) = \begin{cases} a(\varepsilon, \xi) & \text{for } 0 \leq t \leq \varepsilon, \\ a(t, \xi) & \text{for } \varepsilon \leq t \leq T. \end{cases}$$

In the case *i)* we obtain (19) for ε sufficiently small and for some constants C and K. From this the conclusion follows taking $\varepsilon = |\xi|^{-1}$. In the case *ii)* we have

$$\int_0^T \frac{|a_\varepsilon'(t, \xi)|}{a_\varepsilon(t, \xi)} dt \leq C\varepsilon^{1-q}, \qquad \int_0^T \frac{|a(t, \xi) - a_\varepsilon(t, \xi)|}{(a_\varepsilon(t, \xi))^{1/2}} dt \leq C\varepsilon,$$

and it is enough to choose $\varepsilon = |\xi|^{-1/q}$ to conclude.

To prove Theorem 6 we take a_ε as in the Theorem 2, but in estimating the two integrals in (15) we will exploit the Hölder-continuity of a in a neighborhood of 0 and the differentiability of a away from 0. More precisely we obtain, with C a positive constant, β a positive constant to be chosen, and ε sufficiently small

$$\int_0^T \frac{|a_\varepsilon'|}{a_\varepsilon} dt \leq \int_0^{\varepsilon^\beta} \frac{|a_\varepsilon'|}{a_\varepsilon} dt + \int_{\varepsilon^\beta}^T \frac{|a_\varepsilon'|}{a_\varepsilon} dt \leq C(\varepsilon^{\beta + \alpha - 1} + \varepsilon^{\beta(1-q)}),$$

$$\int_0^T \frac{|a - a_\varepsilon|}{(a_\varepsilon)^{1/2}} dt \leq \int_0^{\varepsilon^\beta} \frac{|a - a_\varepsilon|}{(a_\varepsilon)^{1/2}} + \int_{\varepsilon^\beta}^T \frac{|a - a_\varepsilon|}{(a_\varepsilon)^{1/2}} \leq C(\varepsilon^{\beta+\alpha} + \varepsilon^{1+\beta(1-q)}).$$

The conclusion follows immediately choosing $\beta = (1-\alpha)/q$ and $\varepsilon = |\xi|^{-1}$.

We remark finally that in the very recent work [23] the hypotheses of Theorem 5,i), have been, in some sense, weakened; in fact the C^∞-well-posedness is proved under the assumption

$$|\partial_t a_{ij}(t)| \leq C|t - \bar{t}|^{-1} \log |t - \bar{t}|^{-1} \qquad \text{for } |t - \bar{t}| < 1/2,$$

which is obviously less restrictive than (25) with $q = 1$, but in this last result we have to suppose that $a_{ij} \in C^2([0, T] \setminus \{\bar{t}\})$ and moreover

$$|\partial_t^2 a_{ij}(t)| \leq C|t - \bar{t}|^{-2}(\log |t - \bar{t}|)^2 \qquad \text{for } |t - \bar{t}| < 1/2.$$

4. Counter examples

Starting from the counter example of [27], as referred in the Paragraph 2, many other examples have been constructed, no more for sequences L^k of operators of type (1), but for a single equation, showing that all the results stated in Paragraphs 2 and 3 are, in general, optimal. In all these examples a crucial role is played by the functions α_ε as given in (10) or, for some more sophisticated non existence examples (see [30]), by a refinement of them defined as follows: we consider a real non-negative 2π–periodic C^∞ function ρ defined on $\mathbf{R}$ such that $\rho(\tau) = 0$ for all τ in a neighborhood of 0 and

$$\int_0^{2\pi} \rho(\tau) \cos^2 \tau \, d\tau = \pi.$$

We define, for all $\tau \geq 0$ and $\varepsilon \in \,]0, \bar{\varepsilon}]$,

$$\alpha_\varepsilon(\tau) = 1 - 4\varepsilon\rho(\tau) \sin 2\tau + 2\varepsilon\rho'(\tau) \cos^2 \tau - 4\varepsilon^2 \rho^2(\tau) \cos^4 \tau,$$

$$w_\varepsilon(\tau) = \cos \tau \exp(-2\varepsilon \int_0^\tau \rho(s) \cos^2 s \, ds),$$

and, for all $\tau < 0$, $\alpha_\varepsilon(\tau) = \alpha_\varepsilon(-\tau)$, $w_\varepsilon(\tau) = w_\varepsilon(-\tau)$. As $\alpha_\varepsilon(\tau) = 1$ and $w_\varepsilon(\tau) = \cos \tau$ in a neighborhood of the origin, α_ε and w_ε are C^∞ functions on $\mathbf{R}$. Moreover, α_ε and w_ε verify the following properties:

$$\begin{cases} w_\varepsilon'' + \alpha_\varepsilon(\tau)w_\varepsilon = 0, \\ w_\varepsilon(0) = 1, \ w_\varepsilon'(0) = 0; \end{cases} \tag{26}$$

there exists $M > 0$, not depending on ε, such that for all $\tau \in \mathbf{R}$,

$$|\alpha_\varepsilon(\tau) - 1| \leq M\varepsilon, \qquad |\alpha_\varepsilon'(\tau)| \leq M\varepsilon;$$

and finally $\alpha_\varepsilon(\tau)$ and $w_\varepsilon(\tau)\exp(\varepsilon\tau)$ are 2π–periodic in $]-\infty,0]$ and $[0,+\infty[$. The importance of this construction resides in the fact that, as $|\tau| \to \infty$, the functions w_ε are exponentially decreasing. Handling suitably these functions it is possible to show the optimality of most part of the results cited here before; we recall, e.g., the following nonexistence theorem ([22, Th. 4]):

Theorem 7 *There exists a positive function* $a \in \mathcal{C}^\infty(\mathbf{R}\setminus\{0\})\cap\mathcal{C}^{0,\alpha}(\mathbf{R})$ *for all* $\alpha \in\,]0,1[$, *with*

$$\sup_{t>0} t^q|a'(t)| < +\infty,$$

for all $q > 1$, *and there exist* u_0, $u_1 \in \mathcal{C}^\infty(\mathbf{R})$ *such that the Cauchy problem*

$$\begin{cases} \partial_t^2 u - a(t)\partial_x^2 u = 0, \\ u(0,x) = u_0(x),\ u_t(0,x) = u_1(x), \end{cases}$$

has no solution in $\mathcal{C}^1([0,r[\,;\mathcal{D}'(\,]-r,r[\,))$, *for all* $r > 0$.

The example in [30] deals with an operator of the form (1) where $n = 1$ and $a \in \cap_{\alpha<1}\mathcal{C}^{0,\alpha}(\mathbf{R}^2)$, which is not solvable in any neighborhood of the origin.

On the other hand in [24], with a constructing technique similar to the previous cited ones, but much more complicate, the following result is shown ([24, Th. A]).

Theorem 8 *Let us consider the equation*

$$Pu = \partial_t^2 u - a(t)\partial_x^2 u + c(t,x)u = 0. \tag{27}$$

There exists a function $a(t)$, *with* $a(t) \geq 1/2$, $a \in \cap_{\alpha<1}\mathcal{C}^{0,\alpha}(\mathbf{R})$ *and there exists a* $\mathcal{C}^\infty$ *function* $c(t,x)$ *such that (27) has a solution* $u \in \mathcal{C}^\infty(\mathbf{R}^2)$ *with* $\operatorname{supp} u = \{(t,x) : t \geq 0\}$.

The proof of this theorem, which is very elaborate, employs some techniques which are not so far, from a logical point of view, from those ones used by De Giorgi [33], Leray [74], Cohen [19], Hörmander [53, pp. 224-230] and Pliś [89] in their examples, together with suitable energy estimates. However, the key point is always given by the functions α_ε and the properties of the solutions to the Cauchy problem (26).

The question of the minimal regularity to be imposed to the coefficients of the operator (27), under the hypothesis (2) of strict hyperbolicity, to ensure the *uniqueness* of the solution to the Cauchy problem has not yet a complete answer: on one side in [25] it is proved that the *LL*-continuity of a which is, as recalled in Paragraph 2, sufficient to the

C^∞-well-posedness, is also almost necessary to the same result ([25, Th. 2.4]). On the other side, as seen before, in [24] a nonuniqueness example is constructed for an operator P as defined in (27) with $a \in \cap_{\alpha<1} C^{0,\alpha}$. Recently in [21] the example of [24] has been refined in the following way. Let μ be a modulus of continuity, i.e. μ is a non-negative function defined on $[0, r]$ for some $r \in (0, 1)$, continuous, strictly increasing, concave and such that $\mu(0) = 0$. We say that a function f defined on an open set Ω is μ-continuous (and we will write $f \in C^\mu(\Omega)$) if for all K compact set in Ω there exists $\varepsilon > 0$ such that

$$\sup_{y,z \in K,\, 0<|y-z|<\varepsilon} \frac{|f(y) - f(z)|}{\mu(|y - z|)} < +\infty.$$

Then we have the following result ([21, Th.1]).

Theorem 9 *Suppose that $\int_0^\tau \mu(s)^{-1} ds < +\infty$ and that the function $s \mapsto -\mu(s)/(s \log s)$ is decreasing on $]0, r]$. Then there exists a function $a(t) \in C^\mu(\mathbf{R})$ such that the conclusion of the Theorem 8 still remains valid.*

The next question arises naturally.
Problem *Suppose that the coefficient $a(t)$ verifies, e.g., the following condition*

$$|a(t') - a(t'')| \le C|t' - t''|\,|\log|t' - t''||^{-1} \log\log|t' - t''|^{-1} \qquad (28)$$

for all $|t' - t''| < 1/4$, is the solution to the Cauchy problem for the operator P given in (27) unique, for all $c(t, x) \in C^\infty$?

Remark that a function verifying (28) is not in general a LL-continuous function, but on the other hand it does not satisfy the hypotheses of the nonuniqueness Theorem 9.

References

[1] d'Alembert, J.: Traité de dynamique. Paris, 1743.

[2] d'Alembert, J.: Recherches sur la courbe que forme une corde tendue, mise en vibration. Hist. Ac. Sci. Berlin, 1747, 3 (1749), 214–219.

[3] d'Alembert, J.: Suites des Recherches sur la courbe que forme une corde tendue, mise en vibration. Hist. Ac. Sci. Berlin, 1747, 3 (1749), 220–229.

[4] d'Alembert, J.: Réflexions sur la cause générale des vents. Paris, 1747.

[5] d'Alembert, J.: Addition au mémoire sur la courbe que forme une corde tendue, mise en vibration, Hist. Ac. Sci. Berlin, 1750, 6 (1752), 355–360.

[6] d'Alembert, J.: Recherches sur les vibrations des cordes sonores. *O-puscules mathématiques*, I, Paris, 1761, 1–64.

[7] Alinhac, S.: Blowup for nonlinear hyperbolic equations. Progress in Nonlinear Differential Equations and their Applications, Birkhäuser, Boston 1995.

[8] Bernoulli, D.: Réflexions et éclaircissemens sur les nouvelles vibrations des cordes. Hist. Ac. Sci. Berlin, 1753, 9 (1755), 147–172.

[9] Bernoulli, D.: Sur le mélange de plusieurs espéces de vibrations simples isochrones, qui peuvent coexister dans un même système de corps. Hist. Ac. Sci. Berlin, 1753, 9 (1755), 173–195.

[10] Bensoussan, A., Lions, J.-L., Papanicolaou, G.: Asymptotic Analysis for Periodic Structures. North-Holland, Amsterdam, 1978.

[11] Bony, J.-M.: Calcul symbolique et propagation des singularités pour les équations aux dérivées partielles non linéaires. Ann. Sci. École Norm. Sup. (4) 14 (1981), 209–246.

[12] Bony, J.-M., Shapira, P.: Existence et prologement des solutions holomorphes des équations aux dérivées partielles. Invent. Math. 17 (1972), 95–105.

[13] Bony, J.-M., Shapira, P.: Existence et prolongement des solutions analytiques des systèmes hyperboliques non stricts. C. R. Acad. Sci. Paris 274 (1972), 86–89.

[14] Bony, J.-M., Shapira, P.: Problème de Cauchy, existence et prolongement pour les hyperfonctions solutions des équations hyperboliques non strictes. C. R. Acad. Sci. Paris 274 (1972), 188–191.

[15] Burgatti, P.: Sull'estensione del metodo d'integrazione di Riemann all'equazioni lineari d'ordine n con due variabili indipendenti. Rend. Reale Accad. Lincei (5) 15 (1906), 602–609.

[16] Calderón, A.P., Zygmund, A.: Singular integral operators and differential equations. Amer. J. Math. 79 (1957), 901–921.

[17] Cauchy, L.-A.: Mémoire sur l'intégration des équations linéaires aux différentielles partielles et à coefficients constants. Oeuvres Complètes, 2e série, Gauthier-Villars, Paris, 1905.

[18] Cicognani, M.: The Cauchy problem for strictly hyperbolic operators with non-absolutely continuous coefficients. to appear in Tsukuba J. Math.

[19] Cohen, P: The non–uniqueness of the Cauchy problem. O. N. R. Techn. Report 93, Stanford 1960.

[20] Colombini, F., De Giorgi, E., Spagnolo, S.: Sur les équations hyperboliques avec des coefficients qui ne dépendent que du temp. Ann. Sc. Norm. Sup. Pisa, 6 (1979), 511–559.

[21] Colombini, F., Del Santo, D.: An example of non-uniqueness for a hyperbolic equation with non-Lipschitz-continuous coefficients. To appear.

[22] Colombini, F., Del Santo, D., Kinoshita, T.: Well-posedness of the Cauchy problem for a hyperbolic equation with non-Lipschitz coefficients, to appear in Ann. Sc. Norm. Sup. Pisa.

[23] Colombini, F., Del Santo, D., Reissig, M.: On the optimal regularity of coefficients in hyperbolic Cauchy problems. To appear.

[24] Colombini, F., Jannelli, E., Spagnolo, S.: Nonuniqueness in hyperbolic Cauchy problems. Ann. of Math. 126 (1987), 495–524.

[25] Colombini, F., Lerner, N.: Hyperbolic operators with non–Lipschitz coefficients. Duke Math. J. 77 (1995), 657–698.

[26] Colombini, F., Spagnolo, S.: Sur la convergence de solutions d'équations paraboliques. J. Math. Pures Appl. 56 (1977), 263–305.

274

[27] Colombini, F., Spagnolo, S.: On the convergence of solutions of hyperbolic e-
quations. Comm. Partial Differential Equations 3 (1978), 77–103.

[28] Colombini, F., Spagnolo, S.: Second order hyperbolic equations with coefficients
real analytic in space variables and discontinuous in time. J. Analyse Math. 38
(1980), 1–33.

[29] Colombini, F., Spagnolo, S.: Hyperbolic equations with coefficients rapidly oscillating in time: a result of nonstability. J. Differ. Equations 52 (1984), 24–38.

[30] Colombini, F., Spagnolo, S.: Some examples of hyperbolic equations without
local solvability. Ann. scient. Éc. Norm. Sup. (4) 22 (1989), 109–125.

[31] Courant, R., Hilbert, D.: Methods of Mathematical Physics. Interscience, New
York, 1, 1953, 2, 1962.

[32] Courant, R., Lax, P.D.: Remarks on Cauchy's problem for hyperbolic partial
differential equations with constant coefficients in several independent variables.
Comm. Pure Appl. Math. 8 (1955), 497–502.

[33] De Giorgi, E.: Un esempio di non unicità della soluzione del problema di Cauchy
relativo ad una equazione differenziale lineare a derivate parziali di tipo parabol-
ico. Rend. Mat. 14 (1955), 382–387.

[34] De Giorgi, E., Spagnolo, S.: Sulla convergenza degli integrali dell'energia per
operatori ellittici del secondo ordine. Boll. Unione Mat. Ital. (4) 8 (1973), 391–
411.

[35] Demidov, S.S.: Création et développement de la théorie des équations différentielles aux dérivées partielles dans les travaux de J. d'Alembert. Rev. His-
toire Sci. Appl. 35 (1982), 3–42.

[36] Egorov, Yu.V., Shubin, M.A.: Linear Partial Differential Equations. Foundations
of the Classical Theory. Encyclopaedia of Mathematical Science 30, Springer,
Berlin, 1991.

[37] Euler, L.: Additamentum ad dissertationem de infinitis curvis ejusdem generis.
Comm. Ac. Sci. Petrop. 1734–35. 7 (1740), 184–200.

[38] Euler, L.: De vibratione chordarum exercitatio Hist. Ac. Sci. Berlin, 1748, 4
(1750), 68–85 (= Id., Opera omnia, Ser. II, vol. 10, 63–77).

[39] Euler, L.: Remarques sur les mémoires précédents de M. Bernoulli. Hist. Ac. Sci.
Berlin, 1753, 9 (1755), 196–222 (= Id., Opera omnia, Ser. II, vol. 10, 232–254).

[40] Euler, L.: De chordis vibrantibus disquisitio ulterior. Novi comm. Ac. Sci.
Petrop., 1772, 17 (1773), 381–409 (= Id., Opera omnia, Ser. II, vol. 11/1, 62-80).

[41] Friedrichs, K.O.: Symmetric hyperbolic linear differential equations, Comm.
Pure Appl. Math. 7 (1954), 345–392.

[42] Friedrichs, K.O., Lewy, H.: Über die Eindeutigkeit und das Abhängigkeitsgebiet
der Lösungen beim Anfangswertproblem linearer hyperbolischer Differentialgleichungen. Mat. Ann. 98 (1928), 192–204.

[43] Fourier, J.: Œuvres. 2 voll., Gauthier-Villars, Paris, 1888–1890.

[44] Gårding, L.: Linear hyperbolic partial differential equations with constant coefficients. Acta Math. 85 (1950), 1–62.

[45] Gårding, L.: Solution directe du problème de Cauchy pour les équations hyperboliques. Colloques Internationaux du Centre Nat. de la Recherche Scient. 71
(1956), 71-90.

[46] Gårding, L.: L'inegalité de Friedrichs et Lewy pour les équations hyperboliques linéaires d'ordre superieur. C. R. Acad. Sci. Paris 239 (1954), 849–850.

[47] Gelfand, I.M.: Some questions of analysis and differential equations. Uspehi Mat. Nauk 14 (3) (1959), 3–19; Amer. Math. Soc. Transl. 26 (2) (1963), 201–219.

[48] Hadamard, J.: Le problème de Cauchy et les équations aux dérivées partielles linéaires hyperboliques. Hermann et Cie, Paris 1932.

[49] Helmholtz, H.: Theorie der Luftschwingungen in Röhren mit offenen Enden. J. Reine Angew. Math. 57 (1860), 1–72.

[50] Hilbert, D.: Grundzüge einer allgemeinen Theorie der linearen Integralgleichungen. Sechste Mitteilung, Gött. Nachr. (1910), 355–419.

[51] Holmgren, E.: Om Cauchys Problem vid de lineära partiella differentialekvationerna of 2:dra ordningen. Arkiv f. Mat., Astr. och Fys. 2 (1905), 1–13.

[52] Holmgren, E.: Sur les systèmes linéaires aux dérivées partielles du premier ordre à charactéristiques réelles et distinctes. Arkiv f. Mat., Astr. och Fys. 6 (1909), 1–10.

[53] Hörmander, L.: Linear Partial Differential Operators. Springer, Berlin 1963.

[54] Hörmander, L.: The Analysis of Linear Partial Differential Operators, I-IV. Springer, Berlin 1983–1985.

[55] Hörmander, L.: Lectures on Nonlinear Hyperbolic Differential Equations. Math. et Appl. 26, Springer, Berlin 1997.

[56] Hurd, A.E., Sattinger, D.H.: Questions of existence and uniqueness for hyperbolic equations with discontinuous coefficients. Trans. Amer. Math. Soc. 132 (1968), 159–174.

[57] Kirchhoff, G.: Zur Theorie der Lichtstrahlen. Sitzungsber. Akad. Wiss. zu Berlin 1882, 641–669.

[58] Ivriĭ, V.Ja.: Linear Hyperbolic Equations. Encyclopaedia of Mathematical Science, 33, Springer, Berlin 1993.

[59] Jannelli, E.: Regularly hyperbolic systems and Gevrey classes. Ann. Mat. Pura. Appl. 140 (1985), 133–145.

[60] John, F.: Plane Waves and Spherical Means Applied to Partial Differential Equations. Interscience, New York 1955.

[61] John, F.: Hyperbolic and Parabolic Equations. L. Bers, F. John and M. Schechter, Partial differential equations. Proceedings of the Summer Seminar, Boulder, Colorado, 1957. Lectures in Applied Mathematics, III. Interscience Publishers John Wiley & Sons, Inc. New York-London-Sydney 1964.

[62] Juškevič, A.P.: Istoriya matematiki v Rossii do 1917 goda. [History of mathematics in Russia up to 1917] Izdat. Nauka, Moscow 1968 (Russian).

[63] Kline, M.: Mathematical thought from ancient to modern times. Oxford University Press, New York 1972.

[64] Kleinerman, S., Majda, A.: Formation of singularities for wave equations including the nonlinear vibrating string. Comm. Pure Appl. Math. 33 (1980), 241–263.

[65] Kubo, A., Reissig, M.:Construction of parametrix for hyperbolic equations with slow oscillations in non-Lipschitz coefficients. To appear.

[66] Kubo, A., Reissig, M.: C^∞-well posedness of the Cauchy problem for quasilinear hyperbolic equations with coefficients non-Lipschitz in t and smooth in x. To appear.

[67] Lagrange, J.-L.: Nouvelles recherches sur la nature et la propagation du son. Misc. Taur., 1760-1761, 2 (1762), 11-172 (= Id., Œuvres, t. I, Paris, 1867, 151-316).

[68] Lagrange, J.-L.: Misc. Taur., t. 1_3, 1759, i-x, 1-112 (= Id., Œuvres, t. I, Paris, 39-148).

[69] Lax, P.D.: Asymptotics solutions of oscillatory initial value problems. Duke Math. J. 24 (1957), 627-646.

[70] Lax, P.D.: Differential equations, difference equations and matrix theory. Comm. Pure Appl. Math. 11 (1958), 175-194.

[71] Lax, P.D.: Lectures on hyperbolic differential equations. Stanford University, 1963.

[72] Leray, J.: Lectures on hyperbolic equations with variable coefficients. Inst. Advanced Study, Princeton, 1952.

[73] Leray, J.: On linear hyperbolic differential equations with variable coefficients on a vector space. Ann. Math. Studies 33 (1954), 201-210.

[74] Leray, J.: Équations hyperboliques non-strictes: contre-exemples, du type De Giorgi, aux théorèmes d'existence et d'unicité. Mat. Ann. 162 (1966), 228-236.

[75] Levi, E.E.: Sul problema di Cauchy. Rend. Reale Accad. Lincei (5) 16 (1907), 105-112.

[76] Levi, E.E.: Sulle equazioni lineari totalmente ellittiche alle derivate parziali. Rend. Circ. Mat. Palermo 24 (1907), 275-317.

[77] Levi, E.E.: Sul problema di Cauchy per le equazioni lineari in due variabili a caratteristiche reali. Rend. R. Ist. Lomb. di Scienze e Lettere (2) 41 (1908), 408-428.

[78] Magnus, W., Winkler, S.: Hill's Equation. Wiley, New York 1966.

[79] Martineau, A.: Sur les functionnelles analytiques et la transformation de Fourier-Borel. J. Analyse Math. 11 (1963), 1-164.

[80] Mizohata, S.: Le problème de Cauchy pour les systèmes hyperboliques et paraboliques. Mem. Coll. Sci. Univ. Kyoto, Ser. A, 32 (1959), 181-212.

[81] Mizohata, S.: Systèmes hyperboliques. J. Math. Soc. Japan 11 (1959), 205-233.

[82] Mizohata, S.: The theory of partial differential equations. University Press, Cambridge 1973.

[83] Nicoletti, O.: Sulla estensione dei metodi di Picard e di Riemann ad una classe di equazioni a derivate parziali. Atti della R. Accademia delle scienze fisiche e mat. di Napoli (2) 8 (1897), 1-22.

[84] Nishitani, T.: Sur les équations hyperboliques à coefficients höldériens en t et de classe de Gevrey en x. Bull. Sci. Math. (2) 107 (1983), 113-138.

[85] Petrowski, I.G.: Über das Cauchysche Problem für Systeme von partiellen Differentialgleichungen. Mat. Sbornik 2 (44) (1937), 815-866.

[86] Petrowski, I.G.: Über das Cauchysche Problem für ein System linearer partieller Differentialgleichungen im Gebiete der nichtanalytischen Funktionen. Bull. U-niv. Moscou, Sér. Int., Sect. A, 1 (7) (1938), 1-74.

[87] Petrowski, I.G.: Lectures on partial differential equations. 2nd ed. Gos. Izd. Tekh. Teor. Lit., Moscow 1953. English translation, Interscience, New York 1954.

[88] Picard, É.: Mémoires sur la théorie des équations aux dérivées partielles et la méthode des approximations successives. J. Math. Pures Appl. (4) 6 (1890), 145–210.

[89] Pliś, A.: The problem of uniqueness for the solutions of a system of partial differential equations. Bull. Acad. Pol. Sci. 2 (1954), 55–57.

[90] Poisson, D.: Mémoire sur la théorie du son. Journal de l'École Polytechn. 7 (1807), 319–392.

[91] Poisson, D.: Mém. de l'Acad. des Sci., Paris, (2) 3 (1818), 121–176.

[92] Rellich, F.: Verallgemeinerung der Riemannschen Integrationsmethode auf Differentialgleichungen *n*-ter Ordnung in zwei Veränderlichkeiten. Math. Ann. 103 (1930), 249–278.

[93] Riemann, B.: Über die Fortpflanzung ebener Luftwellen von endlicher Schwingungsweite. Abhandl. Königl. Ges. Wiss. Göttingen 8 (1850).

[94] Rubinowicz, A.: Herstellung von Lösungen gemischter Randwertprobleme bei hyperbolischen Differentialgleichungen zweiter Ordnung durch Zusammenstückelung aus Lösungen einfacherer gemischter Randwertaufgaben. Monatsh. Math. 30 (1920), 65–79.

[95] Rubinowicz, A.: Eindeutigkeit der Lösungen der Maxwellschen Gleichungen. Phyzik. Z. 27 (1926), 707–710.

[96] Schauder, J.: Das Anfangswertproblem einer quasilinearen hyperbolischen Differentialgleichung zweiter Ordnung. Fund. Math. 24 (1935), 213–246.

[97] Spagnolo, S.: Sulla convergenza di soluzioni di equazioni paraboliche ed ellittiche. Ann. Sc. Norm. Sup. Pisa, 22 (1968), 571–597.

[98] Spagnolo, S.: Convergence in energy for elliptic operators. Proc. 3rd Symp. numer. Solut. partial Differ. Equat., College Park 1975, (1976), 469–498.

[99] Truesdell, C.: The rational mechanics of flexible or elastic bodies. 1638–1788, dans L. Euler, Opera omnia, Ser. II, 11, Turici, 1960.

[100] Wallner, C.R.: Totale und partielle Differentialgleichungen. M. Cantor, Vorlesungen über Geschichte der Mathematik, Bd. 4, Leipzig, 1908, 871–1074.

[101] Wieleitner, H.: Geschichte der Mathematik. II. Teil, Leipzig, 1911.

[102] Zaremba, S.: Sopra un teorema d'unicità relativo alla equazione delle onde sferiche. Rend. Reale Accad. Lincei (5) 24 (1915), 904–908.

MULTIPLE HIGHLY OSCILLATORY SHOCK WAVES

Ya-Guang Wang
Department of Mathematics
Shanghai Jiao Tong University
Shanghai 200030, China
ygwang@online.sh.cn

Dedicated to Prof. K. Kajitani on his 60'th birthday

Abstract   This note is devoted to the rigorous theory of nonlinear geometric optics for multiple shocks to a general $N \times N$ conservation law in one space variable. For the problem of multiple weak shocks perturbed by small amplitude, high frequency oscillatory waves, we obtain that the leading profiles of oscillatory shocks are solutions to an integro-differential system with free boundaries, the leading terms of shock fronts don't oscillate, and oscillations only appear in the leading terms of shock speeds

1.   Introduction

Consider the following Cauchy problem for nonlinear hyperbolic equations with highly oscillatory initial data:

$$\begin{cases} \partial_t u^\epsilon + \sum_{j=1}^{n} A_j(t, x, u^\epsilon) \partial_{x_j} u^\epsilon = f(t, x, u^\epsilon), & t > 0, x \in I\!\!R^n \\ u^\epsilon(0, x) = \underline{u}_0(x) + \epsilon u_0^\epsilon(x) \end{cases} \tag{1}$$

where $L = \partial_t + \sum_{j=1}^{n} A_j(t, x, u) \partial_{x_j}$ is a first

order hyperbolic system with respect to t for any $u \in I\!\!R^N$, and

$$u_0^\epsilon(x) = U_0(x, \frac{\varphi_0(x)}{\epsilon}) + o(1)$$

with $U_0(x, \theta)$ being (almost/quasi-)periodic in θ

279

H.G.W. Begehr et al. (eds.), Analysis and Applications - ISAAC 2001, 279–291.
© *2003 Kluwer Academic Publishers.*

The topic of nonlinear geometric optics is to construct a solution u^ϵ of (1) in the form of

$$u^\epsilon(t, x) = \underline{u}(t, x) + \epsilon U(t, x; \frac{\varphi_1(t, x)}{\epsilon}, \dots, \frac{\varphi_m(t, x)}{\epsilon}) + o(\epsilon) \qquad (2)$$

where $U(t, x; \Theta)$ is (almost/quasi-)periodic in $\Theta \in I\!\!R^m$. The most important ingredients of this problem include:

- the existence of solution u^ϵ to (1) for a time interval $0 \leq t < T_0$ independent of $\epsilon \in (0, \epsilon_0]$;

- the derivation of problems for the profile U and phase functions $\varphi_j(t, x)$, and the existence of solutions $(U, \varphi_1, \dots, \varphi_m)$ to the corresponding problems;

- the rigorous justification of the asymptotics (2).

There has been rich literature on this field for the case of smooth background state $\underline{u}(t, x)$ (cf. [4] and references therein).

An important problem is to study the interaction between highly oscillatory waves and singular waves. Recently, some interesting results in this aspect were obtained. In [7], a formal analysis was developed for the problem of a single multidimensional shock, the papers [2, 9] were devoted to the rigorous justification for the problem of a single shock in one space variable, and Williams ([12]) established a rigorous theory on the nonlinear geometric optics for a single multidimensional shock. The case of two shock waves was studied by the author in [10] for the 2×2 conservation law, Corli in [3] studied the interaction between highly oscillatory waves and a one-dimensional contact discontinuity while we had considered the problem of delta-waves.

This note is devoted to the theory of nonlinear geometric optics for multiple shocks to a general $N \times N$ conservation law in one space variable, which has many prototypes in physics and mechanics such as gas dynamical equations. There are two motivations at least for studying this problem, one is to obtain the stability of weak shocks under the perturbation of high frequency oscillations, and the other one is to study the interaction phenomena between multiple shocks and oscillatory waves. Similar to [10], for the problem of multiple weak shocks perturbed by small amplitude, high frequency oscillatory waves, we shall obtain that the leading profiles of oscillatory shocks are solutions to an integro-differential system with free boundaries, the leading terms of shock fronts don't oscillate, and oscillations only appear in the leading terms of shock speeds.

In next section, we formulate the problems and state the main results. Then, in Section 3, we sketch the proof of the main results briefly, which gives our nonlinear geometric optics.

2. Formulations of Problems and Main Results

Suppose that the following $N \times N$ conservation law in one space variable:

$$\partial_t u + \partial_x f(u) = 0 \tag{3}$$

is strictly hyperbolic with respect to t for each $u \in \mathbb{R}^N$, let $\lambda_k(u)$ ($k = 1, \ldots, N$) be the eigenvalue of $f'(u)$ with $\lambda_1(u) < \ldots < \lambda_N(u)$, and $r_k(u)$ ($l_k(u)$ resp.) be the right (left resp.) eigenvector of $f'(u)$ with respect to $\lambda_k(u)$ satisfying the normalization $l_j(u) \cdot r_k(u) = \delta_{jk}$.

Without loss of generality, we assume that the Cauchy problem for (3) with piecewise constant initial data:

$$u(0, x) = \begin{cases} u_-, & x < 0 \\ u_+, & x > 0 \end{cases} \tag{4}$$

admits only two shocks

$$\underline{u} = \begin{cases} u_-, & -\infty < x < \sigma_1 t \\ u_\star, & \sigma_1 t < x < \sigma_2 t \\ u_+, & \sigma_2 t < x < \infty \end{cases} \tag{5}$$

with $\sigma_1 < \sigma_2$ and $u_\star$ being constants, and $\underline{u}$ satisfies the Rankine-Hugoniot condition

$$\sigma_1(u_\star - u_-) = f(u_\star) - f(u_-), \quad \sigma_2(u_+ - u_\star) = f(u_+) - f(u_\star) \tag{6}$$

and the Lax entropy condition:

$$\begin{cases} \lambda_i^\star < \sigma_1 < \lambda_i^-, & \lambda_{i-1}^- < \sigma_1 < \lambda_{i+1}^\star \\ \lambda_j^+ < \sigma_2 < \lambda_j^\star, & \lambda_{j-1}^\star < \sigma_2 < \lambda_{j+1}^+ \end{cases} \tag{7}$$

for fixed i, j with $1 \leq i < j \leq N$, where $\lambda_k^\natural = \lambda_k(u_\natural)$ for any $k \in \{1, \ldots, N\}$ and $\natural \in \{\pm, \star\}$. It shall be seen that our discussion also works for the general case that the Cauchy problem (3),(4) determines m shocks with $2 \leq m \leq N$ as in [1].

Denote by $r_k^\natural = r_k(u_\natural)$ and $l_k^\natural = l_k(u_\natural)$ for $\natural \in \{\pm, \star\}$. We impose two stability conditions on the shocks (5):

(S1) the matrices

$$\begin{cases} M_1 = (r_1^-, \ldots, r_{i-1}^-, u_\star - u_-, r_{i+1}^\star, \ldots, r_N^\star) \\ M_2 = (r_1^\star, \ldots, r_{j-1}^\star, u_+ - u_\star, r_{j+1}^+, \ldots, r_N^+) \end{cases}$$

are nonsingular;

(S2) let $C_0 = \max(|u_+|, |u_\star|, |u_-|)$, there is a small constant $\eta(C_0) \in (0,1)$ depending only on C_0, such that

$$\sum_{l=j}^{N} \left| \sum_{k=1}^{i} (\vec{e}_p^T M_1^{-1} r_k^\star)(\vec{e}_k^T M_2^{-1} r_l^\star) \right| < \eta(C_0) < 1$$

for any $p \in \{j, \ldots, N\}$, where $\vec{e}_k = (0, \ldots, 0, 1, 0, \ldots, 0)^T$ is the standard unit vector.

Remark 1 (1) When $N = 2$, $i = 1$, $j = 2$, which means that the conservation law (3) is a 2×2 system, let

$$u_\star - u_- = ar_1^\star + br_2^\star, \qquad u_+ - u_\star = cr_1^\star + dr_2^\star,$$

the condition (S2) is equivalent to

$$\left| \frac{bc}{ad} \right| < \eta(C_0) < 1$$

which is similar to that one given in [8, 10].

(2) When the shocks (5) are weak, the eigenvalues $\lambda_i(u)$ and $\lambda_j(u)$ of $f'(u)$ are genuinely nonlinear (i.e. $\nabla \lambda_k(u) \cdot r_k(u) \neq 0$ for $k = i, j$), the hypotheses (S1), (S2) hold always, and the condition (S1) is equivalent to the one-dimensional stability condition given by Majda in [6].

Given a small closed neighborhood $\omega \subset \{t = 0\}$ of the origin, suppose Ω is the closure of a determinacy domain of ω for the Cauchy problem of (3) when $|u - \underline{u}| < \delta$ for a fixed small constant $\delta > 0$. Equip the space $C^k(\Omega)$ of the k-th continuously differentiable functions in Ω with a family of norms

$$\|u\|_{k,\Omega}^\epsilon = \sum_{|\alpha| \leq k} \epsilon^{|\alpha|} \|\partial^\alpha u\|_{L^\infty(\Omega)}.$$

We say that a family $u^\epsilon \in C^k(\Omega)$ is *bounded in* $C_\epsilon^k(\Omega)$ if norms $\|u^\epsilon\|_{k,\Omega}^\epsilon$ are bounded, and $\phi^\epsilon \in C^k([0,T])$ *bounded in* $\tilde{C}_\epsilon^k([0,T])$ if $\|d_t\phi^\epsilon\|_{k-1,[0,T]}^\epsilon$ are bounded for $k \geq 1$. Let $C_p^0(\Omega : I\!R^q)$ be the space of continuous functions in $(t,x) \in \Omega$ valued as a continuous almost periodic function in $\theta \in I\!R^q$. For $k \in I\!N$, define the space $C_p^k(\Omega : I\!R^q)$ as the set of all $U \in C_p^0(\Omega : I\!R^q)$ whose derivatives $\partial_{(t,x;\theta)}^\alpha U$ belong to $C_p^0(\Omega : I\!R^q)$ for any $|\alpha| \leq k$.

Consider the following Cauchy problem of small perturbations of $\underline{u}$:

$$\begin{cases} \partial_t U^\epsilon + \partial_x f(U^\epsilon) = 0, \quad t > 0, \ x \in I\!R \\ U^\epsilon(0,x) = \begin{cases} u_+ + \epsilon u_{+,0}^\epsilon(x), & x > 0 \\ u_- + \epsilon u_{-,0}^\epsilon(x), & x < 0 \end{cases} \end{cases} \tag{8}$$

where $\epsilon > 0$ is small enough, and $u_{\pm,0}^\epsilon \in C^1(\omega^\pm)$ with $\omega^+ = \omega \cap \{x > 0\}$ and $\omega^- = \omega \cap \{x < 0\}$, satisfying that there are $U_{\pm,0} \in C_p^1(\omega^\pm : I\!R)$ such that

$$\|u_{\pm,0}^\epsilon(x) - U_{\pm,0}(x, \frac{x}{\epsilon})\|_{1,\omega^\pm}^\epsilon = o(1) \tag{9}$$

when ϵ goes to zero. Obviously, from (9) we have the boundedness of $u_{\pm,0}^\epsilon$ in $C_\epsilon^1(\omega^\pm)$.

We suppose that the problem (8) determines two shocks:

$$U^\epsilon(t,x) = \begin{cases} u_- + \epsilon u_-^\epsilon(t,x), & x < \sigma_1 t + \epsilon \varphi_1^\epsilon(t) \\ u_\star + \epsilon u_\star^\epsilon(t,x), & \sigma_1 t + \epsilon \varphi_1^\epsilon(t) < x < \sigma_2 t + \epsilon \varphi_2^\epsilon(t) \\ u_+ + \epsilon u_+^\epsilon(t,x), & x > \sigma_2 t + \epsilon \varphi_2^\epsilon(t). \end{cases} \tag{10}$$

The purpose of this paper is to study the local existence of U^ϵ in the form of (10) to the problem (8), and to analyse the asymptotic properties of $(u_\sharp^\epsilon, \varphi_1^\epsilon, \varphi_2^\epsilon)$ with respect to ϵ for $\sharp \in \{\pm, \star\}$.

Without loss of generality, we consider the general case

$$1 < i < j < N$$

and the special case $i = 1$ or $j = N$ is easier as in [10].

Denote by $\tilde{u}_{-,0}^\epsilon \in C_\epsilon^1(\omega)$ ($\tilde{U}_{-,0} \in C_p^1(\omega : I\!R)$ resp.) the extension of $u_{-,0}^\epsilon$ ($U_{-,0}(x,\theta)$ resp.) in $x \in \omega$ with the asymptotic property:

$$\|\tilde{u}_{-,0}^\epsilon(x) - \tilde{U}_{-,0}(x, \frac{x}{\epsilon})\|_{1,\omega}^\epsilon = o(1), \quad \text{when } \epsilon \to 0 \tag{11}$$

is valid.

By employing the result of Joly, Métivier and Rauch [5, Theorem 2.9.1] for the following Cauchy problem:

$$\begin{cases} \partial_t U_L^\epsilon + \partial_x f(U_L^\epsilon) = 0, \quad t > 0, \ x \in I\!R \\ U_L^\epsilon(0,x) = u_- + \epsilon \tilde{u}_{-,0}^\epsilon(x) \end{cases} \tag{12}$$

we obtain T, $\epsilon_0 > 0$ such that on $\Omega_T = \Omega \cap \{t \leq T\}$, there is a unique solution $U_L^\epsilon(t,x) = u_L + \epsilon \tilde{u}_L^\epsilon(t,x)$ of (12) with $\tilde{u}_L^\epsilon(t,x)$ bounded in $C_\epsilon^1(\Omega_T)$ for any $\epsilon \in (0, \epsilon_0]$; moreover, we have the asymptotic expansion:

$$\tilde{u}_L^\epsilon(t,x) = \tilde{U}_L(t,x; \frac{t}{\epsilon}, \frac{x}{\epsilon}) + o(1) \quad \text{in } C_\epsilon^1(\Omega_T) \tag{13}$$

when $\epsilon \to 0$, where $\tilde{U}_L \in C_p^1(\Omega_T : I\!\!R^2)$ is the unique solution to the initial value problem of an integro-differential system, see Joly et al. [5] for the detail.

Denote by $\{x = \lambda_1^- t + \epsilon \gamma_-^\epsilon(t)\}$ and $\{x = \lambda_N^+ t + \epsilon \gamma_+^\epsilon(t)\}$ the first and last characteristic curves of (8) respectively. Obviously, $\Omega^L = \{x < \lambda_1^- t + \epsilon \gamma_-^\epsilon(t)\}$ and $\Omega^R = \{x > \lambda_N^+ t + \epsilon \gamma_+^\epsilon(t)\}$ are the determinacy domains of ω_- and ω_+ respectively, which implies

$$u_L^\epsilon(t,x) = \tilde{u}_L^\epsilon(t,x)|_{\Omega^L}, \quad U_L(t,x; \tau, \theta) = \tilde{U}_L|_{(t,x) \in \Omega^L} \tag{14}$$

being independent of the above extension for initial data $(u_{-,0}^\epsilon, U_{-,0})$.

Clearly, $\gamma_-^\epsilon(t)$ should satisfy:

$$\begin{cases} d_t \gamma_-^\epsilon = (\lambda_1(u_- + \epsilon u_L^\epsilon(t, \lambda_1^- t + \epsilon \gamma_-^\epsilon(t))) - \lambda_1^-)/\epsilon \\ \gamma_-^\epsilon(0) = 0 \end{cases} \tag{15}$$

where $u_L^\epsilon \in C_\epsilon^1(\Omega_T^L)$ is given in (14).

Employing the results of [5] for the problem (15) of an ordinary differential equation, we deduce:

Lemma 1 *There is a unique solution* $\gamma_-^\epsilon \in \tilde{C}_\epsilon^2[0,T]$ *to the problem* (15), *and the following asymptotics*

$$\begin{cases} \gamma_-^\epsilon(t) = \Gamma_-(t) + o(1) \quad \text{in } L^\infty[0,T] \\ d_t \gamma_-^\epsilon(t) = \chi_-(t, \frac{t}{\epsilon}) + o(1) \quad \text{in } C_\epsilon^1[0,T] \end{cases} \tag{16}$$

holds, where

$$\begin{cases} \Gamma_-(t) = \int_0^t \nabla_u \lambda_1(u_-) \overline{U}_L(s, \lambda_1^- s) ds \\ \chi_-(t, \tau) = \nabla_u \lambda_1(u_-) U_L(t, \lambda_1^- t; \tau, \lambda_1^- \tau + \Gamma_-(t)) \end{cases} \tag{17}$$

with $U_L \in C_p^1$ *given in* (14), *and* $\overline{U}_L(t,x)$ *the mean value*

$$\overline{U}_L(t,x) = I\!\!E_0 U_L = \lim_{\rho \to \infty} \frac{1}{(2\rho)^2} \int_{-\rho}^{\rho} \int_{-\rho}^{\rho} U_L(t,x; \tau, \theta) d\tau d\theta. \tag{18}$$

In the same way, we obtain the solution $U_R^\epsilon(t,x) = u_+ + \epsilon u_R^\epsilon(t,x)$ to (8) in $\Omega^R = \{x > \lambda_N^+ t + \epsilon \gamma_+^\epsilon(t)\}$ and the asymptotic expansion:

$$u_R^\epsilon(t,x) = U_R(t,x; \frac{t}{\epsilon}, \frac{x}{\epsilon}) + o(1) \quad \text{in} \quad C_\epsilon^1(\Omega_T^R), \tag{19}$$

where $U_R \in C_p^1(\Omega_T^R, I\!\!R^2)$ satisfies an integro-differential system. Furthermore, we have the existence of $\gamma_+^\epsilon \in \tilde{C}_\epsilon^2[0,T]$, and the asymptotics:

$$\begin{cases} \gamma_+^\epsilon(t) = \Gamma_+(t) + o(1) & \text{in} \quad L^\infty[0,T] \\ d_t \gamma_+^\epsilon(t) = \chi_+(t, \frac{t}{\epsilon}) + o(1) & \text{in} \quad C_\epsilon^1[0,T] \end{cases} \tag{20}$$

holds, where

$$\begin{cases} \Gamma_+(t) = \int_0^t \nabla_u \lambda_N(u_+) \overline{U}_R(s, \lambda_N^+ s) ds \\ \chi_+(t,\tau) = \nabla_u \lambda_N(u_+) U_R(t, \lambda_N^+ t; \tau, \lambda_N^+ \tau + \Gamma_+(t)) \end{cases} \tag{21}$$

Denote by $\Omega^- = \{\lambda_1^- t + \epsilon \gamma_-^\epsilon(t) < x < \sigma_1 t + \epsilon \varphi_1^\epsilon(t)\}$, $\Omega^\star = \{\sigma_1 t + \epsilon \varphi_1^\epsilon(t) < x < \sigma_2 t + \epsilon \varphi_2^\epsilon(t)\}$ and $\Omega^+ = \{\sigma_2 t + \epsilon \varphi_2^\epsilon(t) < x < \lambda_N^+ t + \epsilon \gamma_+^\epsilon(t)\}$ respectively. It remains to study the existence of $(u_-^\epsilon, u_\star^\epsilon, u_+^\epsilon, \varphi_1^\epsilon, \varphi_2^\epsilon)$, which makes (10) to be a shock wave solution to the problem (8). From the Rankine-Hugoniot condition, we know that $(u_-^\epsilon, u_\star^\epsilon, u_+^\epsilon, \varphi_1^\epsilon, \varphi_2^\epsilon)$ satisfy the following problem:

$$\begin{cases} \partial_t u_\sharp^\epsilon + f'(u_\sharp + \epsilon u_\sharp^\epsilon) \partial_x u_\sharp^\epsilon = 0, \quad (t,x) \in \Omega^\sharp \; (\sharp = -, \star, +) \\ (\sigma_1 + \epsilon d_t \varphi_1^\epsilon)(\epsilon u_\star^\epsilon - \epsilon u_-^\epsilon + u_\star - u_-) = f(u_\star + \epsilon u_\star^\epsilon) - f(u_- + \epsilon u_-^\epsilon), \\ \qquad\qquad\qquad\qquad\qquad\qquad \text{on } x = \sigma_1 t + \epsilon \varphi_1^\epsilon(t) \\ (\sigma_2 + \epsilon d_t \varphi_2^\epsilon)(\epsilon u_+^\epsilon - \epsilon u_\star^\epsilon + u_+ - u_\star) = f(u_+ + \epsilon u_+^\epsilon) - f(u_\star + \epsilon u_\star^\epsilon), \\ \qquad\qquad\qquad\qquad\qquad\qquad \text{on } x = \sigma_2 t + \epsilon \varphi_2^\epsilon(t) \\ u_-^\epsilon = u_L^\epsilon, \quad \text{on } x = \lambda_1^- t + \epsilon \gamma_-^\epsilon(t) \\ u_+^\epsilon = u_R^\epsilon, \quad \text{on } x = \lambda_N^+ t + \epsilon \gamma_+^\epsilon(t) \\ \varphi_1^\epsilon(0) = \varphi_2^\epsilon(0) = 0 \end{cases} \tag{22}$$

which is a Goursat problem with free boundaries, where $u_{R,L}^\epsilon(t,x)$ is given above.

As in [8, 10], we introduce the transformations:

$$T^- : \begin{cases} y = t \dfrac{x - (\lambda_1^- t + \epsilon \gamma_-^\epsilon(t))}{(\sigma_1 - \lambda_1^-)t + \epsilon(\varphi_1^\epsilon(t) - \gamma_-^\epsilon(t))} \\ z = t \dfrac{\sigma_1 t + \epsilon \varphi_1^\epsilon(t) - x}{(\sigma_1 - \lambda_1^-)t + \epsilon(\varphi_1^\epsilon(t) - \gamma_-^\epsilon(t))} \end{cases}$$

for $u_-^\epsilon(t,x)$ in Ω^-,

$$T^\star : \begin{cases} y = t\dfrac{\sigma_2 t + \epsilon\varphi_2^\epsilon(t) - x}{(\sigma_2 - \sigma_1)t + \epsilon(\varphi_2^\epsilon(t) - \varphi_1^\epsilon(t))} \\[2mm] z = t\dfrac{x - (\sigma_1 t + \epsilon\varphi_1^\epsilon(t))}{(\sigma_2 - \sigma_1)t + \epsilon(\varphi_2^\epsilon(t) - \varphi_1^\epsilon(t))} \end{cases}$$

for $u_\star^\epsilon(t,x)$ in $\Omega^\star$, and

$$T^+ : \begin{cases} y = t\dfrac{x - (\sigma_2 t + \epsilon\varphi_2^\epsilon(t))}{(\lambda_N^+ - \sigma_2)t + \epsilon(\gamma_+^\epsilon(t) - \varphi_2^\epsilon(t))} \\[2mm] z = t\dfrac{\lambda_N^+ t + \epsilon\gamma_+^\epsilon(t) - x}{(\lambda_N^+ - \sigma_2)t + \epsilon(\gamma_+^\epsilon(t) - \varphi_2^\epsilon(t))} \end{cases}$$

for $u_+^\epsilon(t,x)$ in Ω^+, then, the problem (22) is equivalent to the following boundary value problem on $\Omega = \{y, z > 0\}$:

$$\begin{cases} L_\sharp^\epsilon(u_\sharp^\epsilon, \varphi_1^\epsilon, \varphi_2^\epsilon)u_\sharp^\epsilon = 0, \quad \sharp = +, -, \star \\ \mathcal{F}_1^\epsilon(t, \gamma_1 u_-^\epsilon, \gamma_1 u_\star^\epsilon, d_t\varphi_1^\epsilon, \varphi_1^\epsilon) = 0, \quad \text{on } z = 0 \\ \mathcal{F}_2^\epsilon(t, \gamma_2 u_\star^\epsilon, \gamma_2 u_+^\epsilon, d_t\varphi_2^\epsilon, \varphi_2^\epsilon) = 0, \quad \text{on } y = 0 \\ \gamma_1 u_+^\epsilon(t) = known \in C_\epsilon^1[0,T] \\ \gamma_2 u_-^\epsilon(t) = known \in C_\epsilon^1[0,T] \\ \varphi_1^\epsilon(0) = \varphi_2^\epsilon(0) = 0 \end{cases} \quad (23)$$

where $L_\sharp^\epsilon u_\sharp^\epsilon = 0$ are the first order quasilinear hyperbolic systems derived from those in (22) by using the transformations $T^\sharp$ for $\sharp = -, \star, +$, $\gamma_1 u(t) = u(t,0)$, $\gamma_2 u(t) = u(0,t)$, $\mathcal{F}_1^\epsilon = 0$ and $\mathcal{F}_2^\epsilon = 0$ are the Rankine-Hugoniot conditions given in (22).

Let us discuss the problem (23) in formal analysis.

For $\sharp \in \{-, \star, +\}$, denote by

$$R_\sharp = (r_1^\sharp, \dots, r_N^\sharp), \qquad L_\sharp = (l_1^\sharp, \dots, l_N^\sharp)^T$$

and

$$\Lambda_\sharp = diag[\lambda_1^\sharp, \dots, \lambda_N^\sharp].$$

Obviously, we have $L_\sharp R_\sharp = I$ and $L_\sharp f'(u_\sharp)R_\sharp = \Lambda_\sharp$. Suppose that $(u_\sharp^\epsilon, \varphi_k^\epsilon)$ are the form of

$$\begin{cases} u_\sharp^\epsilon(y,z) = R_\sharp U_\sharp(y,z; \frac{y}{\epsilon}, \frac{z}{\epsilon}) + O(\epsilon), \quad \sharp = -, \star, + \\ \varphi_k^\epsilon(t) = \varphi_k(t, \frac{t}{\epsilon}) + \epsilon\phi_k(t, \frac{t}{\epsilon}) + o(\epsilon), \quad k = 1, 2 \end{cases} \quad (24)$$

where $U_\sharp(y,z; \xi, \eta)$ and $(\varphi_k(t,\tau), \phi_k(t,\tau))$ are almost periodic in $(\xi, \eta) \in \mathbb{R}^2$ and $\tau \in \mathbb{R}$ respectively.

By plugging the expression (24) into (23), it follows

$$\partial_\tau \varphi_k(t,\tau) = 0, \quad k = 1,2$$

which means that the leading terms of shock fronts don't oscillate.

In order to formulate the problem of $U_\sharp$, we define $\mathbb{E}_\sharp = diag[\mathbb{E}_1^\sharp, \dots, \mathbb{E}_N^\sharp]$, where

$$\begin{cases} \mathbb{E}_k^- u(y,z;\xi,\eta) = \lim_{\rho\to\infty} \frac{1}{2\rho} \int_{-\rho}^{\rho} u(y,z;\xi - \frac{\lambda_1^- - \lambda_k^-}{\sigma_1 - \lambda_k^-} s, \eta + s) ds \\ \mathbb{E}_k^\star u(y,z;\xi,\eta) = \lim_{\rho\to\infty} \frac{1}{2\rho} \int_{-\rho}^{\rho} u(y,z;\xi - \frac{\lambda_k^\star - \sigma_2}{\lambda_k^\star - \sigma_1} s, \eta + s) ds \\ \mathbb{E}_k^+ u(y,z;\xi,\eta) = \lim_{\rho\to\infty} \frac{1}{2\rho} \int_{-\rho}^{\rho} u(y,z;\xi + s, \eta - \frac{\lambda_k^+ - \lambda_N^+}{\lambda_k^+ - \sigma_2} s) ds \end{cases} \quad (25)$$

are mean value operators for any $u \in C_p^0(\Omega : \mathbb{R}^2)$.

Denote by

$$\begin{cases} P_-(\partial_y, \partial_z) = (\Lambda_- - \lambda_1^- I)\partial_y + (\sigma_1 I - \Lambda_-)\partial_z \\ P_\star(\partial_y, \partial_z) = (\sigma_2 I - \Lambda_\star)\partial_y + (\Lambda_\star - \sigma_1 I)\partial_z \\ P_+(\partial_y, \partial_z) = (\Lambda_+ - \sigma_2 I)\partial_y + (\lambda_N^+ I - \Lambda_+)\partial_z, \end{cases}$$

then, by formal analysis, we obtain the following problem:

$$\begin{cases} \mathbb{E}_\sharp U_\sharp = U_\sharp, \quad \sharp = -, \star, + \\ \mathbb{E}_\pm (P_\pm(\partial_y, \partial_z)U_\pm + L_\pm f''(u_\pm)(R_\pm(\partial_\xi - \partial_\eta)U_\pm, R_\pm U_\pm) \\ \qquad\qquad + (d_\pm I + h_\pm \Lambda_\pm)(\partial_\xi - \partial_\eta)U_\pm) = 0 \\ \mathbb{E}_\star (P_\star(\partial_y, \partial_z)U_\star + L_\star f''(u_\star)(R_\star(\partial_\eta - \partial_\xi)U_\star, R_\star U_\star) \\ \qquad\qquad + (d_\star I + h_\star \Lambda_\star)(\partial_\eta - \partial_\xi)U_\star) = 0 \\ \chi_1(u_\star - u_-) + (\sigma_1 I - \Lambda_\star)R_\star \gamma_1 U_\star - (\sigma_1 I - \Lambda_-)R_- \gamma_1 U_- = 0, \\ \qquad\qquad\qquad\qquad\qquad\qquad \text{on } z = \eta = 0 \\ \chi_2(u_+ - u_\star) + (\sigma_2 I - \Lambda_+)R_+ \gamma_2 U_+ - (\sigma_2 I - \Lambda_\star)R_\star \gamma_2 U_\star = 0, \\ \qquad\qquad\qquad\qquad\qquad\qquad \text{on } y = \xi = 0 \\ \gamma_1 U_+(t,\tau) = U_+(t,0;\tau,0) = known \\ \gamma_2 U_-(t,\tau) = U_-(0,t;0,\tau) = known \end{cases}$$

$$(26)$$

where $\chi_k(t,\tau) = d_t \varphi_k(t) + \partial_\tau \phi_k(t,\tau)$ $(k = 1,2)$ are leading terms of shock speeds $d_t \varphi_k^\varepsilon(t)$, $(d_\sharp, h_\sharp) \in C^1$ is calculated from $(\varphi_1, \varphi_2, \Gamma_+, \Gamma_-)$, e.g.

$$d_\star = \frac{(z\sigma_1 + y\sigma_2)(\varphi_2 - \varphi_1)}{(\sigma_2 - \sigma_1)(y + z)^2} - \frac{y d_t \varphi_2 - z d_t \varphi_1}{y + z}, \qquad h_\star = \frac{\varphi_1 - \varphi_2}{(\sigma_2 - \sigma_1)(y + z)}.$$

Remark 2 If $U_\natural = E_\natural U_\natural$ for $\natural \in \{-, \star, +\}$, then the oscillation phases of $U_{-,k}$, $U_{\star,k}$ and $U_{+,k}$ are $\varphi_k^-(y,z) = (\sigma_1 - \lambda_k^-)y - (\lambda_k^- - \lambda_1^-)z$, $\varphi_k^\star(y,z) = (\lambda_k^\star - \sigma_1)y - (\sigma_2 - \lambda_k^\star)z$ and $\varphi_k^+(y,z) = (\lambda_N^+ - \lambda_k^+)y - (\lambda_k^+ - \sigma_2)z$ respectively Thus, $E_\natural U_\natural = U_\natural$ implies that the oscillation phases are solutions to eikonal equations, and the problem (26) is a boundary value problem for an integro-differential system.

To solve (26), we should first determine (φ_1, φ_2). By acting the mean value operator

$$E_0 u(y,z) = \lim_{\rho \to \infty} \frac{1}{(2\rho)^2} \int_{-\rho}^{\rho} \int_{-\rho}^{\rho} u(y,z;\xi,\eta) d\xi d\eta$$

on (26), it follows that $\overline{U}_\natural = E_0 U_\natural$ and (φ_1, φ_2) satisfy the following linear problem:

$$\begin{cases} P_\natural(\partial_y, \partial_z)\overline{U}_\natural = 0, \quad \natural = -, \star, + \\ d_t\varphi_1(u_\star - u_-) + (\sigma_1 I - \Lambda_\star)R_\star \gamma_1 \overline{U}_\star - (\sigma_1 I - \Lambda_-)R_- \gamma_1 \overline{U}_- = 0, \\ \qquad\qquad\qquad\qquad\qquad\qquad\qquad\qquad\qquad\qquad \text{on } z = 0 \\ d_t\varphi_2(u_+ - u_\star) + (\sigma_2 I - \Lambda_+)R_+ \gamma_2 \overline{U}_+ - (\sigma_2 I - \Lambda_\star)R_\star \gamma_2 \overline{U}_\star = 0, \\ \qquad\qquad\qquad\qquad\qquad\qquad\qquad\qquad\qquad\qquad \text{on } y = 0 \\ \gamma_1 \overline{U}_+(t) = \overline{U}_+(t,0) = known, \quad \gamma_2 \overline{U}_-(t) = \overline{U}_-(0,t) = known \\ \varphi_1(0) = \varphi_2(0) = 0. \end{cases}$$

$$(27)$$

The main results of this paper are stated as follows:

Theorem 1 *Denote by $\Omega_T = \{y, z > 0, y + z < T\}$. Under the assumptions (S1), (S2),*

$$u_{\pm,0}^\epsilon(x) - U_{\pm,0}(x, \tfrac{x}{\epsilon}) = o(1) \quad \text{in } C_\epsilon^1(\omega_\pm)$$

and the compatibility conditions for problems (23) and (26) of order one are satisfied, we have

(1) there are $T, \epsilon_0 > 0$ such that the problem (23) has unique solutions $(u_-^\epsilon, u_\star^\epsilon, u_+^\epsilon)$ and $(\varphi_1^\epsilon, \varphi_2^\epsilon)$ bounded in $C_\epsilon^1(\Omega_T)$ and $\tilde{C}_\epsilon^2[0,T]$ for any $\epsilon \in (0, \epsilon_0]$;

(2) there are unique solutions $(U_-, U_\star, U_+) \in C_p^1(\Omega_T, \mathbb{R}^2)$, $(\chi_1, \chi_2) \in C_p^1([0,T], \mathbb{R})$ to (26), and $(\varphi_1, \varphi_2) \in C^2[0,T]$ to (27);

(3) when ϵ goes to zero, the following asymptotic expansions hold:

$$\begin{cases} u_\natural^\epsilon(y,z) = R_\natural U_\natural(y,z; \tfrac{y}{\epsilon}, \tfrac{z}{\epsilon}) + o(1) \quad \text{in } C_\epsilon^1(\Omega_T) \\ d_t\varphi_k^\epsilon(t) = \chi_k(t, \tfrac{t}{\epsilon}) + o(1) \quad \text{in } C_\epsilon^1[0,T] \\ \varphi_k^\epsilon(t) = \varphi_k(t) + o(1) \quad \text{in } L^\infty[0,T] \end{cases}$$

$$(28)$$

for $\natural = -, \star, +$ *and* $k = 1, 2$.

3.   Sketch of The Proof

Mainly, Theorem 1 shall be proved by the following four steps:

STEP 1 Construct the first approximate solutions $(u_-^{\epsilon,0}, u_\star^{\epsilon,0}, u_+^{\epsilon,0}) \in C_\epsilon^1$ (Ω_T) and $(\varphi_1^{\epsilon,0}, \varphi_2^{\epsilon,0}) \in \tilde{C}_\epsilon^2[0,T]$ satisfying the compatibility conditions of (23), and the first approximate solutions $(U_-^0, U_\star^0, U_+^0) \in C_p^1(\Omega_T, I\!\!R^2)$ and $(\chi_1^0, \chi_2^0) \in C_p^1([0,T], I\!\!R)$ satisfying the compatibility conditions of (26). Moreover, the following asymptotics

$$\begin{cases} u_\natural^{\epsilon,0}(y,z) = R_\natural U_\natural^0(y,z;\frac{y}{\epsilon},\frac{z}{\epsilon}) + o(1) & \text{in } L^\infty(\Omega_T) \quad (\natural = -, \star, +) \\ d_t\varphi_k^{\epsilon,0}(t) = \chi_k^0(t,\frac{t}{\epsilon}) + o(1) & \text{in } L^\infty[0,T] \quad (k = 1,2) \end{cases}$$

(29)

holds while $\epsilon \to 0$.

STEP 2 Solve the problem (23) by the following iteration scheme:

$$\begin{cases} L_\natural^\epsilon(u_\natural^{\epsilon,\nu}, \varphi_1^{\epsilon,\nu}, \varphi_2^{\epsilon,\nu})u_\natural^{\epsilon,\nu+1} = 0, \quad y, z > 0 \quad (\natural = \pm, \star) \\ F_{1,(\gamma_1 u_-^{\epsilon,\nu}, \gamma_1 u_\star^{\epsilon,\nu}, d_t\varphi_1^{\epsilon,\nu})}^\epsilon (\gamma_1 u_-^{\epsilon,\nu+1}, \gamma_1 u_\star^{\epsilon,\nu+1}, d_t\varphi_1^{\epsilon,\nu+1}) = \\ \qquad -\mathcal{F}_1^\epsilon(t, \gamma_1 u_-^{\epsilon,\nu}, \gamma_1 u_\star^{\epsilon,\nu}, d_t\varphi_1^{\epsilon,\nu}, \varphi_1^{\epsilon,\nu}) \\ \qquad +F_{1,(\gamma_1 u_-^{\epsilon,\nu}, \gamma_1 u_\star^{\epsilon,\nu}, d_t\varphi_1^{\epsilon,\nu})}^\epsilon (\gamma_1 u_-^{\epsilon,\nu}, \gamma_1 u_\star^{\epsilon,\nu}, d_t\varphi_1^{\epsilon,\nu}) \\ F_{2,(\gamma_2 u_\star^{\epsilon,\nu}, \gamma_2 u_+^{\epsilon,\nu}, d_t\varphi_2^{\epsilon,\nu})}^\epsilon (\gamma_2 u_\star^{\epsilon,\nu+1}, \gamma_2 u_+^{\epsilon,\nu+1}, d_t\varphi_2^{\epsilon,\nu+1}) = \\ \qquad -\mathcal{F}_2^\epsilon(t, \gamma_2 u_\star^{\epsilon,\nu}, \gamma_2 u_+^{\epsilon,\nu}, d_t\varphi_2^{\epsilon,\nu}, \varphi_2^{\epsilon,\nu}) \\ \qquad +F_{2,(\gamma_2 u_\star^{\epsilon,\nu}, \gamma_2 u_+^{\epsilon,\nu}, d_t\varphi_2^{\epsilon,\nu})}^\epsilon (\gamma_2 u_\star^{\epsilon,\nu}, \gamma_2 u_+^{\epsilon,\nu}, d_t\varphi_2^{\epsilon,\nu}) \\ \gamma_1 u_+^\epsilon(t) = known, \quad \gamma_2 u_-^\epsilon(t) = known \\ \varphi_1^{\epsilon,\nu+1}(0) = \varphi_2^{\epsilon,\nu+1}(0) = 0 \end{cases}$$

(30)

where $(u_-^{\epsilon,0}, u_\star^{\epsilon,0}, u_+^{\epsilon,0}, \varphi_1^{\epsilon,0}, \varphi_2^{\epsilon,0})$ are given in the first step, and F_k^ϵ is the Fréchet derivative of $\mathcal{F}_k^\epsilon$ with respect to the arguments. That is to say that we use the usual Picard iteration for the equation, and the Newton iteration for the boundary conditions in (23).

By studying a linear Goursat problem similar to (30), we can obtain

(1) there are $T, \epsilon_0 > 0$ such that the sequences $\{u_-^{\epsilon,\nu}, u_\star^{\epsilon,\nu}, u_+^{\epsilon,\nu}\}_{\nu \geq 0}$ and $\{\varphi_1^{\epsilon,\nu}, \varphi_2^{\epsilon,\nu}\}_{\nu \geq 0}$ are bounded in $C_\epsilon^1(\Omega_T^\star)$ and $\tilde{C}_\epsilon^2([0,T])$ respectively when $\epsilon \in (0, \epsilon_0]$;

(2) when ν tends to infinity, $(u_-^{\epsilon,\nu}, u_\star^{\epsilon,\nu}, u_+^{\epsilon,\nu})$ $((\varphi_1^{\epsilon,\nu}, \varphi_2^{\epsilon,\nu})$ resp.) is convergent in $C^0(\Omega_T^\star)$ $(C^1([0,T])$ resp.) uniformly in $\epsilon \in (0, \epsilon_0]$;

(3) for any $\epsilon \in (0, \epsilon_0]$ fixed, the family $\{\nabla u_-^{\epsilon,\nu}, \nabla u_\star^{\epsilon,\nu}, \nabla u_+^{\epsilon,\nu}, d_t^2\varphi_1^{\epsilon,\nu}, d_t^2\varphi_2^{\epsilon,\nu}\}_{\nu \geq 0}$ is equicontinuous,

from which we have that the convergence of $(u_-^{\epsilon,\nu}, u_\star^{\epsilon,\nu}, u_+^{\epsilon,\nu})$ and $(\varphi_1^{\epsilon,\nu}, \varphi_2^{\epsilon,\nu})$ also holds in $C^1(\Omega_T)$ and $C^2([0,T])$ respectively as $\nu \to \infty$; moreover, their limits $(u_-^\epsilon, u_\star^\epsilon, u_+^\epsilon, \varphi_1^\epsilon, \varphi_2^\epsilon)$ are unique solutions of (23).

STEP 3 Noting that the problem (27) is a linear Goursat one, we can immediately obtain $(\varphi_1, \varphi_2) \in C^2([0,T])$.

Solve the nonlinear problem (26) by the following iteration scheme:

$$
\begin{cases}
\mathbb{E}_\sharp U_\sharp^{\nu+1} = U_\sharp^{\nu+1}, \qquad \sharp = -, \star, + \\
\mathbb{E}_\pm(P_\pm(\partial_y, \partial_z)U_\pm^{\nu+1} + L_\pm f''(u_\pm)(R_\pm(\partial_\xi - \partial_\eta)U_\pm^{\nu+1}, R_\pm U_\pm^\nu) \\
\qquad\qquad + (d_\pm I + h_\pm \Lambda_\pm)(\partial_\xi - \partial_\eta)U_\pm^{\nu+1}) = 0 \\
\mathbb{E}_\star(P_\star(\partial_y, \partial_z)U_\star^{\nu+1} + L_\star f''(u_\star)(R_\star(\partial_\eta - \partial_\xi)U_\star^{\nu+1}, R_\star U_\star^\nu) \\
\qquad\qquad + (d_\star I + h_\star \Lambda_\star)(\partial_\eta - \partial_\xi)U_\star^{\nu+1}) = 0 \\
\chi_1^{\nu+1}(u_\star - u_-) + (\sigma_1 I - \Lambda_\star)R_\star\gamma_1 U_\star^{\nu+1} - (\sigma_1 I - \Lambda_-)R_-\gamma_1 U_-^{\nu+1} = 0, \\
\qquad\qquad\qquad \text{on } z = \eta = 0 \\
\chi_2^{\nu+1}(u_+ - u_\star) + (\sigma_2 I - \Lambda_+)R_+\gamma_2 U_+^{\nu+1} - (\sigma_2 I - \Lambda_\star)R_\star\gamma_2 U_\star^{\nu+1} = 0, \\
\qquad\qquad\qquad \text{on } y = \xi = 0 \\
\gamma_1 U_+^{\nu+1}(t, \tau) = known, \quad \gamma_2 U_-^{\nu+1}(t, \tau) = known
\end{cases}
$$
$$(31)$$

with $(U_-^0, U_\star^0, U_+^0, \chi_1^0, \chi_2^0)$ being given from the first step.

By studying a linear problem similar to (31), in the same way as in the second step, we can get that there is $T > 0$ such that the sequences $\{U_-^\nu, U_\star^\nu, U_+^\nu\}_{\nu \geq 0}$ and $\{\chi_1^\nu, \chi_2^\nu\}_{\nu \geq 0}$ are convergent in $C_p^1(\Omega_T : \mathbb{R}^2)$ and $C_p^1([0,T] : \mathbb{R})$ respectively, and the limits $(U_-, U_\star, U_+, \chi_1, \chi_2)$ are unique solutions to (26).

STEP 4 At this step, we explain the idea of the justification of the asymptotic properties (28).

From the problems (23) and (26), we know that the second line of (28) can be implied by its first line. The third line of (28) can be easily obtained from its second line by using

$$
d_t\varphi_k(t) = \mathbb{E}_0\chi_k(t) = \lim_{\rho \to \infty} \frac{1}{2\rho} \int_{-\rho}^{\rho} \chi_k(t, \tau)d\tau. \tag{32}
$$

The proof for the first line of (28) in the L^∞-norm is given by using the technique of simultaneous Picard iteration: for each $\nu \geq 0$, we establish the asymptotics

$$u_{\natural}^{\epsilon,\nu}(y,z) = R_{\natural}U_{\natural}^{\nu}(y,z;\frac{y}{\epsilon},\frac{z}{\epsilon}) + o(1) \quad \text{in} \quad L^{\infty}(\Omega_T) \tag{33}$$

which immediately implies the part of the first line of (26) in the L^{∞}-norm by using the uniformity of the L^{∞}-convergence of $(u_{-}^{\epsilon,\nu}, u_{*}^{\epsilon,\nu}, u_{+}^{\epsilon,\nu})$ in $\epsilon \in (0, \epsilon_0]$.

The asymptotic property of $\epsilon \nabla_{(y,z)} u_{\natural}^{\epsilon}$ is obtained by studying the nonlinear problem (23) directly as in [5, 10].

Acknowledgement The author would like to express his gratitude to Michael Reissig for his invitation to participate at the third ISAAC Congress in Berlin. The research of this paper was partially supported by the NSF and the Education Ministry of China.

References

[1] Bui, A.T., Li, D.: The double shock front solutions for hyperbolic conservation laws in multidimensional space. Trans. Amer. Math. Soc. 316(1989), 233–250.

[2] Corli, A.: Weakly nonlinear geometric optics for hyperbolic systems of conservation laws with shock waves. Asymptotic Anal. 10(1995), 117–172.

[3] Corli, A.: Asymtotic analysis of contact discontinuities. Ann. Mat. Pura Appl. 173(1997), 163–202.

[4] Joly, J.L., Métivier, G., Rauch, J.: Several recent results in nonlinear geometric optics. Partial Differential Equations and Mathematical Physics, Birkhäuser, Boston 1996, 181–206.

[5] Joly, J.L., Métivier, G., Rauch, J.: Resonantly one dimensional nonlinear geometric optics. J. Funct. Anal. 114(1993), 106–231.

[6] Majda, A.: The stability of multi-dimensional shock fronts. Memoirs of AMS 275(1983), 1–95.

[7] Majda, A., Artola, A.: Nonlinear geometric optics for hyperbolic mixed problems. Analyse Mathematique et Applications, Gauthier-Villars, Paris 1988, 319–356.

[8] Métivier, G.: Interaction de deux chocs pour un systeme de deux lois de conservation, en dimension deux d'espace. Trans. Amer. Math. Soc. 296(1986), 431–479.

[9] Wang, Y.G.: Nonlinear geometric optics for shock waves, II: system case. Z. Anal. Anwendungen 16(1997), No. 4, 857–918.

[10] Wang, Y.G.: Nonlinear geometric optics for two shock waves. Comm. PDE 23(1998), 1621–1692.

[11] Wang, Y.G., Oberguggenberger, M.: Semilinear geometric optics for generalized solutions. Z. Anal. Anwendungen 19(2000), No. 4, 913–926.

[12] Williams, M.: Highly oscillatory multidimensional shocks. Comm. Pure Appl. Math. 52(1999), 129–192.

EXPONENTIAL TIME DECAY SOLUTIONS OF SCHRÖDINGER EQUATIONS AND OF WAVE EQUATIONS IN EVEN DIMENSIONAL SPACES

Yuya Dan

Kunihiko Kajitani
Institute of Mathematics
University of Tsukuba
kajitani@math.tsukuba.ac.jp

Abstract   We investigate the sufficient conditions on the initial data in order that the solutions of the Cauchy problem for Schrödinger equations with some potentials and also for even dimentional wave equations decay exponentially in time.

1.   Introduction

In [1] and [2] we have constructed the solutions to the Cauchy problem for the Schrödinger equation without potentials, which have the analytically smoothing effect and the exponential time decay. In this paper we shall investigate the Schrödinger equation with potentials and moreover we shall construct exponential time decay solutions of the wave equation in even dimensional spaces if the initial data belong to the image of a Fourier integral operator.

We consider the following Cauchy problem:

$$\begin{cases} \partial_t u(t,x) = (i\triangle + V(x,D))u(t,x) & \text{for } t \in (0,\infty), \ x \in \mathbb{R}^n, \\ u(0,x) = u_0(x) & \text{for } x \in \mathbb{R}^n, \end{cases} \tag{1}$$

where $i = \sqrt{-1}$, $\partial_t = \dfrac{\partial}{\partial t}$ and $\triangle$ is Laplacian in $\mathbb{R}^n$ which is defined by

$$\triangle = \sum_{j=1}^{n} \frac{\partial^2}{\partial x_j^2} \tag{2}$$

293

H.G.W. Begehr et al. (eds.), Analysis and Applications - ISAAC 2001, 293–301.
© 2003 *Kluwer Academic Publishers.*

294

and

$$V(x, D) = \sum_{j=1}^{n} V_j(x, D)D_j + V_0(x, D).$$

We investigate sufficient conditions on u_0 and V for the solution u of (1) to decay exponentially with respect to time t, that is, for a compact set $K \subset \mathbb{R}^n$ there are positive constants ε and C_0 such that the solution u of (1) satisfies

$$\sup_{x \in K} |u(t, x)| \leqq C_0 e^{-\varepsilon t}$$

for $t > 0$. We can see in [3] that in general the solutions of (1.1) does not necessarily decay exponentially, even if its initial datum $u_0(x)$ decays exponentially. For example, when $V = 0$, for $u_0(x) = e^{-|x|^2}$ we have the solution

$$u(t, x) = \frac{1}{\sqrt{1 + it}^n} e^{-\frac{|x|^2}{4(1+it)}},$$

which has no exponential decay. We shall prove that the solutions decay exponentially with respect to t if the initial value u_0 belongs to the image of a Fourier integral operator from Sobolev space.

For $\mu \in \mathbb{C}$, denote

$$\phi(x, \xi) = x(\xi - \frac{i\mu\xi}{|\xi|}) \tag{3}$$

and define

$$I_\phi(x, D)u(x) = \frac{1}{\sqrt{2\pi}^n} \int e^{i\phi(x,\xi)} \hat{u}(\xi)d\xi, \tag{4}$$

for u from the Sobolev space H^s, where $\hat{u}$ stands for the Fourier transform of u. Denote by B^k the set of functions which are bounded in R^n together with it's kth derivatives.

We can prove the following result.

Theorem 1. *Let* $\psi \in H^{[\frac{n}{2}]+1}(\mathbb{R}^n)$ *and* $V_j(x) \in B^{[\frac{n}{2}]+1}(\mathbb{R}^n)(j = 0, \cdots, n)$. *If* $u_0 = I_\phi(x, D)\psi$ *and* $V_j(x, D) = I_\phi(x, D)V_j(x)(j = 0, 1, \cdots, n)$, *and if* $C := \sup_{x \in R^n} ReV_0(x) < Re\mu Im\mu$ *and* $Re\mu \geq \sup_{x \in R^n} \sum_{j=1}^{n} |ReV_j(x)|$, *then for any* $\delta > 0$ *there is* $C_0 > 0$ *such that the solution* u *of (1) satisfies*

$$|u(t, x)| \leqq C_0 e^{(-2 Re\mu Im\mu + C)t + (|Re| + \delta)\langle x \rangle} \tag{5}$$

for $t > 0$ and $x \in \mathbb{R}^n$.

Next we consider the following Cauchy problem,

$$\begin{cases} \partial_t^2 u(t,x) = \triangle u(t,x) & \text{for } t \in (0, \infty), \ x \in \mathbb{R}^n, \\ u(0,x) = u_0(x), \quad u_t(0,x) = u_1(x), & \text{for } x \in \mathbb{R}^n. \end{cases} \tag{6}$$

It is well-known that in odd dimensional spaces the solutions of (6) decay exponentially with respect to t if the initial data decay exponentially with respect to space variables but in even dimensional spaces, for example $n = 2$, we can give a solution of (6) with $u_0 = 0$ and $u_1(x) = e^{-|x|^2}$ as follows,

$$u(t,x) = \frac{1}{2\pi} \int_{|x-y|<t} \frac{e^{-|y|^2}}{\sqrt{t^2 - |x-y|^2}} dy,$$

which does not decay exponentially. We investigate sufficient conditions on $u_i (i = 0, 1)$ in order that the solutions of the Cauchy problem for the wave equation (6) decay exponentially with respect to t, that is, we can prove the following theorem.

Theorem 2. *Let $\psi \in H^{[\frac{n}{2}]+1}(\mathbb{R}^n)$. If $u_0 = I_\phi(x, D)\psi$ and $u_1(x) = - sign(\Re\mu)I_\phi(x, D)(\mu + i|D|)\psi$, then for any $\delta > 0$ there is $C > 0$ such that the solution u of (6) satisfies*

$$|u(t,x)| \leqq Ce^{-2|Re\mu|t + (|Re|\delta)\langle x \rangle} ||\psi||_{[\frac{n}{2}]+1}, \tag{7}$$

for $t > 0$ and $x \in \mathbb{R}^n$.

2.   Notation

For $x = (x_1, \ldots, x_n) \in \mathbb{R}^n$ and $\xi = (\xi_1, \ldots, \xi_n) \in \mathbb{R}^n$, an inner product and a norm of vectors is denoted by

$$x \cdot \xi = x_1\xi_1 + \cdots + x_n\xi_n \qquad |x| = \sqrt{x \cdot x}$$

respectively, and we also use $\langle x \rangle = \sqrt{1 + |x|^2}$. Throughout this paper,

$$\tilde{\xi} := \frac{\xi}{|\xi|} \quad \text{for } \xi \neq 0.$$

We define the Fourier transform of $u \in L^2(\mathbb{R}^n)$ as follows :

$$\hat{u}(\xi) = \int_{\mathbb{R}^n} e^{-ix\cdot\xi} u(x) \, dx,$$

where $dx = (2\pi)^{-n/2}dx$. Under this notation, the inverse Fourier transform can be written as :

$$u(x) = \int_{\mathbb{R}^n} e^{ix\cdot\xi}\hat{u}(\xi)\,d\xi.$$

For a set Ω in $\mathbb{R}^n$, $L^2(\Omega)$ denotes a set of functions u satisfying

$$\int_\Omega |u(x)|^2\,dx < \infty$$

and $\|u\|_{L^2(\Omega)}$ denotes the norm of $L^2(\Omega)$ which is defined the above quantity. We also use $L^2 = L^2(\mathbb{R}^n)$ and $\|u\| = \|u\|_{L^2(\mathbb{R}^n)}$ for simplicity.

For a real number s denote $\|u\|_s = \|\langle\xi\rangle^s\hat{u}(\xi)\|$ and H^s is the set of functions whose Sobolev norms $\|\cdot\|_s$ are finite.

Let $m \in \mathbb{R}$, $1 \geq \rho \geq \delta$ and $\delta \neq 1$. Then $S^m_{\rho,\delta}$ means the symbol class of pseudo-differential operators which is defined by

$$\begin{aligned}
S^m_{\rho,\delta} &= \{p(x,\xi) \in C^\infty(\mathbb{R}^n \times \mathbb{R}^n); \; |p^{(\alpha)}_{(\beta)}(x,\xi)| \\
&\leqq C_{\alpha,\beta}\langle\xi\rangle^{m-\rho|\alpha|+\delta|\beta|}, \forall\alpha,\beta \in Z_\pm\},
\end{aligned}$$

where

$$p^{(\alpha)}_{(\beta)}(x,\xi) = \partial^\alpha_\xi D^\beta_x p(x,\xi)$$

and

$$\partial^\alpha_\xi = \partial^{\alpha_1}_{\xi_1}\cdots\partial^{\alpha_n}_{\xi_n}, D^\beta_x = (-i)^{|\beta|}\partial^{\beta_1}_{x_1}\cdots\partial^{\beta_n}_{x_n},$$

for both $\alpha = (\alpha_1,\ldots,\alpha_n)$ and $\beta = (\beta_1,\ldots,\beta_n)$ are in $\mathbb{Z}^n_+ = (\mathbb{N}\bigcup\{0\})^n$.

3.   Fourier Integral Operators

In this section, we shall explain a Fourier integral operator introduced in Dan-Kajitani [1] and we mention the properties of these tools, so that we prove our theorems. To begin with, we give the definition of the function space on which Fourier integral operators act.

Definition 1. For a real number δ and a non-negative number s, we define weighted Sobolev spaces as follows :

$$H^s_\delta = \{u \in L^2_{\text{loc}}(\mathbb{R}^n); \; e^{\delta\langle x\rangle}u(x) \in H^s\}.$$

At the next stage, we define a Fourier integral operator I_φ, which is applied to transform the original equation (1) to another equation with respect to a new unknown function.

Definition 2. Let $\varphi(x,\xi) = x \cdot \xi - i\mu x \cdot \tilde{\xi}$ for a complex number $\mu \in \mathbb{C}$. For $v \in H^s$, we denote by I_φ the Fourier integral operator

$$I_\varphi v(x) = \int_{\mathbb{R}^n} e^{i\varphi(x,\xi)} \hat{v}(\xi) \, d\xi,$$

where $d\xi = (2\pi)^{-n/2} d\xi$ and $\hat{v}$ stands for the Fourier transform of v.

Then we obtain the following result. The proof is given in [1] but to be self-contained we give a proof here.

Lemma 1. *Let s be a non-negative integer and $\varphi(x,\xi) = x \cdot \xi - i\mu x \cdot \tilde{\xi}$ for $\mu \in \mathbb{C}$. Then, I_φ operates continuously from H^s to $H^s_{-(|Re\mu|+\delta)}$ for all $\delta > 0$ and satisfies*

$$\|I_\varphi v\|_{H^s_{-(|Re\mu|+\delta)}} \leqq C\|v\|_s, \tag{8}$$

for any $v \in H^s$.

Proof It suffices to show (8). In the first place, we have from the definition of I_φ,

$$e^{-(|Re\mu|+\delta)\langle x \rangle} I_\varphi v(x) = e^{-(|Re\mu|+\delta)\langle x \rangle} \int_{\mathbb{R}^n} e^{ix\cdot\xi + \mu x \cdot \tilde{\xi}} \hat{v}(\xi) \, d\xi$$

$$= \int_{\mathbb{R}^n} e^{ix\cdot\xi} e^{\mu x \cdot \tilde{\xi} - (|Re\mu|+\delta)\langle x \rangle} \hat{v}(\xi) \, d\xi$$

$$= \int_{\mathbb{R}^n} e^{ix\cdot\xi} p(x,\xi) \hat{v}(\xi) \, d\xi, \tag{9}$$

where we put $p(x,\xi) := e^{\mu x \cdot \tilde{\xi} - (|Re\mu|+\delta)\langle x \rangle}$. We may denote the last integral by $Pv(x)$ using the corresponding capital letter $P = p(x,D)$ as pseudo-differential operators. Remember in mind for the further discussion that $p(x,\xi)$ decays exponentially with respect to the space variable $x \in \mathbb{R}^n$, that is,

$$|p(x,\xi)| = e^{(Re\mu)x\cdot\tilde{\xi} - (|Re\mu|+\delta)\langle x \rangle}$$
$$\leqq e^{-\delta\langle x \rangle},$$

because the estimate $x \cdot \tilde{\xi} \leqq |x| \leqq \langle x \rangle$ holds for every $\xi \in \mathbb{R}^n \setminus \{0\}$.

For the purpose of avoiding effects of the singularity at the origin $\xi = 0$, we should divide integral region $\mathbb{R}^n$ into two parts, namely the unit ball in $\mathbb{R}^n$ and the complement. Let us take $\chi_0 \in C_0^\infty(\mathbb{R}^n)$ such that $\text{supp}\chi_0 \subset \{\xi \in \mathbb{R}^n; |\xi| \leqq 1\}$, $0 \leqq \chi_0(\xi) \leqq 1$ for all $\xi \in \mathbb{R}^n$, and $\chi_0(\xi) = 1$ in some neighborhood of $\xi = 0$. Then, we can divide p as follows :

$$\begin{aligned} p(x,\xi) &= p(x,\xi)\chi_0(\xi) &+& p(x,\xi)(1 - \chi_0(\xi)) \\ &=: p_0(x,\xi) &+& p_1(x,\xi). \end{aligned}$$

First, we shall examine $p_0(x, \xi)$ which is supported in the neighborhood of $\xi = 0$. Denote

$$P_0 v(x) = \int_{\mathbb{R}^n} e^{ix \cdot \xi} p_0(x, \xi) \hat{v}(\xi) \; d\xi.$$

Hence,

$$
\begin{aligned}
|D_x^\alpha P_0 v(x)| &\leqq \int_{|\xi| \leqq 1} \left| D_x^\alpha \left[e^{ix \cdot \xi} p(x, \xi) \right] \hat{v}(\xi) \right| \; d\xi \\
&\leqq \sum_{\alpha' \leqq \alpha} \binom{\alpha}{\alpha'} \int_{|\xi| \leqq 1} |\xi^{\alpha - \alpha'} D_x^{\alpha'} p(x, \xi) \hat{v}(\xi)| \; d\xi.
\end{aligned}
$$

Since

$$
\begin{aligned}
|D_x^{\alpha'} p(x, \xi)| &= \left| D_x^{\alpha'} \left[e^{\mu x \cdot \bar{\xi} - (|\mathrm{Re}\,\mu| + \delta)\langle x \rangle} \right] \right| \\
&\leqq C_{\alpha'} e^{-\delta \langle x \rangle},
\end{aligned}
$$

we obtain

$$
\begin{aligned}
|D_x^\alpha P_0 v(x)| &\leqq C_\alpha e^{-\delta \langle x \rangle} \int_{|\xi| \leqq 1} 1 \cdot |\hat{v}(\xi)| \; d\xi \\
&\leqq \tilde{C}_\alpha \|v\| e^{-\delta \langle x \rangle}
\end{aligned}
$$

for $|\alpha| \leqq s$. Thus, it is proved that

$$\|P_0 v\|_s \leqq C_0 \|v\| \tag{10}$$

for some positive constant C_0. Next, we shall examine p_1. Recall that we discuss the integral (9) on the region except for the neighborhood of $\xi = 0$, then we have

$$
\begin{aligned}
\left| p_1{}_{(\beta)}^{(\alpha)}(x, \xi) \right| &\leqq \tilde{C}_{\alpha, \beta} \langle \xi \rangle^{-|\alpha|} \langle x \rangle^{|\alpha|} e^{-\delta \langle x \rangle} \\
&\leqq C_{\alpha, \beta} \langle \xi \rangle^{-|\alpha|}.
\end{aligned}
$$

Thus, p_1 belongs to the symbol class $S_{1,0}^0$. Therefore the boundedness theorem for pseudo-differential operators implies

$$\|P_1 v\|_s \leqq C_1 \|v\|_s. \tag{11}$$

Summing up (10) and (11), we obtain after all (8). This completes the proof of Lemma 1.

As an example let $\psi(x) = e^{-|x|^2}$. Then if $n = 1$ we can calculate

$$I_\phi(x, D)\psi(x) = \frac{e^{-|x|^2}}{2}\left\{e^{\mu x} + e^{-\mu x} + \frac{i\int_0^t e^{\tau^2}d\tau}{\sqrt{2\pi}}(e^{\mu x} - e^{-\mu x})\right\}.$$

4. Schrödinger Equation

Let $\varphi(x, \xi) = x \cdot \xi - i\mu x \cdot \tilde{\xi}$. Transform the unknown function u to a new one v by $u = I_\varphi v$ in the original equation (1), then

$$i\Delta u = i\Delta I_\varphi v$$

$$= i\int_{\mathbb{R}^n} \Delta e^{ix\cdot(\xi - i\mu\tilde{\xi})}\hat{v}(\xi)\ d\xi$$

$$= -i\int_{\mathbb{R}^n} e^{i\varphi(x,\xi)}\sum_{j=1}^n (\xi_j - i\mu\tilde{\xi}_j)^2\hat{v}(\xi)\ d\xi$$

$$= -i\int_{\mathbb{R}^n} e^{i\varphi(x,\xi)}(|\xi| - i\mu)^2\hat{v}(\xi)\ d\xi$$

$$= -iI_\varphi(|D| - i\mu)^2 v(t, x). \tag{12}$$

Hence we obtain the following equation

$$\partial_t v(t, x) = \{-i\,(|D| - i\mu)^2 + \sum_{j=1}^n V_j(x)D_j + V_0(x)\}v(t, x), \tag{13}$$

$$v(0, x) = \psi(x). \tag{14}$$

We can find easily the solution $v \in C^1([0, \infty) : H^{s-2}) \cap C^0([0, \infty) : H^s)$ of this Cauchy problem (13)-(14), if we give an initial datum $\psi(x) \in H^s$. This equation implies the following energy estimate

$$\partial_t \|v(t, \cdot)\|_s^2 = 2\operatorname{Re}(\partial_t v(t, \cdot), v(t, \cdot))_{H^s}$$

$$= 2\operatorname{Re}(\{-i(|D| - i\mu)^2 + \sum_{j=1}^n V_j(x)D_j + V_0(x)\}v(t, \cdot), v(t, \cdot))_{H^s}$$

$$= 2\operatorname{Re}((-2\mu|D| + \sum_{j=1}^n V_j(x)D_j + V_0(x))v(t, \cdot), v(t, \cdot))_{H^s}$$

$$- 4\operatorname{Re}\mu\operatorname{Im}\mu\,\|v(t, \cdot)\|_s^2$$

$$\leq (-4\operatorname{Re}\mu\operatorname{Im}\mu + C)\,\|v(t, \cdot)\|_s^2,$$

if $2\operatorname{Re}\mu \geq \sup_{x \in R^n} \sum_{j=1}^n |\operatorname{Re}V_j(x)|$, where $C = \sup_{x \in R^n} 2\operatorname{Re}V_0(x)$. Moreover we obtain by Gronwall's inequality

$$\|v(t, \cdot)\|_s \leq e^{(-2\operatorname{Re}\mu\operatorname{Im}\mu + C)t}\|\psi\|_s. \tag{15}$$

This estimate shows that v decays exponentially in t if $\mathrm{Re}\mu\,\mathrm{Im}\mu > C$. We put $u(t,x) = I_\varphi v(t,x)$. Then we can see that u satisfies (1) with $u_0 = I_\phi(x,D)\psi$. On the other hand, for $s = [n/2]+1$, Sobolev's lemma implies

$$e^{-(|\mathrm{Re}\mu|+\delta)\langle x\rangle}|u(t,x)| \leqq C_s \|e^{-(|\mathrm{Re}\mu|+\delta)\langle x\rangle}u(t,x)\|_s$$
$$= C_s\|u(t,x)\|_{H^s_{-(|\mathrm{Re}\mu|+\delta)}}$$
$$= C_s\|I_\varphi v(t,x)\|_{H^s_{-(|\mathrm{Re}\mu|+\delta)}}$$
$$\leqq \tilde{C}_s\|v(t,x)\|_s \qquad (16)$$

with some positive constants C_s and $\tilde{C}_s$, and we apply here Lemma 1 in the last inequality. After all, we conclude from (4.4) and (4.5) that taking $\varepsilon = 2\,\mathrm{Re}\mu\,\mathrm{Im}\mu - C > 0$ and $\delta_0 = \delta + |\mathrm{Re}\mu|$ we obtain (5). This completes the proof of Theorem 1.

5.   Wave Equation

We seek a solution u of the Cauchy problem (6) in the following form,

$$u(t,x) = I_\phi(x,D)v(t,x).$$

Then it follows from (12) and from the assumption of Theorem 2 that v satisfies

$$\frac{\partial^2 v(t,x)}{\partial t^2} = -(|D|-i\mu)^2 v(t,x), \quad t \in (0,\infty),\ x \in$$
$$\qquad (17)$$
$$v(0,x) = \psi(x),\ v_t(0,x) = -sign(\Re\mu)I_\phi(x,D)(\mu+i|D|)\psi, \quad x \in \mathbb{R}^n.$$
$$\qquad (18)$$

We can find a solution v of the above equation. Put

$$v_1(t,x) = (\partial_t - i|D| - \mu)v, \quad v_2(t,x) = (\partial_t + i|D| + \mu)v.$$

Then if v satisfies (17) we get the following equations

$$(\partial_t + i|D| + \mu)v_1(t,x) = 0, \quad (\partial_t - i|D| - \mu)v_2(t,x) = 0.$$

Solving the above equations we get

$$v_1(t,x) = e^{-(i|D|+\mu)t}v_1(0,x), \quad v_2(t,x) = e^{(i|D|+\mu)t}v_2(0,x).$$

Since v satisfies the initial condition (18), if $\mathrm{Re}\mu > 0$ we have the initial datum $v_2(0,x) = 0$ and if $\mathrm{Re}\mu < 0$ we have the initial datum $v_1(0,x) = 0$. Hence if $\Re\mu > 0$ we get the solution of (17)-(18) as follows,

$$v(t,x) = \frac{v_1(t,x) - v_2(t,x)}{2(i|D|+\mu)} = \frac{e^{-(i|D|+\mu)t}v_1(0,x)}{2(i|D|+\mu)},$$

and, consequently,

$$\|v(t,\cdot)\|_{[\frac{n}{2}]+1} \leq \frac{C}{\mathrm{Re}\mu} e^{-\mathrm{Re}\mu t} \|v_1(0,\cdot)\|_{[\frac{n}{2}]+1}.$$

By use of Sobolev's lemma and Proposition 2.1 we get

$$|e^{-(\mathrm{Re}\mu+\delta)<x>}u(t,x)| \leq C\|e^{-(\mathrm{Re}\mu+\delta)<x>}u(t)\|_{[\frac{n}{2}]+1} \leq \|v(t)\|_{[\frac{n}{2}]+1}$$

$$\leq \frac{C}{\mathrm{Re}\mu} e^{-\mathrm{Re}\mu t} \|v_1(0,\cdot)\|_{[\frac{n}{2}]+1}.$$

If $\mathrm{Re}\mu < 0$, we obtain analogously

$$|e^{-(-\mathrm{Re}\mu+\delta)<x>}u(t,x)| \leq \frac{C}{-\mathrm{Re}\mu} e^{\mathrm{Re}\mu t} \|v_2(0,\cdot)\|_{[\frac{n}{2}]+1}.$$

Thus we have proved Theorem 2.

References

[1] Dan, Y., Kajitani, K.: Smoothing effect and exponential Time decay of Solutions of Schrödinger equations. Preprint.

[2] Kajitani, K.: Analytically smoothing effect for Schrödinger equations. Proceedings of the International Conference on Dynamical Systems and Differential Equations in Southwest Missouri State University (1996). An added Volume I to Discrete and Continuous Dynamical Systems, 1998, 350–352.

[3] Rauch, J.: Local decay of scattering solutions to Schrödinger's equation. Commun. Math. Phys. 61 (1978), 149–168.

INVERSE SCATTERING FOR A SMALL NONSELFADJOINT PERTURBATION OF THE WAVE EQUATION

Kiyoshi MOCHIZUKI

Department of Mathematic
Tokyo Metropolitan University
Hachioji, Tokyo 192-0397, Japan
mochizuk@comp.metro-u.ac.jp

Abstract   In this paper we consider the wave equation $\Box u + b(x)u_t = 0$ in $\mathbf{R}^n$, $n \geq 3$, with $b(x)$ which is small and decays exponentially as $|x| \to \infty$. We show that the scattering operator exists for each given $b(x)$ and inversely the scattering amplitude at fixed nonzero energy determines the coefficient $b(x)$.

1.   Introduction

We consider the wave equation of the form

$$w_{tt} + b(x)w_t - \Delta w = 0, \quad (x,t) \in \mathbf{R}^n \times \mathbf{R}, \tag{1}$$

where $n \geq 3$ and $b(x)$ is a complex valued "small" function which decays sufficiently fast at infinity. Thus, (1.1) is a perturbed equation of the free wave equation

$$w_{0tt} - \Delta w_0 = 0, \quad (x,t) \in \mathbf{R}^n \times \mathbf{R}. \tag{2}$$

In this paper we discuss the three fundamental problems of scattering between these two equations:

1) to show the existence of the scattering operator,
2) to obtain the expression of the scattering amplitude,
3) to develop a reconstruction procedure of $b(x)$ from the scattering amplitude.

The existence of the scattering state for (1) is shown in Mochizuki [9] in the dissipative case $b(x) \geq 0$. But the completeness of the Moller wave operator is not proved there. In this note we require a smallness

H.G.W. Begehr et al. (eds.), Analysis and Applications - ISAAC 2001, 303–316.

304

of $b(x)$. Then the small perturbation theory of Kato [6] is applicable to obtain the completeness.

An expression of the scattering amplitude in the momentum space is obtained based on the spectral representation of the perturbed operator. Many works have been done for the Schrödinger operator, and the results are known to be useful for the conservative wave equation (see e.g., Mochizuki [8],[10]). In this note we shall show that the argument is also applicable to our nonselfadjoint problem.

The inverse scattering problem is studied mainly for the Schrödinger operator

$$L = -\Delta + V(x) \text{ in } \mathbf{R}^n. \tag{3}$$

The so-called Gelfand-Levitan theory is developed for the 1-dimensional problem or the spherically symmetric potential $V = V(|x|)$. The 3-dimensional problem is initially studied by Faddeev [2], where a uniqueness and a reconstruction procedure are given by use of the scattering amplitude for all high energies. He requires that the potential $V(x)$ is real, $\in C^1$ and behaves like $O(|x|^{-3-\delta})$ as $|x| \to \infty$. This results are extended by Mochizuki [7] for a complex valued, C^0 −potential satisfying the same decay-rate at infinity. The method is however not applicable to our wave equation since the behavior at high energy of the scattering amplitude is not so good as in the Schrödinger case.

Faddeev has published several papers to develop the Gelfand-Levitan theory for multi-dimensional Schrödinger opeartors (see [3] and its references). The results seem not to be completed yet, but these works gave important tools to develop the scattering inverse problems with fixed energy. Weder [11] and Eskin-Ralston [1] gave results for the Schrödinger equations, and Isozaki [4, 5] for the Dirac equations and for the wave equations with stratified media.

Note that Isozaki reduces the problem to the Schrödinger evolution equation by use of the invariance principle of the scattering operator. We can not use the invariance principle to our wave equation, but the asymptotic behavior of the meromorphic extension of the Faddeev resolvent can be directly used to obtain a reconstruction procedure with fixed energy.

2. A survey of the Schrödinger scattering

First we shall give a survey of the Schrödinger scattering developed by [2] and [7].

Let us consider the Schrödinger operator (3) in the Hilbert space $L^2 = L^2(\mathbf{R}^n)$, where

(A) $V(x)$ is a complex valued, locally L^2 function decaying like $O(|x|^{-1-\delta})$

($\delta > 0$) at infinity. If $n \geq 4$, we further require the Stummel condition which restricts the order of singularities.

Then L becomes a closed operator with the domain $D(L) = H^2$, where H^j ($j = 1, 2, \cdots$) is the Sobolev space. On the other hand, the free operator $L_0 = -\Delta$ is a selfadjoint operator with the same domain. For $\zeta \in \mathbf{C} \backslash \mathbf{R}$ we put

$$R_0(\zeta) = (L_0 - \zeta)^{-1} \quad \text{and} \quad R(\zeta) = (L - \zeta)^{-1}.$$

They are the resolvents of L_0 and L, respectively. Note that L may have nonreal eigenvalues, and these points become poles of $R(\zeta)$.

We denote the norm and innerproduct of L^2 by $\| \cdot \|$ and $(\cdot, \cdot)$, respectively. Further, we introduce the weighted L^2–space L_s^2, $s \in \mathbf{R}$, with norm

$$\|f\|_s^2 = \int (1 + |x|^2)^s |f(x)|^2 dx.$$

Here and in the following we simply write $\int$ for the integral over $\mathbf{R}^n$.

Let $\lambda > 0$ and $\epsilon > 0$. Then as an operator in $\mathcal{B}(L_s^2, L_{-s}^2)$, where $s > 1/2$, $R_0(\lambda^2 \pm i\epsilon)$ is continuously extended to $\epsilon = 0$, and we have

$$\|R_0(\lambda^2 \pm i0)\|_{\mathcal{B}(L_s^2, L_{-s}^2)} \leq \frac{C}{\lambda} \quad \text{as } \lambda \to \infty. \tag{4}$$

This estimate plays a crucial role in our theory. As for the proof see [9].

We choose $1/2 < s < (1 + \delta)/2$ in (4). Then this and the resolvent equation

$$\{1 + R_0(\zeta)V\}R(\zeta) = R_0(\zeta)$$

imply the following lemma ((6) is essentially due to Kato [6]).

Lemma 1 *There exists $\lambda_0 > 0$ such that $R(\lambda^2 \pm i\epsilon) \in \mathcal{B}(L_s^2, L_{-s}^2)$ with $\lambda \geq \lambda_0$ is extended continuously to $\epsilon = 0$, and we have*

$$\|R(\lambda^2 \pm i0)\|_{\mathcal{B}(L_s^2, L_{-s}^2)} \leq \frac{C}{\lambda} \quad as \ \lambda \to \infty. \tag{5}$$

Moreover, there exists $C > 0$ such that for any $\epsilon > 0$ and $f \in L^2$ we have

$$\int_J \|R(\lambda^2 \pm i\epsilon)f\|_{-s}^2 \, 2\lambda d\lambda \leq C\|f\|^2, \tag{6}$$

where $J = [\lambda_0, \infty)$.

It follows from this lemma that for any $f, g \in L^2$, the limit

$$(E(J)f, g) = \lim_{\epsilon \to 0} \frac{1}{2\pi i} \int_J (\{R(\lambda^2 + i\epsilon) - R(\lambda^2 - i\epsilon)\}f, g) 2\lambda d\lambda$$

exists and gives a projection (not necessarily orthogonal) which commutes with L. The orthogonal projection $E_0(J)$ corresponding to L_0 is defined by

$$(E_0(J)f, g) = \lim_{\epsilon \to 0} \frac{1}{2\pi i} \int_J (\{R_0(\lambda^2 + i\epsilon) - R_0(\lambda^2 - i\epsilon)\}f, g) 2\lambda d\lambda.$$

The Fourier transform

$$[F_0(\lambda)f](\omega) = (2\pi)^{-n/2} \lambda^{(n-1)/2} \int e^{-i\lambda\omega \cdot x} f(x) dx, \tag{7}$$

$(\lambda, \omega) \in \mathbf{R}_+ \times S^{n-1}$, gives the spectral representation of L_0. A generalized Fourier transform corresponding to L is given by

$$F_\pm(\lambda) = F_0(\lambda)\{I - VR(\lambda^2 \mp i0)\}.$$

We put for $f \in L_s^2$,

$$(F_0 f)(\lambda, \omega) = [F_0(\lambda)f](\omega),$$

$$(F_\pm f)(\lambda, \omega) = [F_\pm(\lambda)f](\omega).$$

Then F_0 is extended to a unitary operator from $E_0(J)L^2$ to $L^2(J; S^{n-1})$. Also one can show that $F_\pm$ is extended to a bijection from $E(J)L^2$ to $L^2(J; S^{n-1})$ with the inverse represented as

$$(F_\pm^{(*)} \hat{f})(x) = \int_J F_\pm(\lambda)^{(*)} \hat{f}(\lambda, \cdot) d\lambda,$$

$$F_\pm(\lambda)^{(*)} = \{I - R(\lambda^2 \pm i0)V\} F_0(\lambda)^*.$$

By use of these operators, we obtain

$$E_0(J) = F_0^* F_0 \quad \text{and} \quad E(J) = F_\pm^{(*)} F_\pm. \tag{8}$$

Moreover, $F_\pm^{(*)} F_0$ has the intertwining property

$$F_\pm^{(*)} F_0 e^{iL_0 t} = e^{iLt} F_\pm^{(*)} F_0. \tag{9}$$

Lemma 2 *For any $f \in E_0(J)L^2$, the limit*

$$\lim_{t \to \pm\infty} e^{iLt} e^{-iL_0 t} E_0(J)f = F_\pm^{(*)} F_0 f$$

exists in $E(J)L^2$.

Proof By means of (8) and (9),

$$e^{iLt} e^{-iL_0 t} E_0(J)f - F_\pm^{(*)} F_0 f = e^{iLt}\{E_0(J) - F_\pm^{(*)} F_0\} e^{-iL_0 t} f.$$

So, to complete the proof, we have only to show that the property

$$\|\{E_0(J) - F_\pm^{(*)} F_0\}e^{-iL_0 t}f\| \to 0 \ \text{ as } t \to \pm\infty \tag{10}$$

holds for f belonging in a dense set of $E_0(J)L^2$.

We choose f to satisfy $[F_0 f](\lambda, \omega) \in C_0^\infty(\mathbf{R}_+; L^2(S^{n-1}))$, and put

$$G_{f,\pm i\epsilon}(t) = \int_J R(\lambda \mp i\epsilon)V F_0(\lambda)^* e^{-i\lambda^2 t}[F_0 f](\lambda, \cdot)d\lambda.$$

Let $\epsilon \to 0$. Then for any fixed $t > 0$, it converges to

$$G_{f,\pm i0}(t) = \int_J R(\lambda \mp i0)V F_0(\lambda)^* e^{-i\lambda^2 t}[F_0 f](\lambda, \cdot)d\lambda$$

$$= \{E_0(J) - F_\pm^{(*)} F_0\}e^{-iL_0 t}f$$

weakly in L_s^2. On the other hand, since

$$R(\lambda \mp i\epsilon) = \mp i \int_0^\infty e^{\pm i(L-\lambda^2 \pm i\epsilon)s}ds$$

and

$$\left\|V \int_J e^{\pm i\lambda^2 t} F_0(\lambda)^*[F_0 f](\lambda, \cdot)d\lambda\right\| \le C(1 + |t|)^{-1-\gamma}$$

for some $\gamma > 0$, it follows that

$$\|G_{f,\pm i\epsilon}(t)\| \le C \int_0^\infty \|e^{\pm i(L \pm i\epsilon)s}V \int_J e^{\pm i\lambda^2(t\pm s)}$$

$$\times F_0(\lambda)^*[F_0 f](\lambda, \cdot)d\lambda\|ds \le C|t|^{-\gamma}.$$

These imply (10) to hold. □

The Moller wave operator is defined by

$$W_\pm(J) = s - \lim_{t\to\pm\infty} e^{iLt}e^{-iL_0 t}E_0(J)$$

The above lemma shows not only the existence of the Moller wave operator but also its completeness. Namely, the ranges of $W_+(J)$ and $W_-(J)$ are coincide and are equal to $E(J)L^2$. Thus, one can define the scattering operator

$$S(J) = W_+^{-1}(J)W_-(J).$$

Substituting the expression $W_\pm(J) = F_\pm^{(*)} F_0$ in this formula, we have

$$S(J) = F_0^* F_+ F_-^{(*)} F_0 = I - F_0^* F_+(F_+^{(*)} - F_-^{(*)})F_0.$$

Thus it follows that

$$2\pi i A(\lambda^2) \equiv F_0(\lambda)\{I - S(\lambda)\}F_0^*(\lambda) = 2\pi i F_+(\lambda) V F_0^*(\lambda) \qquad (11)$$

$$= 2\pi i F_0(\lambda)\{V - VR(\lambda^2 - i0)V\}F_0^*(\lambda).$$

Here $A(\lambda^2)$ is called the scattering amplitude.

Now, we can prove the following theorem.

Theorem 1 Let $n \geq 3$. Other than (A) assume $V(x) \in L^1 \cap L_s^2$. Then all the information of the high-energy scattering amplitude uniquely determine $V(x)$.

Proof The kernel of $A(\lambda^2)$ is given by

$$a(\lambda, \omega, \omega') = (2\pi)^{-n} \lambda^{n-1} \left[\int e^{i\lambda(\omega - \omega') \cdot x} V(x) dx \right. \qquad (12)$$

$$\left. + \int e^{i\lambda\omega \cdot x} V(x) R(\lambda^2 - i0)(V e^{-i\lambda\omega' \cdot})(x) dx \right].$$

Here the second integral is estimated by

$$\left| \int \cdots \right| \leq \|R(\lambda^2 - i0)\|_{\mathcal{B}(L_s^2, L_{-s}^2)} \|V\|_s^2.$$

For each $\xi \in \mathbf{R}^n \backslash \{0\}$, we choose $\eta \in S^{n-1}$ to satisfy $\xi \cdot \eta = 0$, and put

$$\omega(\lambda) = (1 - |\xi|^2/4\lambda^2)^{1/2}\eta + \xi/2\lambda,$$

$$\omega'(\lambda) = (1 - |\xi|^2/4\lambda^2)^{1/2}\eta - \xi/2\lambda.$$

Then $\omega(\lambda), \omega'(\lambda) \in S^{n-1}$ and $\lambda(\omega(\lambda) - \omega'(\lambda)) = \xi$. Thus, letting $\lambda \to \infty$ and taking (5) into account, we conclude

$$\lim_{\lambda \to \infty} (2\pi)^{n/2} \lambda^{-n+1} a(\lambda, \omega(\lambda), \omega'(\lambda)) = (2\pi)^{-n/2} \int e^{-i\xi \cdot x} V(x) dx = \hat{V}(\xi),$$

which gives a uniqueness and a reconstruction procedure. $\qquad \square$

3.　Faddeev resolvent and Eskin-Ralston resolvent

Note that the distorted plane wave

$$\varphi(x, k) = e^{ik \cdot x} - R(|k|^2 - i0)(V e^{ik \cdot})(x),$$

where $k \in \mathbf{R}^n \backslash \{0\}$ and $\lambda^2 = |k|^2$, is used in the expression (12). This is a solution of the equation

$$(-\Delta + V(x) - |k|^2)\varphi = 0. \qquad (13)$$

The inverse scattering with fixed energy will be based on another solution of the form

$$\varphi = e^{ik \cdot x}\{1 + v(x,k)\}.$$

It follows from (13) that

$$-\Delta v - 2ik \cdot \nabla v + Vv = -V.$$

For $k \in \mathbf{R}^n$ we choose $\gamma \in S^{n-1}$ to satisfy $t = k \cdot \gamma \geq 0$ and put $\eta = k - t\gamma$. We replace t by $z \in \mathbf{C}_+$ and put

$$G_{\gamma,0}(|\mu|,z) = (2\pi)^{-n} \int \frac{e^{i(x-y)\cdot\xi}}{\xi^2 + 2z\gamma \cdot \xi - |\mu|^2}\,d\xi.$$

Then Faddeev unperturbed resolvent depending on the direction γ is defined as

$$R_{\gamma,0}(|k|,t) = e^{it\gamma \cdot x}G_{\gamma,0}(|\eta|,t)e^{-it\gamma \cdot x}. \tag{14}$$

The following proposition is due to Weder [11].

Proposition 1 (1) $G_{\gamma,0}(|\mu|,z) \in \mathcal{B}(L_s^2, L_{-s}^2)$, $s > 1/2$, *is continuous in* $|\mu| \in \bar{\mathbf{R}}_+$, $\gamma \in S^{n-1}$ *and* $z \in \bar{\mathbf{C}}_+$ *except* $(|\mu|,z) = (0,0)$.
(2) $G_{\gamma,0}(|\mu|,z)$ *is analytic in* $z \in \bar{\mathbf{C}}_+$.
(3) *For each* $\epsilon_0 > 0$ *there exists* $C > 0$ *such that*

$$\|G_{\gamma,0}(|\mu|,z)\|_{\mathcal{B}(L_s^2, L_{-s}^2)} \leq C(|\mu| + |z|)^{-1}$$

if $|\mu| + |z| \geq \epsilon_0$.
(4) *For* $t \in \mathbf{R}$,

$$(-\Delta - |\mu|^2 - t^2)R_{\gamma,0}(|k|,t) = I.$$

The statement (3) corresponds to (4). But in order to make use of this estimate, we need to obtain an analytic continuation of $G_{\gamma,0}(\sqrt{|k|^2 - t^2}, t)$ in t, which is established by Eskin-Ralston [1]. For $a \in \mathbf{R}$ let

$$\mathcal{H}_a = \{f; e^{a|x|}f(x) \in L^2\},$$

and for $\epsilon > 0$ let

$$D_\epsilon = \{z \in \mathbf{C}_+; |z| < \epsilon/2\}.$$

Then there exists an operator valued function $U_{\gamma,0}(\lambda, z)$ satisfying the following proposition.

Proposition 2 (1) *For any* $\delta > 0$ *there exists* $\epsilon > 0$ *such that* $U_{\gamma,0}(\lambda, z) \in \mathcal{B}(\mathcal{H}_\delta, \mathcal{H}_{-\delta})$ *and depends analytically on* $z \in D_\epsilon$.
(2) $U_{\gamma,0}(\lambda, z)$ *has a boundary value when* $z \to (-\epsilon/2, \epsilon/2)$ *and*

$$U_{\gamma,0}(\lambda, t) = G_{\gamma,0}(|\eta|, t) = G_{\gamma,0}(\sqrt{\lambda^2 - t^2}, t).$$

(3) *For $\tau > 0$,*

$$U_{\gamma,0}(\lambda, i\tau) = G_{\gamma,0}(\sqrt{\lambda^2 + \tau^2}, i\tau).$$

The perturbed Faddeev resolvent can be defined for a.e. $t \in (-\epsilon/2, \epsilon/2)$ as follows:

$$R_\gamma(\lambda, t) = \{I - R_{\gamma,0}(\lambda, t)V\}^{-1} R_{\gamma,0}(\lambda, t). \tag{15}$$

In the expression (2.9) of the scattering amplitude, we replace $R(\lambda^2 - i0)$ by this resolvent, and call it the Faddeev scattering amplitude.

The perturbed Eskin-Ralston resolvent is defined by

$$U_\gamma(\lambda, t) = e^{-it\gamma\cdot x} R_\gamma(\lambda, t) e^{it\gamma\cdot x}. \tag{16}$$

If

$$V(x) = O(e^{-\delta_0|x|}) \quad \text{as } |x| \to \infty \tag{17}$$

for some $\delta_0 > 0$, by use of the above propositions with $\delta < \delta_0$, we see that $U_\gamma(\lambda, t)$ has a unique meromorphic continuation on D_ϵ and

$$\|U_\gamma(\lambda, i\tau)\|_{B(L^2_s, L^2_{-s})} \leq C/\tau \tag{18}$$

for large $\tau > 0$.

These properties of $U_\gamma(\lambda, z)$ can be used to show a reconstruction procedure with fixed energy. Note that the Faddeev scattering amplitude $A_\gamma(\lambda, t)$ has a relation with the physical scattering amplitude $A(\lambda)$.

The next statement holds due to Isozaki [4].

Proposition 3 *The Faddeev resolvent is given by*

$$R_{\gamma,0}(\lambda, t) = R_0((\lambda + i0)^2) - 2\pi i F_0(\lambda)^* \Phi_\gamma(t) F_0(\lambda),$$

where $\Phi_\gamma(t) = \chi(\gamma \cdot \theta \geq t/\lambda)$ (characteristic function of $\theta \in S^{n-1}$).

So, if we put $K = 2\pi i A(\lambda) \Phi_\gamma(t)$, then the Faddeev scattering amplitude is obtained by

$$A_\gamma(\lambda, t) = (I - K)^{-1} A(\lambda), \tag{19}$$

except a null set of t near $t = 0$.

4. Existence of the scattering operator

We return to the wave equation (1). In the following we require

$(A.1)$ $b(x)$ is a complex valued measurable function satisfying

$$|b(x)| \leq b_0(1 + |x|)^{-1-\delta}, \quad 0 < \delta < 1, \text{ in } x \in \mathbf{R}^n,$$

where b_0 is a constant satisfying the smallness condition

$$0 < b_0 < (\sqrt{5} - 1)\delta/4. \tag{20}$$

Defining $K = \sqrt{-\Delta}$, this operator is positive selfadjoint in L^2 with domain $D(K) = H^1$. Put $v = \{Kw, w_t\}$. Then (1) is rewritten in the form ([8, 10])

$$\tfrac{1}{i} v_t = \Lambda v = \Lambda_0 v + V^b v, \tag{21}$$

where

$$\Lambda_0 = \frac{1}{i} \begin{pmatrix} 0 & K \\ -K & 0 \end{pmatrix}, \quad V^b = \frac{1}{i} \begin{pmatrix} 0 & 0 \\ 0 & -b(x) \end{pmatrix}.$$

We consider (21) as an evolution equation in the Hilbert space $[L^2]^2 = L^2 \times L^2$.

Then Λ_0 is a selfadjoint operator with domain $D(\Lambda_0) = [H^1]^2$. On the other hand V is nonselfadjoint, but bounded in $[L^2]^2$. Moreover, noting $n \geq 3$, we can have ([9])

Lemma 3 *Under (A.1) we have for any* $\kappa \in \mathbf{C} \setminus \mathbf{R}$,

$$b_0 \|(\Lambda_0 - \kappa I)^{-1}\|_{B([L_s^2]^2, [L_{-s}^2]^2)} \leq c_0 < 1.$$

The smallness assumption (20) is required only to show $c_0 < 1$.

With this lemma both $(\Lambda_0 - \kappa)^{-1}$ and $(\Lambda - \kappa)^{-1}$ are extended to $\kappa = \lambda \pm i0$ ($\lambda \neq 0$) as maps of $[L_s^2]^2$ to $[L_{-s}^2]^2$. Here we have chosen s to satisfy $1/2 < s < (1 + \delta)/2$.

Moreover, we can follow the argument of Kato [6] to obtain

Theorem 2 (1) *The matrix operators* Λ *and* Λ_0 *are similar to each other. More precisely,*

$$(\mathcal{W}_\pm f, g)_{[L^2]^2} = (f, g)_{[L^2]^2}$$

$$\mp \frac{1}{2\pi i} \int_{-\infty}^{\infty} (V^b (\Lambda_0 - \lambda \mp 0)^{-1} f, (\Lambda - \lambda \pm 0)^{-1} g)_{[L^2]^2} d\lambda$$

defines a bijection on $[L^2]^2$ *and*

$$\Lambda = \mathcal{W}_\pm \Lambda_0 \mathcal{W}_\pm^{-1}.$$

(2) $\mathcal{W}_\pm$ *coincides with the Moller wave operator:*

$$\mathcal{W}_\pm = s - \lim_{t \to \pm\infty} e^{-i\Lambda t} e^{i\Lambda_0 t},$$

$$\mathcal{W}_\pm^{-1} = s - \lim_{t \to \pm\infty} e^{-it\Lambda_0} e^{it\Lambda}.$$

(3) *The scattering operator S is defined as*

$$S = W_+^{-1} W_-,$$

which is a bijection on $[L^2]^2$ and commutes with Λ_0.

Remark To the above theorem we can weaken the assumption $(A.1)$ as follows:

$(A.1)'$ $b(x)$ is a complex valued measurable function satisfying

$$|b(x)| \leq a(|x|), \quad \text{in } x \in \mathbf{R}^n,$$

where $a(r)$ is a positive, continuous L^1-function of $r > 0$ such that

$$\int_r^\infty a(r)dr \geq ra(r),$$

$$\int_0^\infty a(r)dr < \frac{\sqrt{5}-1}{4}.$$

5. Spectral representations of Λ

We put

$$\mathcal{F}_0(\lambda) = \begin{cases} F_0(\lambda) \begin{pmatrix} 1 & -i \\ i & 1 \end{pmatrix} & \text{for } \lambda > 0, \\[3mm] F_0(-\lambda) \begin{pmatrix} 1 & i \\ -i & 1 \end{pmatrix} & \text{for } \lambda < 0, \end{cases}$$

where $F_0(\lambda)$ is the Fourier transform defined by (7).

Lemma 4 (1) *Let $\Lambda_0 = \displaystyle\int_{-\infty}^\infty \lambda d\mathcal{E}_0(\lambda)$. Then for $f = \{f_1, f_2\} \in [L_s^2]^2$ and $\lambda \neq 0$ we have*

$$\frac{d}{d\lambda}\mathcal{E}_0(\lambda)f = \mathcal{F}_0(\lambda)^* \mathcal{F}_0(\lambda)f.$$

(2) *Put*

$$(\mathcal{F}_0 f)(\lambda, \omega) = [\mathcal{F}_0(\lambda)f](\omega),$$

$$(\mathcal{F}_0^* h)(x) = \int_{-\infty}^\infty [\mathcal{F}_0(\lambda)^* h(\lambda, \cdot)](x)d\lambda.$$

Then $\mathcal{F}_0$ is extended to a unitary operator from $[L^2]^2$ to $[L^2(\mathbf{R}; L^2(S^{n-1}))]^2$ and $\mathcal{F}_0^*$ becomes its inverse, that is,

$$\mathcal{F}_0^* \mathcal{F}_0 = 1 \quad in \quad [L^2]^2.$$

Now define the maps $\mathcal{F}_\pm(\lambda) : [L_s^2]^2 \to \left[L^2(S^{n-1})\right]^2$ and $\mathcal{F}_\pm(\lambda)^{(*)} :$ $\left[L^2(S^{n-1})\right]^2 \to [L_{-s}^2]^2$ respectively by

$$\mathcal{F}_\pm(\lambda) = \mathcal{F}_0(\lambda)\{I - V^b(\Lambda - \lambda \mp 0)^{-1}\},$$

$$\mathcal{F}_\pm(\lambda)^{(*)} = \{I - (\Lambda - \lambda \pm 0)^{-1}V^b\}\mathcal{F}_0(\lambda)^*.$$

Lemma 5 (1) *For $f \in [L_s^2]^2$,*

$$\lim_{\epsilon \to 0} \tfrac{1}{2\pi i}\{(\Lambda - \lambda - i\epsilon)^{-1} - (\Lambda - \lambda + i\epsilon)^{-1}\}f$$

$$= \mathcal{F}_\pm(\lambda)^{(*)}\mathcal{F}_\pm(\lambda)f.$$

(2) *Put*

$$[\mathcal{F}_\pm f](\lambda, \omega) = [\mathcal{F}_\pm(\lambda)f](\omega),$$

$$[\mathcal{F}_\pm^{(*)}g](x) = \int_{-\infty}^{\infty} [\mathcal{F}_\pm(\lambda)^{(*)}g(\lambda, \cdot)](x)d\lambda.$$

Then $\mathcal{F}_\pm$ is extended to a bijection from $[L^2]^2$ onto $\left[L^2(\mathbf{R} \times S^{n-1})\right]^2$ with the inverse $\mathcal{F}_\pm^{-1} = \mathcal{F}_\pm^{(*)}$.

With this lemma we can prove the following

Theorem 3 *The following identities hold:*

$$\mathcal{W}_\pm = \mathcal{F}_\pm^{(*)}\mathcal{F}_0, \tag{22}$$

$$S = \mathcal{W}_+^{-1}\mathcal{W}_- = \mathcal{F}_0^*\mathcal{F}_+^{(*)-1}\mathcal{F}_-^{(*)}\mathcal{F}_0, \tag{23}$$

$$\mathcal{F}_0(I - S)\mathcal{F}_0^* = 2\pi i\mathcal{F}_-V\mathcal{F}_0^*. \tag{24}$$

The relation (24) gives the scattering amplitude with energy $\lambda \in \mathbf{R}$ in the following form:

$$2\pi i\mathcal{A}(\lambda) = \mathcal{F}_0(\lambda)\{V^b - V^b(\Lambda - \lambda - i0)^{-1}V^b\}\mathcal{F}_0(\lambda)^*. \tag{25}$$

6. Inverse scattering with a fixed energy

The scattering amplitude (25) is an operator matrix. But if we note

$$V^b = \begin{pmatrix} 0 & 0 \\ 0 & ib(x) \end{pmatrix},$$

then one can reduce it to the scaler amplitude

$$A^b(\lambda) = F_0(\lambda)\left[ib + \lambda b\{1 + i\lambda R_0((\lambda + i0)^2)b\}^{-1}\right. \tag{26}$$

$$\left. \times R_0((\lambda + i0)^2)b\right]F_0(\lambda)^*.$$

We shall derive the reconstruction procedure of $b(x)$ from this $A^b(\lambda)$ with a fixed energy $\lambda \neq 0$. For this aim we require other than $(A.1)$ the following condition on $b(x)$:

$(A.2)$ $\quad |b(x)| \leq Ce^{-\delta_0|x|}$ for some $\delta_0 > 0$.

In (26) we replace $R_0((\lambda + i0)^2)$ by the Faddeev resolvent $R_{\gamma,0}(\lambda, t)$. Then the Faddeev scattering amplitude is given by

$$A^b_\gamma(\lambda, t) = F_0(\lambda)\left[b - i\lambda b R^b_\gamma(\lambda, t)b\right]F_0(\lambda); \tag{27}$$

$$R^b_\gamma(\lambda, t) = \{I + i\lambda R_{\gamma,0}(\lambda, t)b\}^{-1}R_{\gamma,0}(\lambda, t).$$

As is noted in Section 3, the Faddeev scattering amplitude $A^b_\gamma(\lambda, t)$ is obtained from the physical scattering amplitude $A^b(\lambda)$.

With the use of this fact, we can show the following theorem.

Theorem 4 Let $n \geq 3$ and assume $(A.1)$ and $(A.2)$. Then $A^b(\lambda)$ with a fixed energy $\lambda \neq 0$ uniquely determines the function $b(x)$.

Proof It follows from (27) that the kernel of $A^b_\gamma(\lambda, t)$ is given by

$$a^b_\gamma(\lambda, \omega, \omega'; t) = (2\pi)^{-n}\lambda^{n-1}\left[\int e^{-i\lambda(\theta - \theta')\cdot x}b(x)dx\right. \tag{28}$$

$$\left. -i\lambda\int e^{-i\lambda\theta\cdot x}b(x)R^b_\gamma(\lambda, t)(be^{i\lambda\theta'\cdot})(x)dx\right].$$

We choose $\omega, \omega' \in S^{n-1}$ to satisfy $\omega \cdot \gamma = \omega' \cdot \gamma = 0$ and put

$$\lambda\theta = \sqrt{\lambda^2 - t^2}\omega + t\gamma, \quad \lambda\theta' = \sqrt{\lambda^2 - t^2}\omega' + t\gamma.$$

Then (28) is reduced to

$$(2\pi)^n\lambda^{-n+1}a^b_\gamma(\lambda, \omega, \omega'; t) = \int e^{-i\sqrt{\lambda^2 - t^2}(\omega - \omega')\cdot x}b(x)dx$$

$$-i\lambda\int e^{-i\sqrt{\lambda^2 - t^2}\omega\cdot x}b(x)U^b_\gamma(\lambda, t)(be^{i\sqrt{\lambda^2 - t^2}\omega'\cdot})(x)dx,$$

where

$$U^b_\gamma(\lambda, t) = e^{-it\gamma\cdot x}R^b_\gamma(\lambda, t)e^{it\gamma\cdot x}.$$

Consequently it follows from (18) that

$$(2\pi)^n \lambda^{-n+1} a_\gamma^b(\lambda, \omega, \omega'; i\tau) \simeq \int e^{-i\sqrt{\lambda^2+\tau^2}(\omega-\omega')\cdot x} b(x) dx \qquad (29)$$

as $\tau \to \infty$.

For any $\xi \in \mathbf{R}^n$ we can choose $\gamma, \eta \in S^{n-1}$ to satisfy

$$\xi \cdot \gamma = \xi \cdot \eta = \gamma \cdot \eta = 0$$

since we have required $n \geq 3$. Put

$$\omega(\tau) = \left(1 - |\xi|^2/4\tau^2\right)^{1/2} \eta + \xi/2\tau,$$

$$\omega'(\tau) = \left(1 - |\xi|^2/4\tau^2\right)^{1/2} \eta - \xi/2\tau.$$

Then $\omega(\tau), \omega'(\tau) \in S^{n-1}$ and

$$\sqrt{\lambda^2 + \tau^2}(\omega(\tau) - \omega'(\tau)) = \sqrt{(\lambda/\tau)^2 + 1}\,\xi \simeq \xi \quad (\tau \to \infty).$$

Thus, from (29) it is concluded that

$$\lim_{\tau \to \infty} (2\pi)^n \lambda^{-n+1} a_\gamma^b(\lambda, \omega(\tau), \omega'(\tau); i\tau) = \int e^{-i\xi \cdot x} b(x) dx = (2\pi)^{n/2} \hat{b}(\xi).$$

$\square$

Final remark

If $b(x)$ satisfies (A.2), $(\Lambda_0 - \lambda \mp i0)^{-1}$ becomes analytic in $\lambda \in \mathbf{R} \backslash \{0\}$ with values in $\mathcal{B}(\mathcal{H}_s, \mathcal{H}_{-s})$. So, without the smallness assumption (20), one can show the existence of $(\Lambda - \lambda \mp i0)^{-1}$ except the discrete set $\Sigma_\pm^b$ which consists of the values $\lambda \in \mathbf{R}$ such that

$$\exists \varphi \in [\mathcal{H}_{-s}]^2; \varphi \neq 0 \text{ and } \{1 + (\Lambda_0 - \lambda \mp i0)^{-1} V^b\}\varphi = 0.$$

So, if we choose the fixed energy $\lambda \notin \Sigma_+^b \cup \Sigma_-^b$, then a result similar to Theorem 4 can be obtained without (20). But the characterization of $\Sigma_\pm^b$ is not known yet.

References

[1] Eskin, G., Ralston, J.: Inverse scattering problem fot the Schrödinger equation with magnetic potential with a fixed energy. Comm. Math. Phys. 173 (1995), 199–224.

[2] Faddeev, L.D.: Uniqueness of the inverse scattering problem. Vestnik Leningrad Univ. 11 (1956), 126–130.

[3] Faddeev, L.D.: Inverse problem of quantum scattering theory. J. Sov. Math. 5 (1976), 334–396.

316

[4] Isozaki, H.: Inverse scattering theory for Dirac operators. Ann. Inst. H. Poincaré Physique Théorique 66 (1997), 237–270.

[5] Isozaki, H.: Inverse problem theory for wave equations in stratified media. J. Differential Equations 138 (1997), 19–54.

[6] Kato, T.: Wave operators and similarity for some non-selfadjoint operators. Math. Ann. 162 (1966), 255–279.

[7] Mochizuki, K.: Eigenfunction expansions associated with the Schrödinger operator with a complex potential and the scattering inverse problem. Proc. Japan Acad. 43 (1967), 638–643.

[8] Mochizuki, K.: Spectral and scattering theory for second order elliptic differential operators in an exterior domain. Lecture Notes Univ. Utah, Winter and Spring 1972.

[9] Mochizuki, K.: Scattering theory for wave equations with dissipative term. Publ. Res. Inst. Math. Sci. 12 (1976), 383–390.

[10] Mochizuki, K.: Scattering theory for wave equations. Kinokuniya, 1983 (in Japanese).

[11] Weder, R.: Generalized limiting absorption method and multidimensional inverse scattering theory. Math. Meth. in Appl. Sci. 14 (1991), 509–524.

International Society for Analysis, Applications and Computation

1. H. Florian et al. (eds.): *Generalized Analytic Functions*. Theory and Applications to Mechanics. 1998   ISBN 0-7923-5043-X

2. H.G.W. Begehr et al. (eds.): *Partial Differential and Integral Equations*. 1999
                                                                      ISBN 0-7923-5482-6

3. S. Saitoh, D. Alpay, J.A. Ball and T. Ohsawa (eds.): *Reproducing Kernels and their Applications*. 1999   ISBN 0-7923-5618-7

4. R.P. Gilbert, J. Kajiwara and Y.S. Xu (eds.): *Recent Developments in Complex Analysis and Computer Algebra*. 1999   ISBN 0-7923-5999-2

5. R.P. Gilbert, J. Kajiwara and Y.S. Xu (eds.): *Direct and Inverse Problems of Mathematical Physics*. 2000   ISBN 0-7923-6005-2

6. H.G.W. Begehr, A. Okay Celebi and W. Tutschke (eds.): *Complex Methods for Partial Differential Equations*. 1999   ISBN 0-7923-6000-1

7. H.G.W. Begehr, R.P. Gilbert and J. Kajiwara (eds.): *Proceedings of the Second ISAAC Congress*. Volume 1. 2000   ISBN 0-7923-6597-6

8. H.G.W. Begehr, R.P. Gilbert and J. Kajiwara (eds.): *Proceedings of the Second ISAAC Congress*. Volume 2. 2000   ISBN 0-7923-6598-4

9. S. Saitoh, N. Hayashi and M. Yamamoto (eds.): *Analytic Extension Formulas and their Applications*. 2001   ISBN 0-7923-6950-5

10. H.G.W. Begehr, R.P. Gilbert and M.W. Wong (eds.): *Analysis and Applications – ISAAC 2001*. 2003   ISBN 1-4020-1384-1

KLUWER ACADEMIC PUBLISHERS – DORDRECHT / BOSTON / LONDON